and

LIBREX —

PLANNING AND DESIGN OF PORTS AND MARINE TERMINALS

2nd Edition

Hans Agerschou, Ian Dand, Torben Ernst, Harry Ghoos,
Ole Juul Jensen, Jens Korsgaard, John M. Land, Tom McKay,
Hocine Oumeraci, Jakob Buus Petersen, Leif Runge-Schmidt,
Hanne L. Svendsen

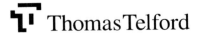 Thomas Telford

Published by Thomas Telford Publishing, Thomas Telford Ltd, 1 Heron Quay, London E14 4JD.
URL: http://www.thomastelford.com

Distributors for Thomas Telford books are
USA: ASCE Press, 1801 Alexander Bell Drive, Reston, VA 20191-4400, USA
Japan: Maruzen Co. Ltd, Book Department, 3–10 Nihonbashi 2-chome, Chuo-ku, Tokyo 103
Australia: DA Books and Journals, 648 Whitehorse Road, Mitcham 3132, Victoria

First edition published 1983 by John Wiley & Sons, Ltd
Reprinted with corrections 1985
Second edition published 2004

Also available from Thomas Telford Books

Breakwaters, coastal structures and coastlines. Edited by N. W. H. Allsop.
ISBN 0 7277 3042 8
Concrete in coastal structures. Edited by T. C. Allen.
ISBN 0 7277 2610 2
Port designer's handbook: Recommendations and guidelines. Carl A. Thoresen.
ISBN 0 7277 3228 5

A catalogue record for this book is available from the British Library

ISBN: 978-07277-3498-3

Contents

ix

Foreword to first edition

The editor and the contributors, who undertook to write this book, have for some time felt the need for a state-of-the-art book on the subject.

Although great strides have been made in the relevant basic and applied sciences, they are but tools needed to practice what we consider an art or a craft, which requires extensive experience for successful results. As both broad and in-depth experience from port planning and engineering practice, applied hydraulic and mathematical research as well as port management practice was indispensable, we believe it would be an impossible task for any individual to write the book we intended to write. The contributors have known each other for many years and have over the years worked closely together in various constellations. Chapters 1, 2, 3 and 5 have been written by Mr Agerschou, Chapter 4 by Mr Sørenson, Chapter 6 by Professor Lundgren and Messrs Agerschou and Sørenson, Chapter 7 by Messrs Korsgaard, Runge Schmidt and Agerschou, Chapter 8 by Mr Ernst and Chapter 9 Messrs Wee and Agerschou. The parts of Chapter 6, which deal exclusively with rubble-mound breakwaters, are based on a monograph by Mr O. J. Jensen of the Danish Hydraulic Institute. The editing has largely been done by Mr Agerschou.

Many improvements in the fields of planning, design, construction and operation of ports have taken place since the Second World War. We have attempted to cover all the more important developments, to some of which we believe we have made smaller or larger contributions.

We are grateful for comments, suggestions and assistance which we have received from the following:

Mr A. Adielson of the Swedish Shipbuilding Standard Centre, Mr G. Bjørk of Hostrup-Schultz and Sørenson, Consulting Engineers, Mr M. S. Chislett of the Danish Ship Research Laboratory, Mr K. G. Clare of the World Bank, Mr D. S. Hill of Dravo Van Houten Inc., Mr R. Kimber of the Victoria Deep Water Terminal Ltd., Messrs. J. Kirkegaard and M. H. Knudsen of the Danish Hydraulic Institute, Messrs S. Langvad and B. Ottesen of Pihl & Son, Contractors, Mr L. Møller of Lloyd's Register

(Copenhagen), Mr U. Møllgaard of Hoff & Overgaard, Planners and Economists (formerly Director General of the Copenhagen Free Port), Messrs P. Mortensen and S. E. Sand of the Danish Hydraulic Institute and Mr P Tryde of the Institute of Hydrodynamics and Hydraulic Engineering, Technical University of Denmark.

The original drawings were made by Mr H. H. Høj, drafts were typed by Ms A. Gram Hansen. The Frank Engelund Foundation graciously provided financial support for the preparation of the manuscript.

The chapters are self-contained to a large extent, but not completely. References made to other chapters are few and generally not absolutely necessary for the understanding of the chapter. Illustrations are labelled in alphabetical order for each section. Literature references are listed alphabetically at the end of each chapter.

We have not intended to cover all subjects associated with ports; for example, design of structures and foundations which is covered in textbooks and other publications is not included. However, we have attempted to deal with all important subjects of particular significance for planning, design and operation of ports and marine terminals, except railways in ports, a field in which we cannot mobilize sufficient experience. We had originally intended to cover fishing ports also and hope to do so in future editions.

Our approach to planning and design aims at rational determination of whatever lends itself to such, and thus attempts to avoid rules of thumb and other solely empirical methods to the extent possible. Where reasonably rational approaches are not yet available, we say so and discuss current practices and their shortcomings.

This is not intended as a textbook for undergraduate engineering students, but it is our hope that it will be useful for graduate students specializing in the field, as well as for practicing engineers and planners, port administrators and operators.

Foreword to second edition

Four of the seven contributors to the first edition have remained. Mr Wee Keng Chi died, at far too young an age, in 1994; perhaps a victim of the cut-throat consulting business. Eight new contributors have participated, while Mr H. Agerschou has remained the editor.

Chapter 9, which was Chapter 8 in the first edition, is basically unchanged due to its general nature, except for its last section on 'Fenders', which has been updated by Mr L. Runge-Schmidt.

The first three sections of Chapter 7 are based on sections 6.0 through 6.2 in the first edition, which were contributed by Professor H. Lundgren. They have been abbreviated extensively as their content is dealt with in more detail in following sections as noted.

The main new areas and topics in this edition are:

- A transparent outline of determination of facilities requirements for container terminals by a hybrid approach which includes a computer simulation program, partly based on original research, in Chapter 3.
- A brief outline of financial feasibility in Chapter 4.
- A detailed analysis of the more important categories of the world fleet ships, based on statistical material acquired specifically for this purpose from Lloyd's Register, in Chapter 5.
- High-speed ferry terminals in Chapter 7.
- Fishing ports in Chapter 7.
- Environmental considerations in Chapter 10.
- Dredging and disposal of contaminated sediments in Chapter 11.
- MARPOL in Chapter 12.
- Mathematical models in Appendix I.
- Case studies concerned with new ports and port extensions in Appendix II.

The author of the 'Fishing ports' section wishes to express his gratitude for information and advice received from Mr Erik Clausen, Chief Engineer and Mr Valdemar V. Leisner, Managing Director of the Port of Esbjerg, both of the Danish State Port Administration.

Chapter 9 (Organisation, Management and Operations) in the first edition has been taken out, as a suitable replacement for Mr Wee Keng Chi was not found. A badly needed update of Chapter 4 ('Physical Planning') in the first edition could unfortunately not be included and thus remains the same as the first edition. This chapter has to some extent been compensated for by Appendix I ('Mathematical Models'), which was kindly contributed by Ms Hanne L. Svendsen, an experienced specialist in the proper usage of this increasingly important tool.

Some topics are dealt with in considerably more detail than in the first edition, in particular breakwaters, in Chapter 7, and computer simulation of container terminals aimed at the improved forecasting of facilities requirements, in Chapter 3.

Our emphasis has shifted according to the major developments, which have taken place since the early 1980s. General cargo throughputs have decreased dramatically as many more commodities are now containerised. Accordingly, general cargo berths are being converted for other usage, such as multi-purpose and container terminals. This involves demolition and/or removal and re-erection for other usage of quayside transit sheds. The increases in DWT sizes of crude oil tankers and dry bulk carriers seem to have stagnated, but the percentages of very large bulk carriers have continued to increase. Carrying capacities measured in TEUs of the largest container ships have continued to increase and now exceed 6500. The beam of container ships is no longer governed by the width of the Panama Canal. This is related to the draught limit of 14 to 15 ms, imposed by currently available or future attainable water depths in access channels, along quays, etc.

CHAPTER I

Introduction

Hans Agerschou

General considerations

A port or a marine terminal is intended to provide facilities for transhipment of ships' cargo and of cargo transported to and from inland locations by rail, road, inland waterway and pipeline. Ships have to be accommodated safely along berths or at anchor, except perhaps under very extreme wind, wave and tidal conditions. Equipment has to be provided for efficient handling of cargo between ship and shore, and storage facilities have to be provided as cargo handling between ship and shore normally takes place at a much higher rate, in particular for bulk cargo, than arrival and removal of cargo by inland carriers. The reason for the much faster loading and unloading of ships is, of course, the high cost of immobilising them.

Storage facilities should be limited to those needed to compensate for the cargo-handling rate differences described above. The demand for storage facilities is reduced by limiting the time during which cargo is stored free of charge, and by sharply increasing the storage charge day by day after the free time has expired. Bulk cargo storage facilities should be provided for at least one shipload, or at large terminals, which can accommodate several bulk carriers simultaneously, for at least the corresponding number of shiploads. In the past, when inland carriers were predominantly trains and barges and service relatively infrequent, more extensive port storage facilities were needed. This is why older ports often have large warehouses, which are now no longer required, due to inland transport services becoming more frequent as trucking has developed and has become continuous by means of pipelines and conveyor belts. Thus the need for port storage has been reduced.

Before the development of rail transport, and before modern road transport became important, a large number of smaller and larger ports were needed because transport by sea was the only possibility or was much cheaper than land transport. Since cheaper and more frequent land transport became available the need for a large number of commercial ports, in particular for the smaller ones, decreased dramatically. Many of them are presently only used to a minor extent, or have been converted for use by pleasure craft.

Planning is clearly concerned not only with physical facilities but also, and equally important, with organisation, management and operations. The latter often become the major problem areas. The design and construction of fixed facilities as well as the choice, design and installation of equipment frequently give fewer and smaller problems than their subsequent efficient usage and maintenance.

General aspects of planning and design

Facilities planning is, by its very nature, concerned with the future and thus has to consider varying degrees of uncertainty. However, it seems to be widely believed that engineering design is normally based on certainty. This is to a high extent true when the properties of materials and the structures or products in which they are used are very well known, and the environment in which they are going to function is not subject to random variation. In certain fields of engineering (electrical, electronics and mechanical) the properties of materials and products are often well known and their environment controlled to a high degree. However, this is far from being the case for harbour and coastal engineering and the associated disciplines of foundation engineering and oceanography. Some of the important construction materials, for use or removal, are soils and rock of widely varying, inhomogeneous, and therefore often not well-known, properties. Further, the facilities and structures are exposed to more or less well-known random variations of wind, tide, currents and waves, which may result in flooding, extreme low water, wind damage, wave damage, erosion and siltation.

Thus, both in regard to planning and to design, we have to live with uncertainty, including more or less well-known probabilities of occurrence of natural phenomena. Similarly, random variations of traffic by ships and land transport occur. This results in variations of the demand for port facilities.

The facilities should be designed for suitably low probabilities of occurrence of the following:

(a) interruption of cargo-handling operations due to flooding and/or wave action;

(b) interruption of ships' access to facilities or the necessity of ships to leave the facilities due to wind, extreme water levels, currents, waves, or combinations of the above;

(c) damage to breakwaters and other structures caused by wave action, currents or wind; and

(d)   damage to quays and berthing platforms caused by ships exposed to wind, currents and wave action.

The chosen probabilities of occurrence should be based upon analyses (to an extent this is realistically possible) of the additional economic and/or financial costs of facilities and the economic and/or financial benefits resulting from reduction of damage and/or of interruption of operations.

The major design parameters, which have to be chosen based upon their probability of occurrence, are thus:

(a)   extreme high water causing flooding of quays and thus interruption of cargo handling as well as damage to cargo and fixed installations;

(b)   extreme low water causing ships to leave berths or not to enter port;

(c)   extreme wind conditions causing:
    (i)   interruption of cargo handling operations;
    (ii)   ships to leave berths or not to come alongside berths; and
    (iii) damage to buildings;

(d)   extreme currents causing:
    (i)   interruption of arrival and departure of ships; and
    (ii)   erosion damage to structures;

(e)   extreme wave action causing:
    (i)   damage to breakwaters;
    (ii)   interruption of cargo-handling operations;
    (iii) ships to leave berths or not to come alongside berths; and
    (iv)   siltation of basins and/or access channels.

The choice of general locations for new ports and port extensions

New ports or port extensions become necessary when traffic increases or when entirely new traffic has to be accommodated. In principle, new ports and also port extensions should be located, if there is a choice, where they will minimise the total sea and land transport costs to the economy of the country for the cargo in question, considering its different origins and destinations. Obviously, this is not exclusively a matter of the geographical location for the port site, but also a matter of different port construction costs for different sites.

Land transport costs should clearly include an appropriate share of the cost of new land transport links or of improvement of existing links. Sea transport costs should consider the effects of the different sizes of ships, which it may be technically and economically feasible to accommodate at different sites. This is, of course, likely to be particularly important for bulk cargo. The total sea and land transport costs in question should not be the first year costs but the sum of the discounted annual transport costs during the useful life of the project. Thus traffic forecasts, and the uncertainties associated with such, are unavoidably introduced into the attempt to determine the optimum port location. However, in some cases it becomes a fairly simple, common-sense matter.

It is sometimes argued that transport projects generate development of production and/or of traffic; for example feeder roads or farm-to-market roads generate agricultural production over and above that required for local consumption. There are hardly any indications that this may be the case for ports. A port has to come into existence as early as connecting land transport facilities and as early as production facilities, which are to be served by the port in question. However, the port has to be part of an economically feasible project containing the above components. The port alone is not feasible but it is an indispensable part of the project.

Factors, which are not readily quantifiable in economic terms, such as overall strategic planning for regional or demographic development, often play an important role in port site selection, which is rarely, if ever, purely an engineering economics exercise.

CHAPTER 2

Facilities requirements

Hans Agerschou

Introduction

Rational determination of facilities requirements, and evaluation of their economic and financial feasibility, are mandatory for projects of any significance. Facilities requirements determination may be based on queuing theory in cases for which exact or approximate analytical solutions are available, or on computer simulation. The economic feasibility of projects is evaluated by application of recognised methodology for comparing the stream of annual economic benefits, normally to the country or the region in question, with the stream of annual costs to the economy in question. Similarly the financial feasibility, from the point of view of the owner or operator of the facilities, is evaluated by application of recognised methodology for analysing the stream of annual net cash flow.

Facilities requirements

General considerations

Rules of thumb for cargo throughput per berth or per unit of quay length are usually quite misleading. They ignore the fact that optimum berth utilisation, which considers the economic cost of having ships wait for berths to become available, varies from less than 0.25 for a single berth to more than 0.75 for a large port. This is clearly important, as the cargo throughput is a linear function of the berth utilisation.

Quay cranes for general cargo have to a large extent been replaced by ships' gear, which has generally seen radical improvement.

General cargo transit sheds have been and are being demolished, as they are no longer needed due to containerisation of the cargo in question, and are replaced by open storage areas.

The number of ships in different categories, which will call at a port during a certain period of time, is determined from throughput forecasts for different commodities and cargo units as well as forecasts of the average tonnage loaded and/or unloaded per ship. The average cargo-handling time for different categories of ships is based on forecast average cargo-handling rates. The average service time is arrived at by adding the forecast average time required for customs' formalities, rigging, idle time, etc. The above forecasts and simple calculations combined with known or forecast statistical distributions for ships' arrival rates and service time are the basic inputs needed to determine berth requirements. The average service time and the statistical distribution of the service time reflect the cargo handling and storage facilities' components of the port system.

The optimum size of covered and open storage facilities should, in principle, be determined by comparison between, on the one hand, the costs of additional areas and/or stacking height and, on the other hand, the benefits from reduction of waiting time and service time for ships and/or land transport, resulting from increased storage facilities, as well as the benefits from reduction of cargo-handling costs and of damage to cargo.

Cargo-handling equipment systems should also in principle be optimised by comparison between, on the one hand, additional capital, operating and maintenance costs for equipment and, on the other hand, benefits from reduction of waiting time and service time for ships and/or land transport, resulting from increased cargo-handling capacity, as well as benefits from reduction of cargo-handling costs and of damage to cargo.

However, the required input data both for storage and for equipment analysis would hardly ever be available for general cargo facilities except perhaps for cases of extreme congestion, whereas they are more likely to be available for container and bulk terminals.

To optimise the entire system, which consists of ships and their cargoes, berths, storage facilities, cargo-handling equipment and land transport it is necessary to employ computer simulation (see Computer simulation later in this chapter and Chapter 3).

Analytical solutions for determination of facilities requirements
Berths

Depending upon the statistical distributions for ships' arrival rates (numbers of arrivals per time unit) and for service time, it is possible to determine berth requirements by using available analytical solutions to queuing theory.

Queuing theory for random arrivals and service time was originally developed for telephone exchange planning by A. N. Erlang and others early in the last century. Exact analytical solutions to queuing theory exist only for two cases, both of which have a Poisson distribution for the arrival rate and either an exponential distribution for service time or

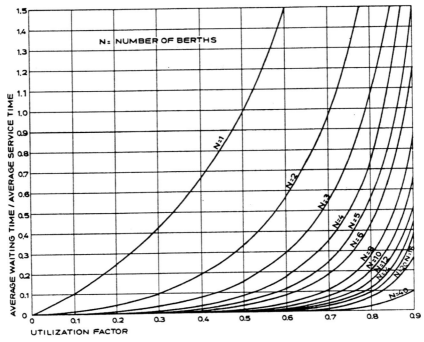

Fig. 2.1 Queuing analysis solutions for exponential service-time distribution. Reproduced by permission of the Dock & Harbour Authority

constant service time (Morse, 1965). For an exponential service-time distribution the ratios between average waiting time and average service time as functions of the berth utilisation factor (utilised berth time divided by available berth time) and the number of berths are shown in Fig. 2.1 (Agerschou and Korsgaard, 1969) and in Table 2.1. Actual arrival rate and service time distributions for many general cargo ports conform sufficiently well to the above distributions according to chi-square goodness of fit tests (Crow *et al.*, 1960). Analytical values for the probability of occurrence of any specified waiting time are also available. However, they are normally of limited importance to the port planner.

The other case for which an exact analytical solution is available has constant service time. The solution includes infinite series, which are conveniently evaluated by computer. It is shown in Fig. 2.2 and in Table 2.2. For only one berth the average waiting time to service-time ratios are exactly one half of the corresponding values for exponential service-time distribution. For two or more berths the average waiting time to service-time ratios varies from 0.5 to 0.6 of those for a corresponding exponential service-time distribution, over the utilisation factor intervals which are normally of interest. Constant or near constant service time is

7

Table 2.1 Queuing analysis solutions for exponential service-time distribution

N =	1	2	3	4	5	6	7	8	9	10
U
0.10	0.111	0.010	0.001	0.000	0.000	0.000	0.000	0.000	0.000	0.000
0.15	0.176	0.021	0.004	0.001	0.000	0.000	0.000	0.000	0.000	0.000
0.20	0.250	0.042	0.010	0.003	0.001	0.000	0.000	0.000	0.000	0.000
0.25	0.333	0.067	0.020	0.007	0.003	0.001	0.000	0.000	0.000	0.000
0.30	0.429	0.099	0.033	0.013	0.006	0.003	0.001	0.001	0.000	0.000
0.35	0.538	0.140	0.053	0.023	0.011	0.006	0.003	0.002	0.001	0.001
0.40	0.667	0.190	0.078	0.038	0.020	0.011	0.006	0.004	0.002	0.001
0.45	0.818	0.254	0.113	0.058	0.033	0.020	0.012	0.008	0.005	0.003
0.50	1.000	0.333	0.158	0.087	0.052	0.033	0.022	0.015	0.010	0.007
0.55	1.222	0.434	0.217	0.126	0.079	0.053	0.037	0.026	0.019	0.014
0.60	1.500	0.562	0.296	0.179	0.118	0.082	0.059	0.044	0.033	0.025
0.65	1.857	0.732	0.401	0.253	0.173	0.124	0.093	0.071	0.055	0.044
0.70	2.333	0.961	0.547	0.357	0.252	0.187	0.143	0.113	0.091	0.074
0.75	3.000	1.286	0.757	0.509	0.369	0.281	0.221	0.178	0.147	0.123
0.80	4.000	1.778	1.079	0.746	0.554	0.431	0.347	0.286	0.240	0.205
0.85	5.667	2.604	1.623	1.149	0.873	0.693	0.569	0.477	0.408	0.357
0.90	9.000	4.263	2.724	1.969	1.525	1.234	1.029	0.877	0.761	0.669
0.95	19.000	9.256	6.047	4.457	3.511	2.885	2.441	2.110	1.855	1.651
0.98	49.000	24.252	16.041	11.950	9.503	7.877	6.718	5.851	5.178	4.641

N =	11	12	13	14	15	16	17	18	19	20
U
0.10	0.000	0.000	0.000	0.000	0.000	0.000	0.000	0.000	0.000	0.000
0.15	0.000	0.000	0.000	0.000	0.000	0.000	0.000	0.000	0.000	0.000
0.20	0.000	0.000	0.000	0.000	0.000	0.000	0.000	0.000	0.000	0.000
0.25	0.000	0.000	0.000	0.000	0.000	0.000	0.000	0.000	0.000	0.000
0.30	0.000	0.000	0.000	0.000	0.000	0.000	0.000	0.000	0.000	0.000
0.35	0.000	0.000	0.000	0.000	0.000	0.000	0.000	0.000	0.000	0.000
0.40	0.001	0.001	0.000	0.000	0.000	0.000	0.000	0.000	0.000	0.000
0.45	0.002	0.002	0.001	0.001	0.001	0.000	0.000	0.000	0.000	0.000
0.50	0.005	0.004	0.003	0.002	0.002	0.001	0.001	0.001	0.000	0.000
0.55	0.010	0.008	0.006	0.005	0.004	0.003	0.002	0.002	0.001	0.001
0.60	0.020	0.016	0.012	0.010	0.008	0.007	0.005	0.004	0.004	0.003
0.65	0.035	0.029	0.024	0.020	0.016	0.014	0.012	0.010	0.008	0.007
0.70	0.061	0.051	0.043	0.037	0.331	0.027	0.023	0.020	0.018	0.016
0.75	0.104	0.089	0.076	0.066	0.058	0.051	0.045	0.040	0.036	0.032
0.80	0.176	0.154	0.135	0.119	0.106	0.095	0.086	0.077	0.070	0.064
0.85	0.310	0.274	0.245	0.220	0.199	0.181	0.165	0.151	0.139	0.128
0.90	0.594	0.533	0.482	0.439	0.402	0.370	0.342	0.317	0.295	0.275
0.95	1.486	1.348	1.233	1.134	1.049	0.975	0.910	0.853	0.801	0.755
0.98	4.202	3.837	3.529	3.266	3.038	2.839	2.663	2.507	2.368	2.243

Table 2.1 (continued)

N =	21	22	23	24	25	26	27	28	29	30
U										
0.10	0.000	0.000	0.000	0.000	0.000	0.000	0.000	0.000	0.000	0.000
0.15	0.000	0.000	0.000	0.000	0.000	0.000	0.000	0.000	0.000	0.000
0.20	0.000	0.000	0.000	0.000	0.000	0.000	0.000	0.000	0.000	0.000
0.25	0.000	0.000	0.000	0.000	0.000	0.000	0.000	0.000	0.000	0.000
0.30	0.000	0.000	0.000	0.000	0.000	0.000	0.000	0.000	0.000	0.000
0.35	0.000	0.000	0.000	0.000	0.000	0.000	0.000	0.000	0.000	0.000
0.40	0.000	0.000	0.000	0.000	0.000	0.000	0.000	0.000	0.000	0.000
0.45	0.000	0.000	0.000	0.000	0.000	0.000	0.000	0.000	0.000	0.000
0.50	0.000	0.000	0.000	0.000	0.000	0.000	0.000	0.000	0.000	0.000
0.55	0.001	0.001	0.001	0.000	0.000	0.000	0.000	0.000	0.000	0.000
0.60	0.003	0.002	0.002	0.001	0.001	0.001	0.001	0.001	0.001	0.001
0.65	0.006	0.005	0.005	0.004	0.003	0.003	0.003	0.002	0.002	0.002
0.70	0.014	0.012	0.011	0.010	0.008	0.008	0.007	0.006	0.005	0.005
0.75	0.029	0.026	0.023	0.021	0.019	0.018	0.016	0.015	0.013	0.012
0.80	0.058	0.054	0.049	0.045	0.042	0.039	0.036	0.033	0.031	0.029
0.85	0.119	0.110	0.103	0.096	0.090	0.084	0.079	0.074	0.070	0.066
0.90	0.258	0.242	0.228	0.215	0.203	0.192	0.182	0.173	0.165	0.157
0.95	0.714	0.676	0.642	0.611	0.583	0.556	0.532	0.510	0.489	0.470
0.98	2.130	2.028	1.934	1.849	1.770	1.698	1.631	1.569	1.511	1.457

N =	31	32	33	34	35	36	37	38	39	40
U										
0.10	0.000	0.000	0.000	0.000	0.000	0.000	0.000	0.000	0.000	0.000
0.15	0.000	0.000	0.000	0.000	0.000	0.000	0.000	0.000	0.000	0.000
0.20	0.000	0.000	0.000	0.000	0.000	0.000	0.000	0.000	0.000	0.000
0.25	0.000	0.000	0.000	0.000	0.000	0.000	0.000	0.000	0.000	0.000
0.30	0.000	0.000	0.000	0.000	0.000	0.000	0.000	0.000	0.000	0.000
0.35	0.000	0.000	0.000	0.000	0.000	0.000	0.000	0.000	0.000	0.000
0.40	0.000	0.000	0.000	0.000	0.000	0.000	0.000	0.000	0.000	0.000
0.45	0.000	0.000	0.000	0.000	0.000	0.000	0.000	0.000	0.000	0.000
0.50	0.000	0.000	0.000	0.000	0.000	0.000	0.000	0.000	0.000	0.000
0.55	0.000	0.000	0.000	0.000	0.000	0.000	0.000	0.000	0.000	0.000
0.60	0.000	0.000	0.000	0.000	0.000	0.000	0.000	0.000	0.000	0.000
0.65	0.002	0.001	0.001	0.001	0.001	0.001	0.001	0.001	0.001	0.001
0.70	0.004	0.004	0.004	0.003	0.003	0.003	0.002	0.002	0.002	0.002
0.75	0.011	0.010	0.010	0.009	0.008	0.007	0.007	0.006	0.006	0.006
0.80	0.027	0.025	0.023	0.022	0.021	0.019	0.018	0.017	0.016	0.015
0.85	0.062	0.059	0.056	0.053	0.050	0.048	0.045	0.043	0.041	0.039
0.90	0.150	0.143	0.137	0.131	0.126	0.120	0.116	0.111	0.107	0.103
0.95	0.452	0.435	0.419	0.404	0.390	0.377	0.365	0.353	0.342	0.332
0.98	1.407	1.359	1.315	1.274	1.235	1.198	1.163	1.130	1.098	1.069

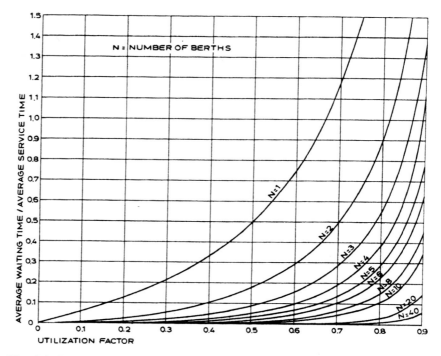

Fig. 2.2 Queuing analysis solutions for constant service time

typical for dry-bulk carriers and especially for tankers. However, bulk-carrier arrivals are often scheduled, non-random events, for which queuing theory is not valid. Thus its use is limited, whereas the use of the exponential service-time solution is widespread.

Exponential service-time distribution and constant service time are special cases of Erlang distributions. The general expression for the probability $P_o(t)$ that the service time exceeds t, is:

$$P_o(t) = e^{-Kut} \sum_{n=0}^{n=K-1} (Kut)^n/n!$$

where u is the average service rate and K is a variable. $K = 1$ corresponds to exponential service time and $K = \infty$ corresponds to constant service time. There is also an application of queuing theory to scheduled arrivals at one berth (Edmond, 1975). It assumes that the ships arrive during time intervals of equal length. If a ship does not arrive during the interval it is assumed not to arrive at all. The service time distribution corresponds to $K = 2$. Analytical solutions are presented for a range of utilisation factors from 0.3 to 0.6 for one berth only, for probabilies of arrival during the interval of 0.9 and 0.75 respectively, and for three trapezoidal distributions of ships' arrivals during the interval. Figures show the probability that the

Table 2.2 Queuing analysis solutions for constant service-time

N =	1	2	3	4	5	6	7	8	9	10
U										
0.10	0.056	0.006	0.001	0.000	0.000	0.000	0.000	0.000	0.000	0.000
0.15	0.088	0.014	0.003	0.001	0.000	0.000	0.000	0.000	0.000	0.000
0.20	0.125	0.024	0.007	0.002	0.001	0.000	0.000	0.000	0.000	0.000
0.25	0.167	0.038	0.012	0.005	0.002	0.001	0.000	0.000	0.000	0.000
0.30	0.214	0.055	0.020	0.008	0.004	0.002	0.001	0.000	0.000	0.000
0.35	0.269	0.077	0.031	0.014	0.007	0.004	0.002	0.001	0.001	0.000
0.40	0.333	0.103	0.045	0.023	0.012	0.007	0.004	0.003	0.002	0.001
0.45	0.409	0.136	0.063	0.034	0.020	0.012	0.008	0.005	0.003	0.002
0.50	0.500	0.177	0.087	0.050	0.031	0.020	0.013	0.009	0.007	0.005
0.55	0.611	0.228	0.118	0.070	0.046	0.031	0.022	0.016	0.012	0.009
0.60	0.750	0.293	0.158	0.098	0.066	0.047	0.034	0.026	0.020	0.015
0.65	0.929	0.378	0.212	0.136	0.095	0.069	0.052	0.040	0.032	0.026
0.70	1.167	0.494	0.286	0.190	0.136	0.102	0.079	0.063	0.051	0.042
0.75	1.500	0.657	0.392	0.267	0.196	0.150	0.119	0.097	0.080	0.068
0.80	2.000	0.903	0.554	0.386	0.289	0.227	0.183	0.152	0.128	0.110
0.85	2.833	1.316	0.827	0.589	0.449	0.359	0.295	0.249	0.213	0.185
0.90	4.500	2.145	1.377	1.000	0.776	0.630	0.526	0.450	0.391	0.345
0.95	9.500	4.631	3.034	2.241	1.769	1.456	1.233	1.067	0.939	0.837
0.98	24.500	12.126	8.020	5.975	4.752	3.938	3.359	2.926	2.592	2.324

N =	11	12	13	14	15	16	17	18	19	20
U										
0.10	0.000	0.000	0.000	0.000	0.000	0.000	0.000	0.000	0.000	0.000
0.15	0.000	0.000	0.000	0.000	0.000	0.000	0.000	0.000	0.000	0.000
0.20	0.000	0.000	0.000	0.000	0.000	0.000	0.000	0.000	0.000	0.000
0.25	0.000	0.000	0.000	0.000	0.000	0.000	0.000	0.000	0.000	0.000
0.30	0.000	0.000	0.000	0.000	0.000	0.000	0.000	0.000	0.000	0.000
0.35	0.000	0.000	0.000	0.000	0.000	0.000	0.000	0.000	0.000	0.000
0.40	0.001	0.000	0.000	0.000	0.000	0.000	0.000	0.000	0.000	0.000
0.45	0.002	0.001	0.001	0.001	0.000	0.000	0.000	0.000	0.000	0.000
0.50	0.003	0.003	0.002	0.001	0.001	0.001	0.001	0.000	0.000	0.000
0.55	0.007	0.005	0.004	0.003	0.002	0.002	0.002	0.001	0.001	0.001
0.60	0.012	0.010	0.008	0.006	0.005	0.004	0.004	0.003	0.002	0.002
0.65	0.021	0.017	0.014	0.012	0.010	0.009	0.007	0.006	0.005	0.005
0.70	0.035	0.029	0.025	0.021	0.019	0.016	0.014	0.012	0.011	0.010
0.75	0.058	0.049	0.043	0.038	0.033	0.029	0.026	0.023	0.021	0.019
0.80	0.095	0.083	0.074	0.065	0.058	0.053	0.047	0.043	0.039	0.036
0.85	0.163	0.145	0.130	0.117	0.106	0.097	0.088	0.081	0.075	0.069
0.90	0.307	0.276	0.250	0.228	0.209	0.192	0.178	0.165	0.154	0.144
0.95	0.753	0.684	0.626	0.577	0.534	0.497	0.464	0.435	0.409	0.386
0.98	2.106	1.924	1.771	1.639	1.526	1.426	1.338	1.261	1.191	1.129

11

Table 2.2 (continued)

N =	21	22	23	24	25	26	27	28	29	30
U										
0.10	0.000	0.000	0.000	0.000	0.000	0.000	0.000	0.000	0.000	0.000
0.15	0.000	0.000	0.000	0.000	0.000	0.000	0.000	0.000	0.000	0.000
0.20	0.000	0.000	0.000	0.000	0.000	0.000	0.000	0.000	0.000	0.000
0.25	0.000	0.000	0.000	0.000	0.000	0.000	0.000	0.000	0.000	0.000
0.30	0.000	0.000	0.000	0.000	0.000	0.000	0.000	0.000	0.000	0.000
0.35	0.000	0.000	0.000	0.000	0.000	0.000	0.000	0.000	0.000	0.000
0.40	0.000	0.000	0.000	0.000	0.000	0.000	0.000	0.000	0.000	0.000
0.45	0.000	0.000	0.000	0.000	0.000	0.000	0.000	0.000	0.000	0.000
0.50	0.000	0.000	0.000	0.000	0.000	0.000	0.000	0.000	0.000	0.000
0.55	0.001	0.001	0.000	0.000	0.000	0.000	0.000	0.000	0.000	0.000
0.60	0.002	0.001	0.001	0.001	0.001	0.001	0.001	0.001	0.000	0.000
0.65	0.004	0.003	0.003	0.003	0.002	0.002	0.002	0.002	0.001	0.001
0.70	0.008	0.008	0.007	0.006	0.005	0.005	0.004	0.004	0.004	0.003
0.75	0.017	0.015	0.014	0.013	0.012	0.011	0.010	0.009	0.008	0.008
0.80	0.033	0.030	0.028	0.026	0.024	0.022	0.021	0.019	0.018	0.017
0.85	0.064	0.060	0.056	0.052	0.049	0.046	0.043	0.041	0.039	0.036
0.90	0.135	0.127	0.120	0.113	0.107	0.102	0.097	0.092	0.087	0.083
0.95	0.365	0.346	0.329	0.313	0.298	0.285	0.273	0.261	0.251	0.241
0.98	1.072	1.021	0.974	0.931	0.892	0.856	0.822	0.791	0.762	0.735

N =	31	32	33	34	35	36	37	38	39	40
U										
0.10	0.000	0.000	0.000	0.000	0.000	0.000	0.000	0.000	0.000	0.000
0.15	0.000	0.000	0.000	0.000	0.000	0.000	0.000	0.000	0.000	0.000
0.20	0.000	0.000	0.000	0.000	0.000	0.000	0.000	0.000	0.000	0.000
0.25	0.000	0.000	0.000	0.000	0.000	0.000	0.000	0.000	0.000	0.000
0.30	0.000	0.000	0.000	0.000	0.000	0.000	0.000	0.000	0.000	0.000
0.35	0.000	0.000	0.000	0.000	0.000	0.000	0.000	0.000	0.000	0.000
0.40	0.000	0.000	0.000	0.000	0.000	0.000	0.000	0.000	0.000	0.000
0.45	0.000	0.000	0.000	0.000	0.000	0.000	0.000	0.000	0.000	0.000
0.50	0.000	0.000	0.000	0.000	0.000	0.000	0.000	0.000	0.000	0.000
0.55	0.000	0.000	0.000	0.000	0.000	0.000	0.000	0.000	0.000	0.000
0.60	0.000	0.000	0.000	0.000	0.000	0.000	0.000	0.000	0.000	0.000
0.65	0.001	0.001	0.001	0.001	0.001	0.001	0.001	0.000	0.000	0.000
0.70	0.003	0.003	0.002	0.002	0.002	0.002	0.002	0.001	0.001	0.001
0.75	0.007	0.006	0.006	0.005	0.005	0.005	0.004	0.004	0.004	0.004
0.80	0.016	0.015	0.014	0.013	0.012	0.011	0.011	0.010	0.010	0.009
0.85	0.034	0.033	0.031	0.029	0.028	0.027	0.025	0.024	0.023	0.022
0.90	0.080	0.076	0.073	0.070	0.067	0.064	0.062	0.060	0.057	0.055
0.95	0.232	0.223	0.215	0.208	0.201	0.194	0.188	0.182	0.177	0.171
0.98	0.710	0.686	0.664	0.643	0.623	0.605	0.587	0.571	0.555	0.540

berth is not occupied at the start of the interval and the probability for a ship to have to wait for access to the berth during the interval. For non-constant time intervals, more complicated ships' arrival distributions, etc., and more than one berth, analytical solutions are not available and computer simulation is required.

Arrival rates for general cargo ports, usually conform to Poisson distributions. The same is the case for container terminals which are used by more than one or a few shipping lines. Although each line schedules its arrivals, late arrivals and lack of co-ordination between the shipping lines result in Poisson distributions. The author has investigated this by means of chi-square, goodness-of-fit tests for several terminals.

In many cases the actual service-time distribution does not fit an exponential distribution sufficiently well or is not sufficiently close to constant service time, but may fit a distribution corresponding to $K = 2$, 3, 4 or more. Service-time distributions may, of course, also be checked by means of goodness-of-fit tests. Approximations of values of the ratio between average waiting time and average service time for K values between 1 and ∞ have been presented (Page, 1972). For $K = 2$ the approximation consists of simple averages of the corresponding $K = 1$ and $K = \infty$ values. For $K = 3$ the approximation consists of weighted averages of the above values, with $K = \infty$ values carrying twice the weight of $K = 1$ values. It is claimed that these approximations are sufficiently accurate for high utilisation factors, but not for low utilisation factors.

The capacity concept

The initial determination of the required number of berths has to be based on an assumed optimum capacity expressed as a berth utilisation. A maximum utilisation of 100 per cent would correspond to infinite average waiting time for the calling ships. The optimum utilisation increases with an increasing number of berths.

In UNCTAD (1986) optimum utilisations are suggested. The only reason given for the chosen values is that they 'are based on a ratio of ship cost to berth cost of 4 to 1'. The author has added to the suggested values the corresponding average waiting-time to service-time ratios, in Table 2.3, assuming $K = 1$ and 2 service-time distributions for general cargo berths, and $K = 4$ or ∞ distributions for multi-user container terminals. A more rational approach would consider the ships' waiting time costs represented by the waiting-time to service-time ratios. The ratios have to be empirical values resulting from economic feasibility studies which have been carried out in the past. The UNCTAD utilisation values for 3 or more berths correspond to ratios varying only from 0.13 to 0.22 for general cargo facilities and from 0.10 to 0.14 for container terminals.

Table 2.3 UNCTAD optimum berth utilisations and corresponding average waiting-time to service-time ratios

Number of berths	Optimum utilisation	Container terminals		General cargo berths	
		$t_{wait}/t_{service}$ $K = 4$	$t_{wait}/t_{service}$ $K = \infty$	$t_{wait}/t_{service}$ $K = 1$	$t_{wait}/t_{service}$ $K = 2$
1	0.40	0.42	0.33	0.67	0.50
2	0.50	0.22	0.18	0.33	0.26
3	0.55	0.14	0.12	0.22	0.17
4	0.60	0.12	0.10	0.18	0.14
5	0.65	0.12	0.10	0.17	0.13
6 or more	0.70	0.12	0.10	0.19	0.14

Note: average waiting-time to service-time ratios shown for '6 or more' are for 6 only.
Source: UNCTAD, 1986 and author.

The lower values for container terminals are reasonable because the value of time for container ships would generally be higher than that for general cargo ships. The ratios for 1 and 2 berths do not seem reasonable, in particular those for 1 berth only, which correspond to waiting time to service time ratios varying from 0.33 to 0.67. According to experience from many economic feasibility studies, reasonable average ratio values for general cargo and container berths would be 0.2 and 0.1 respectively. Based on these assumptions the corresponding utilisation factors shown in Table 2.4 have been prepared.

Comparison of the two tables shows large differences for 1 and 2 berths. For 3 or more berths the differences are smaller, but remain significant. It should be stressed that the above approach is for an initial determination of the required number of berths. This number will be confirmed or modified by means of an economic and/or financial feasibility study.

Table 2.4 Suggested optimum berth utilisations corresponding to empirical average waiting-time to service-time ratios

Number of berths	Container terminals			General cargo berths		
	Optimum $t_{wait}/t_{service}$	Berth utilisation		Optimum $t_{wait}/t_{service}$	Berth utilisation	
		$K = 4$	$K = \infty$		$K = 1$	$K = 2$
1	0.10	0.14	0.17	0.20	0.17	0.21
2	0.10	0.36	0.40	0.20	0.41	0.45
3	0.10	0.49	0.52	0.20	0.54	0.58
4	0.10	0.57	0.60	0.20	0.61	0.65
5	0.10	0.63	0.66	0.20	0.67	0.70
6 or more	0.10	0.67	0.70	0.20	0.71	0.74

Note: berth utilisations for '6 or more' are for 6 only.

Limitations of the analytical solutions

The analytical solutions are imperfect tools as they assume that an infinite number of ships arrive, are serviced and then leave. For determining berth requirements, and economic benefits from providing additional berths, the planner is obviously concerned with finite periods of time, normally one year or one season, during which only a finite number of ships are serviced. Thus the theoretical average waiting time may be expected to deviate somewhat from the actual one. Also queuing theory requires the use of the concept of a nominal number of berths, although, except for bulk terminals, reality consists of one or more stretches of quay along which a varying number of ships may be accommodated simultaneously, depending upon their lengths. The average number of ships, which are accommodated at any one time, would usually be larger than the nominal number of berths. This would, for the same cargo throughput, result in lower ratios of average waiting time to average service time.

Use of the concept of stretches of quay, instead of nominal numbers of berths of a fixed length, requires computer simulation.

Storage facilities

General cargo storage requirements for new ports are difficult to determine; the mix of numerous cargo units, which have different stowage factors (volume-to-weight ratios), is uncertain, the average storage time is uncertain and the short term, peak throughputs are difficult to relate to the average throughput. For existing ports these important parameters should be well known, which makes determination of storage requirements for extensions much more reliable than for new ports.

The required storage area is proportional to the throughput during the period of time in question, the weighted average stowage factor for the forecast cargo mix, and the average storage time. It is inversely proportional to the average stacking height, the length of the period of time in question, and the fraction of the total area that is actually used for storage and not for operating equipment or for giving access to cargo. To accommodate above-average throughput, stowage factor and storage time, as well as below-average stacking height, one or more correction factors have to be employed. Reliable values for such factors are only available in existing ports, where relevant data are properly collected and processed. Specific storage facilities are discussed in Chapter 8.

Storage facilities, which are meant to serve one berth only, should be sufficiently large to hold cargo left by ships which have departed, plus all the cargo to be unloaded from the ship presently alongside that is not directly unloaded to land transport, if delivery from storage does not take place, while cargo is being discharged from the ship. The storage-space requirement is at least theoretically reduced, if delivery does take place, while cargo is still being discharged from the ship. The length of time cargo remains in storage may be influenced by the port operator in the following ways:

(a) the number of days during which no storage charge is levied for cargo, which has been unloaded from ships. For outbound cargo the same purpose is served by refusing to receive cargo for storage until a certain limited number of days before the ship is due to depart;

(b) sharply increasing storage charges day by day after the free time has expired; and

(c) removal of general cargo left in transit sheds a certain limited number of days after the free time has expired. The cargo is moved at the consignees' expense to a warehouse in the port or outside the port.

When delivery takes place from storage during discharge from the ship, and if it is assumed that all cargo has been removed when the free time expires and that both the cargo arrival rate and the removal rate are constant, it can be shown that the required storage capacity is:

(a) when removal and arrival of cargo start on the same day

$$nx - \sum_{p=1}^{p=n} px/n = (n-1)x/2$$

where x is the arrival rate in tonnes per day and n is the free time in days; and

(b) when removal starts one day after the start of arrival, additional storage for cargo arriving during one day has to be provided.

The assumption of constant arrival and removal rates may not be realistic. However, if one examines the storage capacity in an entire port during peak traffic periods a first approximation of storage requirements may be made as indicated if flexibility in the use of storage facilities is feasible. Although this would increase cargo-handling costs it would decrease storage-facilities costs. A trade-off analysis would show which is less costly.

All large quantities of bulk cargo should be handled through storage facilities to permit rapid loading/unloading. For the same reason bulk storage facilities should at least be provided for one shipload, and for terminals which will accommodate several bulk carriers simultaneously, at least for the corresponding number of shiploads. Port or near-port storage requirements, exceeding the above, clearly depend upon inland storage capacity as well as land transport capacity and costs between inland storage and port storage. Without knowledge of these, rational decisions cannot be made. For agricultural commodities, seasonal production and possibly the desire to spread exports or imports over a larger part of the year complicate matters further. Rules of thumb for the required port storage capacity as a fraction of the annual throughput are often quite misleading. Proper determination of port storage capacity necessitates analysis of the entire storage and land-transport system.

If a bulk transport system consists of more than one marine terminal connected by land transport to one or more inland storage facilities, it will normally be economically advantageous to operate it as one integrated system; for example, where grain is exported through several terminals. To make this possible, grain classification has to be sufficiently good to enable loss of identity within categories and classes of grain, and thus permit interchange of grain from different producers, which is loaded at different terminals. This illustrates the complexity of integrated bulk-transport systems.

Equipment

Similar to storage requirements, and partly for the same reasons, equipment requirements for new ports are difficult to ascertain. Downtime and productivity data for equipment are not available, and the efficiency of the future management, cargo-handling workforce and equipment maintenance workforce is somewhat uncertain. There is no doubt that the abilities and attitudes of employees may, on one hand, make a port that is not very well equipped function quite well, and, on the other hand, make a port, which is well equipped, function poorly.

In existing ports where relevant data are properly collected and processed the situation is quite different. Based on downtime and productivity records, requirements for additional equipment similar to existing equipment would be quite well defined. However, for entirely new equipment categories, some, but not all, of the uncertainties associated with new ports would exist.

Cargo-handling equipment for specific purposes is discussed in Chapter 8.

Computer simulation

Simulation is required if analytical solutions are not available, which is typically when:

(a) arrival rates do not conform to a Poisson distribution;
(b) the berth concept is replaced by the total length of quay concept;
(c) all arriving ships cannot be accommodated at all berths;

17

(d) priority is given, e.g. at multi-purpose berths, to ships, which may only be accommodated at such berths;

(e) both alongside handling and lighterage of cargo take place; and

(f) cargo-handling equipment and storage facilities need to be included in analyses of the whole port system.

A large number of simulation programs for alongside ports have been developed over the years. They range from rather simple and not very useful programs to apparently sophisticated, well-thought-out and well-tested programs. Considerable duplication of effort has taken place, as the development work has generally not been co-ordinated between prospective users of simulation programs. Many of the programs are far from transparent in the descriptions made available to prospective users. Programs which claim to simulate the use of container-yard equipment are not realistic and may be quite misleading as discussed in Chapter 3.

In principle it is possible to simulate any stochastic variable including the following:

(a) arrival rates or time between consecutive arrivals of ships;

(b) size of ship;

(c) cargo tonnage loaded and/or unloaded by ship;

(d) availability of quay cranes with known handling-rate distributions;

(e) working days and hours;

(f) time at berth before and after cargo handling;

(g) arrival of cargo in and removal of cargo from storage facilities; and

(h) stoppages of cargo-handling operations, partly or fully, for any reason such as strikes, repairs and extreme weather conditions.

Any desired statistical distribution is generated as a subroutine, based on one input seed number for each subroutine.

It has been demonstrated (Hansen, 1972) that replacing the one ship, at one nominal berth concept, with lengths of ships and with stretches of quay will reduce the required quay length considerably. This assumes that two or more open spaces are aggregated, where this is actually possible, to accommodate larger ships whenever required. The above result is to be expected as it occurs frequently that the average length of ships along a quay is considerably less than the average length of a nominal berth, and thus that more ships than the number of nominal berths are serviced simultaneously. Evidently, this concept is only applicable to certain configurations of quay and storage facilities.

Simulation over shorter and longer periods of time for cases with analytical solutions have been carried out to investigate the reliability of simulation, as illustrated by the following example for five berths, where the simulation in each case was preceded by a running-in period of 12 months. Ten

Table 2.5 Comparison of simulation results with analytical solutions

	Ten simulations of 24-month duration		Ten simulations of 72-month duration		Analytical solution
	Range	Average	Range	Average	
Utilisation factor	0.67–0.83	0.72	0.66–0.75	0.71	0.71
$t_{wait}/t_{service}$	0.17–0.51	0.27	0.16–0.36	0.26	0.27

different seed numbers (first number used in generating desired service-time distribution) were employed. Table 2.5 shows the results.

While simulation did not, in particular for 72-month periods, show large deviations from the analytical utilisation factor, the simulated average waiting-time to service-time ratio in many cases deviated considerably from the analytical value, even when the simulated utilisation factor was exactly the same as the analytical one. This is one of the problems associated with port simulation programs, which points to the need for more than one simulation run, no matter how long, with the same input data. Even ten simulation runs, each of 72-months' duration, do not, in this case, yield an average result, which is exactly equal to the corresponding analytical solution.

The period of time for which simulation should be carried out for a given set of inputs varies with circumstances. If seasonal traffic variations occur, they should of course be reflected in the simulation program. As a general rule, it is recommended to run so many short period (a few months) simulations, that the resulting average waiting-time and service-time values from each period will conform to a normal distribution. It has been shown that although the waiting and service time for individual ships may not conform to a normal distribution, the average values from a number of simulation periods do conform to such a distribution. Therefore, planners may be reasonably confident if their simulation outputs satisfy this requirement. A method exists (Crow et al., 1960) for determining the minimum sample size (number of short periods), to be used with the equal-tails test for goodness of fit. The basic criterion for the minimum duration of the simulation periods is that the queue length at the end of the period must be independent of the queue length at the beginning of the period. A method for investigating correlation between two such sets of variables is available (Crow et al., 1960).

The author has not been able to find any detailed and transparent descriptions in the literature of computer simulation programs for container terminals and therefore outlines in Chapter 3 his own simulation program as part of a hybrid approach and discusses in detail a number of important issues associated with such.

References

Agerschou, H. and Korsgaard, J. (1969) *Systems Analysis for Port Planning*, London: The Dock & Harbour Authority.

Crow, E. L., *et al.* (1960) *Statistics Manual*, New York: Dover Publications.

Edmond, E. D. (1975) *Operating Capacity of Container Berths for Scheduled Services by Queue Theory*, London: The Dock & Harbour Authority.

Hansen, J. B. (1972) 'Optimizing ports through computer simulation. Sensitivity analysis of pertinent parameters', *Operational Research Quarterly*, 23(4).

Morse, P. M. (1965) *Queues, Inventories and Maintenance*, New York and London: John Wiley & Sons Inc.

Page, E. (1972) *Queuing Theory in OR*, London: Butterworths.

UNCTAD (1986) *Port Development. A handbook for planners in developing countries*, New York: United Nations.

CHAPTER 3

Determination of facilities requirements for container terminals by a hybrid approach consisting of computer simulation, analytical and empirical solutions

Hans Agerschou

Introduction

This chapter deals solely with computer simulation, analytical and empirical solutions for planning and optimisation of requirements for new terminal facilities, and for extension of existing facilities. Only numbers and sizes of containers to be handled and stored are considered. The identity of any one single container, which includes its consignee and cargo, cannot possibly be known, when the planning exercise, which includes forecasting for a number of years, takes place. This is *not* concerned with computer programs used by terminal operators for optimising their day-to-day operations, including keeping track of each container as it arrives in the terminal, as it is stored and as it leaves the terminal. For this purpose the specific identity of each container is required before it arrives in the terminal. The input as well as output requirements for the two above-mentioned purposes are very different.

Analytical solutions to queuing theory are limited to determination of an optimum number of nominal berths for assumed numbers and handling rates of container quay cranes and/or ships' gear. Simulation is required for optimisation of quay lengths, numbers of quay cranes and storage-yard sizes.

Exact analytical solutions to queuing theory are available for the following cases:

(a) a nominal number of berths, each of which is assumed to be able to accommodate any arriving ship;

(b) the ship arrival distribution conforms sufficiently well to a random distribution; and

(c)   the service (alongside) time distribution for ships conforms sufficiently well to an exponential distribution, or it is close to constant.

Approximate analytical solutions to queuing theory are available for
service-time distributions, which are found between the exponential
service-time distribution and constant-service time (Page, 1972).

A major problem associated with employment of analytical solutions, in
addition to their limitation to determination of the required numbers of
berths, results from the unavoidable use of a nominal number of berths.
If, instead, stretches of quays, which may accommodate varying numbers
of vessels of different lengths, together with realistic ship length distributions, are used for computer simulation (everything else equal), considerably lower average ship waiting time results for the same berth or quay
utilisation. Conversely, a higher utilisation is possible for the same average waiting time (Hansen, 1972).

Available descriptions of computer simulation programs for container
terminals are, in the author's opinion, usually not sufficiently clear and/
or detailed for a rational evaluation of their utility. Some of them may be
deliberately non-transparent to protect the methodology, which has
been employed, against possible competitors. Some program providers
appear to be more interested in animation, showing movement of ships,
equipment and containers (perhaps for presentations to less knowledge-
able prospective clients) than in realistic and transparent simulation
programs.

Competition, rather than co-operation, between developers of port and
terminal simulation programs has been the rule in the past. Co-operation
would perhaps have yielded more useful results.

In the following sections it is attempted to clearly specify what the
required simulation elements are going to accomplish and how this is
achieved. Where rules and/or inputs for simulation are not available,
analytical and/or empirical solutions and approximations are suggested.

Requirements for a hybrid approach

The containers move through a terminal in the following manner:

(a)   inbound containers with cargo and empties go from ship-side by
transfer equipment, e.g. tractor/trailers, to yard equipment, which
stacks them in separate storage yards from which they are delivered
to land transport by the yard equipment;

(b)   inbound trans-shipment containers go from ship-side by transfer
equipment to yard equipment, which stacks them in a separate
storage yard. From the storage yard they are delivered as outbound
trans-shipment containers, by the yard equipment, to the transfer
equipment, which takes them to the ship-side; and

(c)   outbound containers with cargo and empties arrive by land transport to yard equipment, which stacks them in separate storage yards from which they are delivered to transfer equipment which takes them to ship-side.

If straddle carriers are used as yard equipment, they may also be used as transfer equipment, which modifies the above description.

The approach should in realistic and transparent ways reproduce:

(a)   efficient utilisation of major fixed facilities and major equipment such as stretches of quay, storage-yard areas, container quay cranes, yard and transfer equipment;

(b)   known or assumed statistical distributions and/or discrete values for the following inputs:
   (i)   time of arrivals of ships; and
   (ii)   lengths of arriving ships;

(c)   numbers of inbound containers with cargo and of empties as well as of trans-shipment containers per ship;

(d)   numbers of outbound containers with cargo and of empties as well as of trans-shipment containers per ship;

(e)   container handling rates for quay cranes, yard and transfer equipment, with separate rates for containers with cargo and for empties;

(f)   time in storage (dwell time) for inbound containers with cargo and for empties;

(g)   time in storage (dwell time) for outbound containers with cargo and for empties; and

(h)   numbers of movements (lifts) of containers inside stacks required to give access to other containers for delivery.

Some of these distributions and values are well known, some not so well known, and some have apparently not yet been analysed. Chi-square, goodness-of-fit tests (Crow et al., 1960) of a number of actual distributions are reported below and theoretical distributions recommended for use.

Quays and assignments of berths

The concept of straight-line stretches of quays of known lengths (m) is used for separate categories, e.g. main-line ships and feeder ships. Often there would be only one straight-line stretch of quay in a terminal, but the simulation model should have provisions for four or more stretches.

Any arriving ship would be assigned to the shortest stretch of vacant quay, which will accommodate its length plus space at each end of the ship, with a default value of 15 m.

The first arriving ship would be assigned to a quay space beginning at a variable distance, with a default value of 15 m, from one end of the quay.

The next ship, while the first ship is still alongside, would be assigned to a space beginning at a variable distance, with a default value of 15 m, from the end of the first ship, and so on. When two or more quay spaces are vacant simultaneously, but each is too short to accommodate the next arriving or waiting ship, it is assumed that one or more ships will be moved along the quay as required to create a sufficiently long space, unless one is known to become available within a number of hours, with a default value of 2. To move a ship takes a number of hours, with a default value of 2. While a ship is being moved, its container handling is, of course, interrupted.

The above is based on the rule of first-come, first-served. If it is believed the terminal users accept deviations from this rule, provisions could be made in the program, e.g. to search the line of waiting ships and allow the first one, which could be accommodated at a vacant space, to come alongside.

For the initial series of simulation runs one or more analytically determined quay length(s) will be used. For the next series of runs the length(s) will be increased or reduced, depending upon the results of the first series, and so on.

Ship arrivals, service time and departures
For multi-user terminals the arrivals of ships conform to a random distribution, corresponding to a Poisson distribution for arrival rates. Although each of the users schedule their arrivals, there is apparently no or insufficient co-ordination between them. There is ample proof of the resulting randomness from a number of chi-square, goodness-of-fit tests of actual arrival rate distributions. Some of these tests have been made by the author, using standard methodology. For a single-user terminal, or a terminal with a few users, all arrivals would be scheduled and co-ordinated. Of course delays of arrivals may occur anyway. Thus discrete scheduled arrivals, allowing for some degree of delays, would have to be used.

A number of hours, with a default value of 1, are allowed for formalities and other preparations, before unloading of the containers starts.

Main-line ships are unloaded and loaded by container quay cranes. Feeder ships use their own gear and/or quay cranes. After completion of loading, a number of hours, with a default value of 1, are allowed for formalities and other preparations, before the ship departs.

Delays of arrivals or departures caused by low tide would usually not be acceptable, except under very infrequent extreme low-water conditions. They should be accounted for separately, if significant.

Ships calling at the terminal and the container exchange per ship
Two main categories of ships have to be accounted for. One is main-line ships, which may be broken down into early (smaller), Panamax and

Post-Panamax classes, according to their main dimensions and resulting TEU carrying capacity. Post-Panamax ships are considerably wider than the other categories, and require quay cranes with a very long reach. The second category is feeder ships, which are generally smaller than main-line ships, although some of the older and smaller main-line ships have now become feeder ships.

Based on container throughput forecasts (see 'Container throughput', later in this chapter) and assumed average container exchange (numbers unloaded and loaded) per ship, the number of calling ships of each category is determined (or the other way around). An assumed average container exchange per ship is largely an educated guess, unless some sort of relationship between ship length and container exchange is established.

The writer has carried out analyses of actual distributions of container exchanges per ship and investigated the relationships between the latter and the ship length (see the following section).

For each category of ships a maximum and a minimum length have to be assumed together with a corresponding statistical distribution. The default distribution is assumed to be constant. Actual distributions, if relevant, have to be collected from existing terminals.

The actual length distribution of the world fleet of container ships (Lloyd's Register of Shipping, 31 December, 1999, unpublished) does not conform to any common theoretical distribution.

In case the planned facilities are going to be used exclusively by one or a few known container shipping line(s), their existing fleet(s) and planned additions to the fleet(s) should be used to determine the length distribution for calling ships. Sufficient water depth is assumed to be available along each stretch of quay.

The numbers of inbound and outbound empties per ship will have to be estimated. One way of doing this would be to use percentages of the total numbers of inbound and outbound containers per ship, corresponding to the total forecasts. Another way would be to use statistical distributions (to be determined) with the forecast mean values for the inbound and outbound numbers, respectively, of empties per ship. The latter is more realistic, but also more complicated. It may not yield more useful results, considering the uncertainties associated with forecasts of empties (see 'Container throughput', later in this chapter). If the simple percentage method were used, the number of empties in storage would increase and decrease simultaneously with the number of containers with cargo. This would not adversely affect the results of simulation or of analytically determined yard-equipment requirements (see 'Yard(s) for inbound containers with cargo and internal movement', later in this chapter), because the yards and their equipment are different and separate for containers with cargo and for empties.

The number of trans-shipment (arriving and departing on ships, without leaving the terminal on its landside) containers per ship could be dealt with as described for empties above. The percentage for containers to be loaded would need to be adjusted according to the ratio between the total numbers of loaded containers and of unloaded containers over the simulation period of time, so that the number of loaded trans-shipment containers becomes equal to the number of unloaded trans-shipment containers. However, there are special problems associated with trans-shipment containers (see 'Container throughput').

Statistical analyses of distributions of box exchanges per ship and of lengths of ships for 20 container terminals

General

The analysis is based on data for 5300 calls during the first 7 months of the year 2000 by about 300 different ships at 20 terminals, of which 11 are import/export terminals without trans-shipment, six are exclusively used for trans-shipment and three are mixed import/export and trans-shipment terminals. The data, which have been provided by a major container terminal operator, includes in addition to the number of boxes unloaded and loaded per call, the terminal type and the length overall of the ship.

Chi-square, goodness-of-fit tests have been used to analyse the box-exchange data.

Results for import/export terminals without trans-shipment

For six out of the 11 terminals the box exchange per ship distributions conform sufficiently well or very well to theoretical normal distributions (thus there are no reasons for rejecting this hypotheses). For the other five terminals the fit was poor or very poor. The results of the statistical analyses for the six above-mentioned terminals are shown in Table 3.1 together with some of the inputs.

It would not be unreasonable to think that perhaps there would be some sort of relationship between the mean values for box exchanges per ship and the mean carrying capacities of the ships. Shipping lines would not consider it financially feasible to have large ships call at ports for a relatively small box exchange. The carrying capacity would be largely proportional to the ship length and it is convenient to use the latter as a parameter as it is required anyway as a simulation input.

The mean box exchange values for terminals 3, 16 and 20 are about the same, ranging from 635 to 654. The ship lengths for terminals 3 and 20 are very similar, with a maximum of 194 m. For terminal 16, 87 per cent of the lengths are in the much higher, very narrow, range from 290 to 295 m and

Table 3.1 Normal distributions for box exchanges per ship for import/export terminals

Terminal number	Number of boxes		Ratio SD/mean	Chi-square sum		Ratio Actual/Allowed	Number of calls	Lengths of ships	Percentage within length range of 5 m	Length range in m
	Mean	SD		Allowed	Actual					
3	653.75	286.7	0.44	5.991	1.689	0.28	43	42 between 165 and 194 m minimum: 145 m/maximum: 194 m	35	50
7	393.43	135.96	0.35	3.841	1.32	0.34	51	30 between 175 and 194 m 16 between 145 and 159 m minimum: 145 m/maximum: 194 m	24	50
13	739.08	230.26	0.31	14.07	13.74	0.98	105	61 between 290 and 294 m 25 between 150 and 179 m 14 between 255 and 259 m minimum: 150 m/maximum: 294 m	58	145
15	936.49	372.85	0.40	18.31	17.24	0.94	156	85 between 290 and 294 m 13 between 255 and 274 m 28 between 195 and 209 m minimum: 130 m/maximum: 294 m	54	165
16	642.36	170.28	0.27	9.488	2.779	0.29	87	76 between 290 and 294 m 8 between 255 and 274 m minimum: 190 m/maximum: 294 m	87	105
20	634.98	201.34	0.32	7.815	4.371	0.56	45	All between 160 and 194 m	44	35

Notes: Terminal 13 only for box exchanges of less than 1400. There is a gap from 1400 to 2300.
SD = Standard deviation.

the minimum length is 190 m, which is about the same as the maximum length for terminals 3 and 20. Terminal 7, which has a box-exchange mean value much lower than that for terminal 3, has a somewhat similar length distribution.

Terminals 13 and 15, which have quite different box-exchange mean values, have somewhat similar length distributions. The latter are, however, considerably below those for terminal 16, which has a considerably lower box-exchange mean value than terminals 13 and 15.

It does not appear that any conclusions, in regard to relationships between mean box exchanges and ship lengths, may be drawn from the comparisons described above. This was also the case for terminals in Pusan, Korea, according to Nam *et al.* (2002).

The terminal 13 box-exchange distribution in the upper range of 2300 to 5700 per ship (see below) conforms well to a theoretical constant distribution as determined by linear regression. The corresponding length distribution consists of 22 ships in the 345 to 349 m range, seven in the 315 to 319 m range and 22 in the 290 to 294 m range. The much higher mean box exchange corresponds to the much longer ships, which have a minimum length equal to the maximum length of the ships in Table 3.1.

The goodness-of-fit for box-exchange distributions, defined as the ratio between the actual chi-square sum and the allowed maximum chi-square sum, could reasonably be believed to be related to the narrowness of the ship length distribution.

Terminal 3 has the lowest such goodness-of-fit ratio of 0.28. The length of all calling ships varies within a range of 50 m. However, only one calling ship is outside a narrow range of 30 m.

Terminal 16 has a goodness-of-fit ratio of 0.29 and the length of all calling ships varies within a range of 105 m. However, 87 per cent of them are within an extremely narrow range of 5 m.

Terminal 7 has a goodness-of-fit ratio of 0.34 and the length of all calling ships varies within a range of 50 m. However, 59 per cent of them are within a narrow range of 30 m.

Terminal 20 has a goodness-of-fit ratio of 0.56 and the length of all calling ships varies within a narrow range of 35 m.

Terminal 15 has a rather high goodness-of-fit ratio of 0.94. The length of all calling ships varies within a rather large range of 165 m. However, 54 per cent of them are within an extremely narrow range of 5 m.

Terminal 13 is a special case. The box exchanges per ship are divided into two distinct ranges, one from 150 to 1400 for 105 calling ships and another from 2300 to 5700 for 51 calling ships. There are no box-exchange values in the interval between 1400 and 2300. It is assumed that the terminal may be considered to consist of two separate parts, one of which is perhaps used as a feeder terminal. The distribution for the lower range barely conforms to a theoretical normal distribution with a goodness-of-fit ratio of 0.98. The length of all calling ships varies within a rather large range of 145 m. However, 58 per cent of them are within an extremely narrow range of 5 m.

Thus it looks like there is, to some extent, a relationship between the goodness-of-fit for the normal distributions and the length range of calling ships. The smaller the latter is the better the fit seems to be.

The ratio between the standard deviation and the mean value for the normal distributions for box exchanges per ship varies between 0.27 and 0.44 for the six cases discussed above.

Results for trans-shipment terminals

The box exchanges per ship for the six terminals do not conform to any theoretical distribution known to the writer. They are all far from normal distributions. The distributions have two or more peak values; whereas normal distributions only have one. Why this is so, the author does not know. Perhaps it is caused by a mixture of feeder traffic and main-line traffic.

Results for mixed import/export and trans-shipment terminals

The box exchanges per ship, for the three terminals, do not conform to any theoretical distribution known to the author. One of them is far from a normal distribution. It has two peak values. If it was possible to separate the box exchanges per ship into import/export and trans-shipment, some of the resulting distributions might conform to one or more theoretical distributions.

Container quay cranes

Initially it is assumed that the quay is equipped with one container crane per certain lengths of quay, with a default value of 100 m. The resulting number of cranes is rounded off to a whole number.

Cranes will be assigned to ships on a first-come, first-served basis. An arriving ship will be assigned two available cranes if its length is equal to or less than a certain length, with a default value of 200 m. If its length exceeds the above length but does not exceed another certain length, with a default value of 300 m, three available cranes will be assigned. If its length exceeds the latter length, four or five available cranes will be assigned. If the above numbers of cranes are not available, the ship will have to wait for additional cranes, while the available cranes are being used. After a crane becomes available, it takes a number of hours, with a default value of 2, before it is ready for use at another ship. What actually takes place, when a ship which could use additional cranes is not adjacent to a ship from which a crane becomes available, is that a number of cranes have to be moved along the quay between more than just two ships. A constant handling rate (or a handling-rate distribution) for each crane will be assumed as numbers of containers per crane working hour, with a default value of 30 for containers with cargo and a considerably higher value for empties, during a variable number of hours per day, with a default value of 24. Non-operational days, for any reason such as maintenance, holidays and days with extreme weather conditions, will be generated, with a default value of zero.

Container handling between quay cranes and yard equipment

If tractor-trailer transfer is used between the apron and the yard equipment, containers are placed on and picked up from trailers on the quay apron by the container quay cranes. If straddle carriers are used, the containers are placed on and picked up from the apron.

The cost of the tractor-trailer system constitutes a relatively low percentage of the total costs of the fixed facilities and the equipment. Thus, there is no great need to simulate the transfer of containers by tractor-trailers. The required number of tractor-trailers is determined analytically based on empirical or assumed transfer rates to and from the different yards. Straddle carrier requirements are dealt with as yard equipment below.

Container throughput

Forecasts of annual or seasonal (if required) throughputs of the following container categories should be employed for a number of years ranging from 10 to 20:

(a) main line and feeder line;
(b) inbound, outbound and trans-shipment;
(c) containers with cargo and empties; and
(d) 20 ft and 40 ft containers. Other sizes exist but are at present not significant for overall planning purposes.

The forecast breakdown between main line and feeder line may be quite uncertain.

Throughputs of inbound and outbound containers with cargo are based on established methodology for commodities-flow forecasting and their present and expected future suitability for containerisation. Forecasts of inbound and outbound empties are likely to be quite uncertain. Increasing loss of identity of containers, e.g. by co-operation between shipping lines and other container owners and/or start of local production may reduce the flows dramatically. Trans-shipment throughputs can usually only be forecast by a designated operator, involved at the terminal planning stage.

Container storage yards

Separate container storage yards are provided for each of the following; inbound containers with cargo, outbound containers with cargo, inbound empties, outbound empties and trans-shipment containers.

The area requirements are measured in TEU ground slots (the area required for one 20-ft container) plus operating space for equipment that transfers containers to and from the yards and that stack and deliver containers. Initially the number of TEU ground slots for each yard is determined analytically, based on simple assumptions.

The numbers of containers of different categories moving into storage from ships and from land transport are converted into TEUs by assuming a certain percentage of 40-ft containers. If it is 60, the ratio of numbers of TEUs to numbers of containers is 1.6 ($0.4 \times 1 + 0.6 \times 2$).

Yard(s) for inbound containers with cargo and internal movements

Inbound containers with cargo will have a free storage time of a number of days, usually 3 to 5. After the free time has expired the daily storage charge usually increases steeply day by day. Containers from a certain ship are usually not delivered to land transport until all the containers from that ship have been stacked in the yard. A major operator has promised to provide the author actual distributions of dwell time together with the corresponding free storage time for a number of container terminals.

The dwell-time distribution, applied to the numbers of containers from each ship and aggregated for all ships, would yield the overall distribution of deliveries per time unit.

The overall TEU inflow to storage from ships per time unit, together with the overall TEU deliveries to land transport, would determine the number of TEUs in storage at any one time.

The overall deliveries to land transport per time unit resulting from a series of simulations should probably conform to a normal distribution (see 'Container flows to and from yards by land transport', later in this chapter).

Yard(s) for outbound containers with cargo

Outbound containers will not be received into storage earlier than a certain number of days, usually 4 to 5, before the ship arrives, and they will not be received later than a variable number of hours before the ship arrives, usually 12 to 6 hours. Also they will not be loaded until after unloading is completed.

Table 3.2 Sequence of events for each ship and its containers

Event	Ship arrival/departure	Containers with cargo and empties (separate storage yards) from and to ships	Containers with cargo and empties (separate storage yards) from and to land transport
Start of arrival by land transport to storage yard			From 4 to 5 days before arrival of ship
Arrival by land transport to storage yard			Distributed over interval from start to completion based on empirical dwell time distribution(s) for the two container categories
Completion of arrival by land transport to storage yard			From 6 to 12 hours before arrival of ship
Arrival ready to come alongside	According to Poisson distribution for arrival rates and/or discrete arrivals		
Arrival alongside	Same as above or after waiting for quay space to become available		
Start of unloading		After completion of preparations	
Unloading	According to cranes assigned (may vary over time) and their handling rates		Immediately after completion of unloading of ship
Start of loading of ships and of delivery from storage to land transport		Immediately after completion of unloading of ship	
Loading of ships and delivery from storage to land transport		See unloading	Distributed over time based on empirical dwell-time distribution(s) for the two container categories. Dwell-time distribution would depend upon the free storage time, normally from 3 to 5 days and storage charges (steeply increasing day by day) after the free time ends
Departure	After completion of loading and of preparations		

Notes: The last column is of course not applicable for trans-shipment containers.

It is necessary to run the simulation of alongside arrivals of ships, their unloading and loading ahead of the receipt of containers into storage for a certain ship, to be able to determine the starting time and completion of receipt before arrival. Loops will then have to go back in time to determine the number of TEUs in storage at any one time for the ship in question.

Table 3.2 shows the sequence of events for each ship and its containers. A major operator has promised to provide the writer actual distributions of dwell time together with the time interval, during which containers are received, for a number of container terminals. This would determine their inflow per time unit and, aggregated for all ships, would yield the overall inflow per time unit.

Completion of unloading of a ship triggers the start of its loading. The number of containers loaded per time unit and aggregated for all ships would yield the overall outflow from storage per time unit.

The overall TEU inflow to storage from land transport per time unit, together with the overall TEU outflow to ships per time unit, would determine the number of TEUs in storage at any one time.

The overall inflow per time unit, resulting from a series of simulations, should probably conform to a normal distribution (see 'Container flows to and from yards by land transport', later in this chapter).

Yard(s) for trans-shipment containers

Trans-shipment containers are probably going to have one or a few, close-to-constant, dwell-time values based on scheduled arrivals and departures. The dwell-time value(s) will have to be furnished by the designated operator. Both inbound and outbound flows to and from storage would be according to the cranes assigned to each ship. Loading begins when unloading of all containers (not only those for trans-shipment) has been completed.

Yard(s) for empties

Special rules apply for inbound and outbound empties, respectively. However, in principle they are not different from the above and the approach would be the same.

Container flows to and from yards by land transport

The total numbers of containers received into storage from land transport (road and rail) and delivered to land transport are forecast (see 'Container throughput', earlier in this chapter). The author has carried out chi-square, goodness-of-fit tests of actual distributions of such for a few cases. On a daily basis (data were not available on an hourly basis) the actual distributions conform very well to normal distributions, with standard deviations of very close to 0.4 of their mean values. It is suggested that such distributions be used both for containers with cargo and for empties.

Yard area, layout and equipment

The overall flows of containers of different categories into and out of storage, from and to both ships and land transport, determine the numbers of TEUs in storage and thus their statistical distribution over time. The average stacking height, measured in numbers of containers, is determined by dividing the number of TEUs in storage by the number of assumed TEU ground slots. The width of the stack, measured in numbers of containers, is determined by dividing the number of assumed TEU ground slots per stack by the assumed length of the stack, measured in TEUs. Based on the results, the choice of container-yard equipment, such as rubber-tyred gantry cranes or gantry cranes on rails, with regard to maximum stacking height and span (measured in rows of containers plus width for truck and/or rail-car access) would be made. Also a preliminary layout of the stacks, the roadways and the railtracks would be prepared. The numbers of gantry cranes are determined analytically based on the inflows and outflows of containers (one inbound and one outbound movement), assumed numbers of average internal movements per container (see next section) and average handling rates for the gantry cranes.

With straddle carriers as yard equipment (or other equipment used for empties) the approach would, in principle, be the same as the one described above, modified as required if the equipment is also used for transport between apron and yard. The latter considerably reduces the average handling rate, compared with that of a yard gantry crane.

Based on experience, the number of TEU ground slots and stack lengths would be increased or reduced and the simulation series would be repeated, the container-yard equipment determined and the yard layout prepared. The least-cost storage yard and equipment system would be determined by iteration. Thus a mixture of simulation, analytical and empirical solutions would determine the yard area, layout and equipment requirements.

Stacks for inbound containers with cargo and internal movement (shifting) of containers

The containers unloaded from one ship are to be received by a certain number of consignees varying from one to the number of containers, disregarding LCLs. There are two different categories of containers for one consignee:

(a) those which may be delivered to the consignee in arbitrary order, possibly because their contents are identical; and

(b) those which the consignee wants to have delivered in a certain sequence, possibly because certain goods in the containers are needed in the consignee's manufacturing process before other goods which are in other containers.

The delivery sequences are totally unknown to the port planner. Also he or she is not able to forecast the average number of containers for each consignee or the average number of consignees for each arriving vessel.

If a random location in the stack was assumed for every container, which may be the case if there is only one or a few containers per consignee, the containers would on average be located in a layer at a height of half of the average stacking height. Thus the average number of lifts of other containers, to give access to one to be delivered, would be 0.5 for an average stacking height of two. It would be 1.0 for an average stacking height of three, 1.5 for an average stacking height of four and 2.0 for an average stacking height of five, etc. The random location assumption would be quite wrong and the above numbers too high, if there is considerable block stacking. This assumption would also lead to too high numbers of and average distances of horizontal movement of yard equipment per container. What we are dealing with here are large differences, which are going to lead to too high numbers of yard equipment.

Rules and inputs, for computer simulation of the movements of inbound containers with cargo through the stacks, are simply not available. Thus an analytical approach has to be employed. The latter may use empirical data from existing terminals, e.g. the average number of lifts required to give access to a container to be delivered for a certain average stacking height. If applicable empirical data are not available it is proposed that the average for the two extreme cases is used. One extreme case is zero movements of containers inside the stack. The other extreme case corresponds to random location of containers as discussed above. Thus one half of the values corresponding to random location should be used.

Stacks for outbound containers with cargo and for trans-shipment containers as well as for inbound and outbound empties

The above considerations also apply to the other categories of stacks, which are those for outbound containers with cargo, trans-shipment containers, inbound and outbound empties. However, the details and their significance are different.

Outbound containers with cargo are block stacked for each vessel, according to a stowage plan for that vessel. Thus internal movements per container are usually reduced to 0.2 or less. Trans-shipment containers may reasonably be treated as outbound containers with cargo.

Inbound and outbound empties are block stacked. Internal movements are not free of charge for the terminal users, as is the case for containers with cargo. The numbers of such movements are usually low.

Major simulation outputs

1. The major outputs should include, but not necessarily be limited to:
   (a) waiting time for each category of ships caused by non-availability of quay space and non-operation of container quay cranes, respectively;
   (b) total delay in hours;
   (c) number of occurrences of delayed ships;
   (d) average delay in hours for delayed ships;
   (e) average delay in hours for total number of ships that called during the simulation period; and
   (f) average alongside time in hours for total number of ships that called during the simulation period.

2. Total number of each category of ships that called during the simulation period for comparison with input numbers.

3. Average utilisation of quay(s) equal to sum of ship length plus space between ships multiplied by time alongside in hours for each calling ship, divided by the product of available quay length(s) (minus one space at one end of the quay) and the total number of hours during the simulation period.

4. Average utilisation of container quay cranes equal to sum of utilisation hours for cranes divided by the product of the total number of cranes and the total number of hours during the simulation period.

5. Non-availability of container quay cranes equal to sum of crane hours, when one or more cranes were needed, but not available.

6. Total increase of alongside time in hours for ships of each category, caused by non-availability of container quay cranes, equal to sum of differences between the actual alongside time and the alongside time if the maximum number of cranes had been available for unloading and loading of all containers.

7. Number of occurrences of increase of alongside time for ships of each category, caused by non-availability of cranes.

8. Average increase in alongside time in hours for ships of each category, caused by non-availability of cranes.

9. Average increase in alongside time in hours for all calling ships of each category, caused by non-availability of cranes.

10. Waiting time, similar to number 1, when non-availability of quay cranes does not occur and the number of quay cranes required to achieve this.

11. Number of TEUs in storage in each storage yard at intervals of one hour and the corresponding average stacking height measured in containers.

12. Total throughputs of containers of each category during the simulation period for comparison with the input numbers.

Conclusions

It is only feasible to partly simulate the operations of a container terminal for planning purposes (as defined at the beginning of this chapter). The remainder of the required analyses of the facilities requirements may be done analytically and/or empirically.

For computer simulation the following inputs are suggested:

(a) straight-line stretches of quay rather than nominal numbers of berths;

(b) random arrivals of ships for multi-user terminals;

(c) discrete scheduled arrivals of ships for terminals with one or a few users;

(d) container exchanges per ship call corresponding to a normal distribution with a standard deviation of 0.3 to 0.4 of its mean value for *import/export terminals*;

(e) container exchanges per ship call corresponding to a theoretical distribution or discrete values as specified by the designated operator for a *trans-shipment terminal*. Such a terminal cannot rationally be planned without this and other inputs from the designated operator;

(f) container exchanges per ship call for a *mixed import/export and trans-shipment terminal* with segregated distributions and discrete values as outlined above;

(g) a constant distribution of lengths of calling ships with assumed minimum and maximum values as default value, when no better assumptions are available. It does not seem that there is a relationship between mean container exchange values per ship call and ship lengths; and

(h) dwell-time distributions (to be determined), for the containers in the different categories of storage yards, have to be employed to determine the outflows from and inflows to the storage yards, both per time unit, from the landside. Available results from only a few actual cases seem to indicate that the total outflows and inflows per day conform to normal distributions with a standard deviation of 0.4 of their mean value.

Determination of the required yard equipment is not possible by simulation. For inbound containers with cargo only, the two extreme values are known of the average number of lifts per container required to give access for delivery of other containers. One is zero and the other is the one that corresponds to random location of containers within the stack. The random location case is quite easily treated analytically. Thus an average number of required lifts per container has to be assumed and there is clearly no basis for simulation.

The author intends to complete his own hybrid approach and have it programmed, as soon as he has received promised additional actual statistical data and carried out the required goodness-of-fit tests. He invites co-operation of interested parties, who are in general agreement with the principles and methodology advocated above.

References

Crow, E. L. *et al.* (1960) *Statistics Manual.* New York: Dover Publications.

Hansen, J. B. (1972) 'Optimizing ports through computer simulation. Sensitivity analysis of pertinent parameters', *Operational Research Quarterly*, 23(4).

Nam, K. C. *et al.* (2002) 'Simulation study of container terminal performance', *ASCE Journal of Waterway, Port, Coastal and Ocean Engineering*, May/June.

Page, E. (1972) *Queuing Theory in OR.* London: Butterworths.

CHAPTER 4

Economic and financial feasibility

Hans Agerschou

General considerations

Until approximately 25 years ago decisions concerning new construction, extension and improvement of ports were made without a methodical study of the economic and/or financial feasibility of the project, simply because suitable methodology was not available. Since it became available it has been used extensively and thereby developed further. Although it does not give exact results, and never will, as pointed out below it is the only suitable tool, other than in obvious cases, for separating 'good' projects from 'poor' projects. Economic evaluation of projects aims at answering the three questions:

(a) Why do this at all?
(b) Why do it now?
(c) Why do it this way?

This simple, precise wording has been attributed to the late General J. J. Carty, former chief engineer of the New York Telephone Company.

Evaluation of economic feasibility

Economic feasibility evaluation is basically a determination of the streams of annual economic costs of the project and annual economic benefits resulting from the project, over and above the costs and benefits associated with both the 'without-the-project' situation and with the 'next best' alternative project.

Normally the economic costs and benefits considered are those of the country in question. The cost and benefit streams are developed for a relatively long period of time from the start of construction, through the useful life of the project, considering its likely technological or operational obsolescence. Usually a useful life of at least 20 years is assumed. The streams of annual costs and benefits are discounted using any one year as a basis year and one or more rates of interest. Discounting is accomplished by dividing annual values by the sum of one plus the interest rate, to the power corresponding to the number of years, which the year in question comes after the basis year, and similarly multiplying by the same factor for years before the basis year. It is normally not attempted to consider inflation when estimating future costs and benefits, as general inflation would not influence the economic feasibility indicators.

The rate of interest, which results in the sum of discounted costs and benefits being equal to zero, is called the internal rate of return (IRR). For port projects it normally has one unique value as costs are incurred early and benefits result later. It is an indicator of economic feasibility, which does not require the use of an assumed rate of interest. The two other common indicators are the benefit–cost ratio (BCR) and the net present value (NPV). The former is the ratio between the sum of the discounted benefit stream and the sum of the discounted cost stream, while the latter is the sum of the discounted benefit stream minus the sum of the discounted cost stream. They both require the use of an assumed rate of interest, equal to the opportunity cost of capital. The BCR has another disadvantage in that it depends upon whether gross or net annual costs and benefits, or a mixture of the two, are used. For a more detailed discussion of the three indicators, their relative advantages and shortcomings, see Adler (1971).

During recent years the IRR seems to have become the preferred indicator. The IRR, the BCR and the NPV solutions are all readily available as functions in Excel and other spreadsheet software. The first year return, which is the discounted benefit for the first full year of operation of the project divided by the sum of the discounted stream of annual costs through the year in question, is an indicator of whether the timing of the project is reasonable.

Port facilities may be expected to be extended during the period of 20 to 25 years for which the economic evaluation is carried out. Independent of whether or not a certain project is carried out at present, it normally has to be assumed that extensions will take place during the evaluation period, unless diversion of traffic to other ports would be advantageous. If this assumption is not made, benefits from a project may theoretically become unrealistically high. This is avoided by stopping the increase of benefits when the estimated capacity of the project has been reached and keeping the benefits constant thereafter. So-called shadow pricing is sometimes used both for economic costs and benefits. Shadow foreign exchange rates may be employed if the currency of the country in question is considerably overvalued or undervalued. Similarly, shadow wage rates are used for unskilled labour in countries where considerable unemployment exists. The shadow price theory is presented with detailed guidance for its use by Little and Mirrlees (1969).

The sensitivity of the economic feasibility indicators to the estimates and forecasts, which are required to determine the streams of economic costs and benefits, is analysed by decreasing benefits and/or increasing costs as well as by varying basic forecasts. If the sensitivity to more than two variables is appreciable, in other words the project may be questionable from an economic point of view, it may be advisable to employ risk analysis (Pouliquen, 1970). Risk analysis assigns probability distributions to the important variables. This may be done with more or less confidence depending upon the nature of the variables, as well as upon available information about them. Computer simulation, which generates the probability distributions of the variables as subroutines, is used to determine a large number of values of the indicator, as well as their statistical distribution. The result of the risk analysis is thus a probability distribution for the indicator. The major problem with risk analysis is determination of probability distributions for input variables. For an analysis involving high input uncertainty the process of carrying it out may yield insights, which are more important than the formal end results. IBRD and ADB guidelines give the detailed requirements for projects financed by these lending agencies, e.g. ADB (1997).

Economic costs

Economic costs are the total costs of fixed facilities and equipment, their operation and maintenance (over and above the future costs without the project), less all import duties and taxes associated with the project, such as duties and taxes levied on equipment and on fuel used during construction, operation and maintenance of the project.

Capital costs are the costs of construction and equipment, as well as the costs of replacement required during the life of the project. Normally replacement will only be needed for port operating equipment. Depreciation and interest costs should not be included as the full initial costs and replacement costs are capital costs.

Operating and maintenance costs are the annual costs of staff, fuel, electricity, spare parts and other supplies for additional facilities plus or minus the differences in such costs resulting from modification of existing facilities.

Salvage values for both fixed facilities and equipment at the end of the project life are estimated and entered as negative costs or as benefits. The useful project life employed in the economic evaluation is often considerably shorter than the actual life of fixed facilities such as breakwaters and quay structures.

Economic benefits

Basically all benefits result from avoidance or reduction of costs, which would be incurred if the project in question was not implemented. It is important to attempt to determine who the actual beneficiaries are. This is discussed below for each of the major benefit categories, which are as follows.

Reduction of waiting time and service time for ships and their cargoes

Common reasons for port extension and/or improvement are excessive waiting time before a berth becomes available and/or excessive service time at berth. Obviously, if the average service time is reduced the same number of ships may be serviced and leave the port sooner, which also means that the average waiting time is reduced. The relationships, which govern this, are discussed in Chapter 2, where available analytical solutions are shown and the required methodology outlined for cases without analytical solutions. Thus, in the following we assume that the reductions in waiting time and service time resulting from the port project under consideration have been determined. Introduction of one or two additional shifts also reduces service time and thereby reduces waiting time. However, this is clearly not a result of investment in new or improved facilities and should accordingly be considered separately.

How are these time-savings converted into economic benefits? The basic assumption is that when ships, liners as well as chartered ships, spend less time in port, the resulting benefits are equal to the savings in operating costs for the ships in question plus reduced interest on the value of the cargo. Estimated daily operating costs for stationary ships of different categories and numerous other useful data are made available to subscribers by several organisations. However, the data are for new ships with average operating costs and thus not universally valid. The reduction of interest on the cargo requires determination of the average value of the cargo in question, but is otherwise a rather simple matter.

The approach sketched above is clearly valid to some extent, in particular if only benefits to the global economy are considered. It is obviously advantageous to reduce the use of ships for stationary storage of cargo and to achieve earlier delivery of cargo to consignees. However, it becomes much more complicated when we concern ourselves with benefits to the economy of the country in which the port is located. To be sure that the above savings for ships and their cargo are fully and immediately converted into benefits to the country, the following assumptions have to be made:

(a)   the country owns all the ships that call at the port and operates them efficiently;

(b)   no excess carrying capacity exists;

(c)   traffic, which is scheduled by desired frequency of service, rather than by minimum costs, does not exist; and

(d)   consignees pay for all cargo at time of loading in port of origin.

Under these assumptions time-savings for ships would benefit the economy of the country by making the following possible:

(a) reduction or elimination of acquisition of more ships for present or future traffic; and

(b) sale or release for other traffic of ships presently in service.

The assumptions are all unrealistic. Often more than half of the cargo to and from a country is carried on foreign ships. Domestic ships are sometimes inefficiently operated, and scheduled traffic is not uncommon.

To examine further the problems associated with benefits it is necessary to consider the following categories of traffic:

(a) charter (normally bulk cargo);

(b   liner service (normally general cargo); and

(c) scheduled service (normally general cargo and containers).

Foreign ships, which are chartered by entities of the country in question, are either chartered for a specific number of voyages or a specific period of time. For voyage charter the sailing time as well as the turnaround time (waiting time plus service time) in port are specified in the contract. Thus the average daily charter cost is known. For reductions of turnaround time below the specified time (dispatch) the ship owner normally, according to the contract, returns half of the daily cost of the ship, pro-rated on an hourly basis. Similarly, for increase in the turnaround time (demurrage) the ship owner is paid extra according to the full daily cost of the ship, also pro-rated on an hourly basis. Thus the potential benefits are well known in principle, and are also clearly established for ongoing voyage charter contracts. For future voyage charter contracts the estimated benefits would depend upon forecast charter rates. For time charter, reduction of turnaround time is only significant if it makes possible one or more additional voyages during the charter period and thus results in a lower freight rate per tonne carried. For ongoing time charter contracts the potential benefits of reduced time in port are quite clear. For future time charter, the problems associated with estimates of benefits are the same as for future voyage charter. For ships chartered by foreign ship operators, the charter arrangements are the same as those described above. Here the additional problem is the extent to which benefits from reductions of turnaround time will be passed on to the country in question. Commodity export f.o.b. prices under existing contracts, and similar import c.i.f. prices, would not be affected and thus not be a source of benefits. However, in the future, export f.o.b. prices may increase and import c.i.f. prices may decrease, according to future charter rate reductions.

Foreign operators of ships in liner traffic, which call at ports in two certain geographical areas, e.g. Northern Europe and part of Eastern Africa,

43

generally co-operate with each other and with local operators. This is done both in regard to service and to uniform, so-called conference-freight rates, which are often governed by the least efficient line. Conference-freight rates for several ports in an area, even some which are quite far apart, are often the same. Reduction of freight rates for a more efficient port does generally not take place, whereas surcharges are imposed upon less efficient and/or congested ports. Thus the only immediate and obvious possible benefit from reduction of turnaround time is removal of surcharges. Avoidance of future surcharges is a potential source of benefits.

From the point of view of an efficiently operated shipping line the value of reduction of turnaround time is not the reduction of the operating cost of the ship in port, but is related to potential profits from being able to carry more cargo. This value is clearly not a linear function and not even a continuous function of reduction of the turnaround time.

Scheduled domestic carriers are not affected by reduction of turnaround time if they have idle time in port, unless their schedules are changed to the extent that fewer carriers are needed. The reduction in total carrier costs would be the benefit. Fewer non-scheduled domestic carriers in charter or liner traffic may be required if turnaround time is reduced. Thus smaller fleets, now and in the future, are the benefits, but only if excess ships can be used elsewhere, sold or scrapped.

A study (World Bank, 1977) attempted to determine to which extent benefits from reduced turnaround time were actually captured by the countries in which four different port projects were located. The results, which are however subject to considerable uncertainties, varied from not at all to close to completely.

It is concluded that it is very unlikely that turnaround time reductions can be converted fully into benefits to the economy, and it certainly cannot be expected to take place immediately. Further, the value of such time-savings may vary from zero to more than the full operating cost for a new ship in port, depending upon the categories of ships and of traffic as outlined above. If the above detailed analysis is not carried out, it is recommended that the value of one half of the operating cost for new ships during the reduction of the turnaround time be used as the benefit to the economy. It is assumed to be effective as soon as the reduction of turnaround time takes place. This is more realistic than using the full value of reduction of ships' time in port. The above, recommended approach has yielded apparently reasonable results in numerous economic evaluations of port projects.

Reduction of ocean freight rates by providing facilities for larger ships

This type of benefit is normally associated with bulk-cargo terminals. Very considerable reductions of freight rates have resulted from the enormous increases of sizes of bulk carriers, in particular tankers. Linked to this is generally a reduction in service time per tonne of cargo. Cargo-handling rates are increased so that service time remains largely independent of the size of the bulk carrier. It is generally possible for the country in question to benefit fully from this kind of freight rate reductions.

Reduction of cargo-handling costs

This typically results from:

(a) replacing lighterage with alongside facilities, whereby double handling is eliminated; and

(b) improved cargo-handling equipment.

Reductions of cargo-handling costs will directly benefit the country in question. Better organisation of cargo-handling operations, as well as increased labour and equipment productivity also reduce cargo-handling costs, but this is obviously not a result of extended or improved facilities.

Reduction of cargo damage and/or losses

This kind of benefit normally results from:

(a) replacing lighterage with alongside facilities;

(b) containerisation; and

(c) extended storage facilities which will give better protection of and access to cargo, as well as reduce the average number of times cargo is moved while in storage.

These benefits will materialise as lower cargo insurance premiums. More careful handling of cargo and avoidance of vandalism are benefits attributable to training and improved attitudes of cargo-handling staff.

Diversion of cargo from other ports

This benefit type results from reduction of the sum of ocean transport costs, port related handling and storage costs and land transport costs. The quantity of diverted cargo is determined by transportation costs in the form of user charges, whereas the benefits consist of reduced transportation costs to the economy. Normally the country in question would capture most of such benefits.

Economic benefits and financial benefits

Savings on ocean freight rates and on foreign currency insurance premiums are both economic and financial benefits of equal magnitude, which may be passed on to consumers.

It is sometimes argued that the costs of port improvements may always be recovered by increasing port charges for cargo and ships. Although there may not be any immediate reactions to such increases from shipping-line conferences, they may very well result in earlier increases of conference-freight rates for several ports including the one which increased its charges. Thus from the shipping lines' point of view, losses at one port are recovered at several ports and there need not be any admission of a causal relationship between increase of port charges and increase of freight rates. Many other reasons for the latter may be claimed.

Least-cost solutions

Another approach to facilities optimisation, than the economic feasibility indicator calculation, is valid only under certain conditions. It is determination of the least-cost solution, which will fulfil specified traffic and throughput requirements. This approach may be more readily understood than economic feasibility evaluation. The optimum system is the one which will result in the minimum sum of capital costs and operating and maintenance costs plus the value of time spent by ships and their cargoes while waiting for a berth and while at berth. What is meant by costs is the sum of discounted annual costs, or the average annual cost if applicable, including the cost of capital recovery. Grant and Ireson (1964) give a thorough presentation of appropriate methods and pitfalls to be avoided. The costs of port facilities and the economic value of time-savings for cargo are, in principle, not difficult to evaluate. The economic value of time-savings for ships is a more complex matter, which is discussed earlier in this chapter.

In the following it is attempted to draw some general conclusions based on average annual costs of berths and all related facilities as well as average annual costs to the economy of waiting time and service time for cargo and carriers. Figure 4.1 shows schematically annual costs as functions of the number of berths, if it is assumed that the annual cost of berths, etc. may be approximated by a linear function of the number of berths. It is known that the average waiting time approaches infinity when the utilisation factor approaches 1.0. Thus, the time–cost curve has a vertical asymptote for a number of berths equal to the ratio between the average arrival rate for the system and the average service rate per berth. When the number of berths increases the average waiting time approaches zero. Thus, the time–cost curve has a horizontal asymptote for a time cost equal to the cost of service time only. For each system that has a certain average service time, the cost of service time is constant and therefore irrelevant. However, for comparisons between systems with different average service-time costs it has to be included. The minimum total systems cost and thus the optimum number of berths is found where the

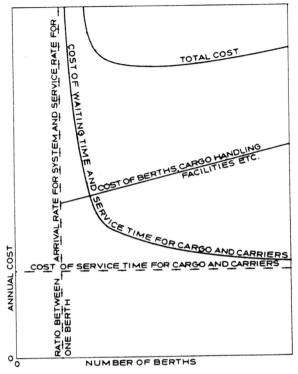

Fig. 4.1 Schematic cost curves for port or terminal system. Reproduced by permission of the Dock & Harbour Authority

inclinations of the two component curves are numerically equal. The total–cost curve is often rather flat around its minimum value. For a range of berths corresponding to relatively high utilisation factors the time–cost curve is much steeper than the berth–cost curve. Thus, it may be advisable to choose a slightly higher number of berths than the calculated optimum number, to avoid sharply increasing costs of ships' waiting time, if the actual future traffic exceeds the forecast traffic.

Financial feasibility

General considerations

There are many similarities between economic analysis and financial analysis of projects. The latter is done from the point of view of the owner of the facilities in question. The main difference is that financial costs are used instead of economic costs, that is taxes, import duties and other levies are included in the costs. A number of years ago the return on net fixed assets was widely used. However, the results were difficult to evaluate, as the benchmarks were to a large extent arbitrary.

Evaluation of financial feasibility

The return on net fixed assets' method has been replaced by the NPV and in particular by the IRR method. The IRR is exclusively concerned with cash flow of revenue and costs. It may be carried out with or *without inclusion of forecast company taxes. Depreciation is clearly not part of it except indirectly when used as a deduction for estimating company taxes as part of the analysis.*

References

ADB (1997) *Guidelines for Economic Analyses of Projects.*

Adler, H. A. (1971) *Economic Appraisal of Transport Projects*, Bloomington, Indiana and London: Indiana University Press.

Grant, E. L. and Ireson, W. (1964) *Principles of Engineering Economy*, New York: The Ronald Press Co.

Little, I. M. D. and Mirrlees, J. A. (1969) *Manual of Industrial Project Analysis in Developing Countries*, Paris: OECD Development Centre.

Pouliquen, L. Y. (1970) 'Risk analysis in project appraisal', *World Bank Staff Occasional Paper No. 11*, Baltimore and London: The Johns Hopkins Press.

World Bank (1977) *'Distributions of Benefits of Port Improvements'*, Case Studies of La Goulette, Pisco Douala and Karachi.

CHAPTER 5

Ships and their influence on port facilities

Hans Agerschou and Jakob Buus Petersen

Introduction

Existing ships, ships being built and ships which are known or expected to be built in the near future are bound to influence plans for improvement and new construction of port and terminal facilities. For very large bulk carriers the specialised terminals for loading and unloading have to be tailored to the requirements of the carriers. On the other hand, existing facilities and physical, economic or financial constraints, in regard to their improvement and extension, will influence the types and sizes of ships which will use the facilities at present and in the future. Transportation costs per tonne–nautical mile decrease with increasing carrying capacity of ships and port handling costs decrease with increased unitisation. Thus small ports and terminals as well as smaller ships and ships not suited for carrying unitised cargo become obsolete, except where short-haul services, in particular roll-on/roll-off, are still feasible.

The relevant characteristics of ships and the port characteristics that they influence are:

(a) main dimensions:
   (i)   length, which governs the length and layout of single-berth terminals, the length of stretches of quay, etc. The length also influences the widths and bends of channels and the size of port basins;
   (ii) beam or breadth, which governs the reach of cargo-handling equipment and influences the width of channels and basins; and
   (iii) draught, which governs the water depth along berths, in channels and in basins.
(b) cargo-carrying capacity, which governs the (minimum) storage requirement for full ship loads and may influence the handling rate for loading/unloading installations;
(c) cargo-handling gear, such as cranes and pumps, which govern cargo-handling rates, in particular for liquid bulk cargo, and which

influence the need for port equipment, such as quay cranes and booster pumps. Cranes, as well as derricks and winches, also restrict the ship motion that may be tolerated without interruption of cargo-handling operations;

(d) types of cargo units, such as bulk, containers, pallets and break-bulk general cargo, which determine requirements for handling equipment (if ship's gear is not used) and storage;

(e) shape, hull strength and motion characteristics:

  (i) size and shape of hull and superstructure, which influence berth and fender system layout, as well as current and wind forces required for design of moorings and fenders;

  (ii) ultimate strength of ships' side shell and supporting members, which influences fender design;

  (iii) ship motions induced by wind waves, swell or long waves, which influence mooring and fender designs; and

  (iv) superstructure configuration, which influences positioning and design of port handling equipment to avoid damage to superstructure and handling equipment.

(f) mooring equipment, such as ropes, wires and constant tension winches, which influences the motion of ships and their mooring forces; and

(g) manoeuvrability at low speed, which influences channel, port entrance and basin layout, as well as the need for harbour tugs.

Main dimensions

General considerations

Important developments have taken place during the past two decades in regard to size, layout, and equipment of ships. The increase of very large super-tanker sizes has stagnated at about 500 000 DWT. There are very few dry bulk carriers larger than 225 000 DWT. The increase in container-ship sizes has continued. Post-Panamax container ships (the beam not limited by the width of the locks of the Panama Canal) have come into being. LASH (lighter aboard ship) carriers have not developed to a significant extent.

Main dimensions

The first edition of this book, published in 1983, included data from the Lloyd's Register of Shipping *Statistical Tables* (1980). The tables showed the relationships between Gross Register Tonnage (GRT) and the three main dimensions for the existing world fleets of the categories of tankers, dry bulk carriers as well as general cargo and passenger ships taken together. Similar tabulations based on Dead Weight Tonnage (DWT), as a more desirable parameter (see next section) were not readily available. The first edition also included original tables, which showed the relationships between Twenty-Foot Equivalent Unit (TEU) carrying capacity and the three main dimensions for cellular container ships.

Tables 5.1, 5.2, 5.3 and 5.4 show the main dimensions for oil tankers, dry-bulk carriers, general cargo ships and cellular container ships, respectively. The number of ships within a certain dimension interval is shown for DWT intervals, except for container ships for which TEU carrying capacity intervals are used. The data, compiled by Lloyd's Register of Shipping, were valid as at 31 December 1999.

The tables demonstrate that there are no well-defined relationships between DWT and TEU capacity, respectively, and each of the main dimensions. Graphs of these relationships would show clouds of plotted points only. We consider tables more useful as they readily yield the statistical distribution of each main dimension over the intervals.

Because of the change from GRT to DWT a comparison of the 1980 and the 1999 end of year world fleets of tankers and dry-bulk carriers cannot be made directly. However, comparison for larger ships, for which GRT is approximately equal to 0.5 DWT (see next section), is mainly of interest.

Back in 1980 there were 879 tankers of about 140 000 DWT and more. This number has decreased to 605. For tankers of about 200 000 DWT and more the corresponding number has decreased from 675 to 448. However, for tankers of about 280 000 DWT and more the number has increased from 142 to more than 200. For the latter category a breakdown above 280,000 DWT was not available in Lloyd's *Statistical Tables* for 1980. At the end of year 1999 there were 10 tankers of 450 000 DWT or more, 25 of 400 000 DWT or more and 126 of 300 000 DWT or more. The 1999 fleet has a pronounced anomaly for tankers from 175 000 to 225 000 DWT, of which there are only 13, compared with about 200 in 1980. The reason is not known to the author.

In 1980 there were 401 dry bulk carriers of about 100 000 DWT or more. This number had increased to 484 by 1999. For carriers of about 140 000 DWT and more the number had increased from 177 to 390 and for carriers of about 200 000 DWT and more it had increased from 34 to 46. However, only 18 of the latter were larger than 225 000 DWT. The 1999 fleet has a pronounced anomaly for dry-bulk carriers from 90 000 to 120 000 DWT, of which there are only 37, compared with more than 150 in 1980. Again, the reason is not known to the author. There are no dry-bulk carriers in the interval from 275 000 to 299 999 DWT.

A similar comparison for general cargo ships is neither possible nor of great interest.

The 1980 container-ship tables were based on a sample of 457 ships. Thus a comparison with the 1999 fleet cannot be made in numbers but

Table 5.1 DWT and dimensions of oil tankers

Oil tankers – number of vessels in divisions of DWT

Overall length (Division in metres)	6000–9999	10 000–19 999	20 000–29 999	30 000–39 999	40 000–49 999	50 000–59 999	60 000–69 999	70 000–79 999	80 000–89 999	90 000–99 999	100 000–109 999	110 000–119 999	120 000–129 999	130 000–139 999	140 000–149 999	150 000–174 999	175 000–199 999	200 000–224 999	225 000–249 999	250 000–274 999	275 000–299 999	300 000–324 999	325 000–349 999	350 000–374 999	375 000–399 999	400 000–424 999	450 000–474 999	475 000 or over
120–124	13	8	–	–	–	–	–	–	–	–	–	–	–	–	–	–	–	–	–	–	–	–	–	–	–	–	–	–
125–129	13	8	–	–	–	–	–	–	–	–	–	–	–	–	–	–	–	–	–	–	–	–	–	–	–	–	–	–
130–134	11	14	–	–	–	–	–	–	–	–	–	–	–	–	–	–	–	–	–	–	–	–	–	–	–	–	–	–
135–139	3	20	1	–	–	–	–	–	–	–	–	–	–	–	–	–	–	–	–	–	–	–	–	–	–	–	–	–
140–144	1	39	–	–	–	–	–	–	–	–	–	–	–	–	–	–	–	–	–	–	–	–	–	–	–	–	–	–
145–149	–	15	–	–	–	–	–	–	–	–	–	–	–	–	–	–	–	–	–	–	–	–	–	–	–	–	–	–
150–154	–	30	14	–	–	–	–	–	–	–	–	–	–	–	–	–	–	–	–	–	–	–	–	–	–	–	–	–
155–159	–	34	8	–	–	–	–	–	–	–	–	–	–	–	–	–	–	–	–	–	–	–	–	–	–	–	–	–
160–164	–	69	21	–	–	–	–	–	–	–	–	–	–	–	–	–	–	–	–	–	–	–	–	–	–	–	–	–
165–169	–	5	113	12	–	–	–	–	–	–	–	–	–	–	–	–	–	–	–	–	–	–	–	–	–	–	–	–
170–174	–	2	79	117	5	–	–	–	–	–	–	–	–	–	–	–	–	–	–	–	–	–	–	–	–	–	–	–
175–179	–	3	12	47	38	–	–	–	–	–	–	–	–	–	–	–	–	–	–	–	–	–	–	–	–	–	–	–
180–184	–	–	6	61	105	1	–	–	–	–	–	–	–	–	–	–	–	–	–	–	–	–	–	–	–	–	–	–
185–189	–	–	3	25	5	–	–	–	–	–	–	–	–	–	–	–	–	–	–	–	–	–	–	–	–	–	–	–
190–194	–	–	–	27	21	6	–	–	–	–	–	–	–	–	–	–	–	–	–	–	–	–	–	–	–	–	–	–
195–199	–	–	1	14	9	3	–	–	–	–	–	–	–	–	–	–	–	–	–	–	–	–	–	–	–	–	–	–
200–204	–	–	–	12	13	3	3	–	–	–	–	–	–	–	–	–	–	–	–	–	–	–	–	–	–	–	–	–
205–209	–	–	–	11	2	22	3	–	–	–	–	–	–	–	–	–	–	–	–	–	–	–	–	–	–	–	–	–
210–214	–	–	–	–	–	19	5	–	–	–	–	–	–	–	–	–	–	–	–	–	–	–	–	–	–	–	–	–
215–219	–	–	–	3	2	8	17	2	–	–	–	–	–	–	–	–	–	–	–	–	–	–	–	–	–	–	–	–
220–224	–	–	–	1	–	2	16	–	2	–	–	–	–	–	–	–	–	–	–	–	–	–	–	–	–	–	–	–
225–229	–	–	–	–	–	11	78	3	–	–	–	–	–	–	–	–	–	–	–	–	–	–	–	–	–	–	–	–
230–234	–	–	–	–	–	2	5	7	–	–	–	–	–	–	–	–	–	–	–	–	–	–	–	–	–	–	–	–
235–239	–	–	–	–	–	2	13	1	1	–	–	–	–	–	–	–	–	–	–	–	–	–	–	–	–	–	–	–
240–244	–	–	–	–	–	2	22	9	18	34	8	–	–	–	–	–	–	–	–	–	–	–	–	–	–	–	–	–
245–249	–	–	–	–	–	–	6	4	50	4	71	–	–	–	–	–	–	–	–	–	–	–	–	–	–	–	–	–
250–254	–	–	–	–	–	–	–	7	81	84	25	6	–	1	–	–	–	–	–	–	–	–	–	–	–	–	–	–
255–259	–	–	–	–	–	–	–	–	48	48	1	9	3	–	–	–	–	–	–	–	–	–	–	–	–	–	–	–
260–264	–	–	–	–	–	–	–	–	2	2	3	3	1	3	–	–	–	–	–	–	–	–	–	–	–	–	–	–
265–269	–	–	–	–	–	–	–	–	2	10	1	4	6	1	–	–	3	–	–	–	–	–	–	–	–	–	–	–
270–274	–	–	–	–	–	–	–	–	2	2	–	3	33	16	22	6	–	–	–	–	–	–	–	–	–	–	–	–
275–279	–	–	–	–	–	–	–	–	–	–	–	–	13	11	49	35	–	2	–	1	1	–	–	–	–	–	–	–
280–284	–	–	–	–	–	–	–	–	–	1	–	3	5	13	6	5	3	–	–	–	–	–	–	–	–	–	–	–
285–289	–	–	–	–	–	–	–	–	–	5	–	–	–	3	6	3	–	1	–	–	–	–	–	–	–	–	–	–
290–294	–	–	–	–	–	–	–	–	–	–	–	–	–	–	1	10	–	–	–	–	–	–	–	–	–	–	–	–
295–299	–	–	–	–	–	–	–	–	–	–	–	–	–	–	–	–	–	–	25	–	–	–	–	–	–	–	–	–
300–304	–	–	–	–	–	–	–	–	–	–	–	–	–	–	–	1	3	–	15	3	–	7	–	–	–	–	–	–
305–309	–	–	–	–	–	–	–	–	–	–	–	–	–	–	–	–	–	–	–	33	13	–	5	–	–	–	–	–
310–314	–	–	–	–	–	–	–	–	–	–	–	–	–	–	–	4	–	–	4	8	28	–	–	–	–	–	–	–
315–319	–	–	–	–	–	–	–	–	–	–	–	–	–	–	–	–	–	–	1	34	46	50	–	–	–	–	–	–
320–324	–	–	–	–	–	–	–	–	–	–	–	–	–	–	–	–	–	1	2	57	11	2	–	–	–	–	–	–
325–329	–	–	–	–	–	–	–	–	–	–	–	–	–	–	–	–	–	–	–	9	20	10	–	–	–	–	–	–
330–334	–	–	–	–	–	–	–	–	–	–	–	–	–	–	–	–	–	–	–	7	–	4	–	–	–	–	–	–
335–339	–	–	–	–	–	–	–	–	–	–	–	–	–	–	–	–	–	–	–	–	–	5	–	2	–	–	–	–
340–344	–	–	–	–	–	–	–	–	–	–	–	–	–	–	–	–	–	–	–	–	–	–	–	12	3	15	4	6
345–349	–	–	–	–	–	–	–	–	–	–	–	–	–	–	–	–	–	–	–	–	–	–	1	–	–	–	–	–
350–354	–	–	–	–	–	–	–	–	–	–	–	–	–	–	–	–	–	–	–	–	–	–	–	–	–	–	–	–
360 or over	–	–	–	–	–	–	–	–	–	–	–	–	–	–	–	–	–	–	–	–	–	–	–	–	–	–	–	–
Totals	41	247	258	330	200	81	168	33	206	190	109	28	61	70	84	64	9	4	47	152	119	78	6	14	3	15	4	6

Table 5.1 (continued)

Oil tankers – number of vessels in divisions of DWT

Extreme breadth / Division in metres	6000–9999	10 000–19 999	20 000–29 999	30 000–39 999	40 000–49 999	50 000–59 999	60 000–69 999	70 000–79 999	80 000–89 999	90 000–99 999	100 000–109 999	110 000–119 999	120 000–129 999	130 000–139 999	140 000–149 999	150 000–174 999	175 000–199 999	200 000–224 999	225 000–249 999	250 000–274 999	275 000–299 999	300 000–324 999	325 000–349 999	350 000–374 999	375 000–399 999	400 000–424 999	450 000–474 999	475 000 or over
15–16	16																											
17–18	18	30																										
19–20	7	48																										
21–22		95	20																									
23–24		41	44	2																								
25–26		28	163	141																								
27–28		5	24	105	17	2																						
29–30			7	46	57	3																						
31–32				36	126	67	137	10																				
33–34						1	2	1	2																			
35–36						8	23	5	3	5																		
37–38							3	11	9	4	2	7	3															
39–40							1	3	17	5	6	13	22	1														
41–42							2	3	67	24	97	6	18	4	1													
43–44									81	125	3	2	4	18	45	15												
45–46									24	17	1		11	28	21	14												
47–48									1	10			1	12	17	16	1	1	1									
49–50									2				1	5		9	3	2	3	19	9							
51–52													1	2		7	5	1	15	54	16	1						
53–54																3			5	23	25	19	5					
55–56																			14	34	57	52	1	11				
57–58																			9	22	12	6		3	3			
59–60																										15		
61 or over																											4	6
Totals	41	247	258	330	200	81	168	33	206	190	109	28	61	70	84	64	9	4	47	152	119	78	6	14	3	15	4	6

Table 5.1 (continued)

Oil tankers – number of vessels in divisions of DWT

Draught (Division in metres)	6000–9999	10 000–19 999	20 000–29 999	30 000–39 999	40 000–49 999	50 000–59 999	60 000–69 999	70 000–79 999	80 000–89 999	90 000–99 999	100 000–109 999	110 000–119 999	120 000–129 999	130 000–139 999	140 000–149 999	150 000–174 999	175 000–199 999	200 000–224 999	225 000–249 999	250 000–274 999	275 000–299 999	300 000–324 999	325 000–349 999	350 000–374 999	375 000–399 999	400 000–424 999	450 000–474 999	475 000 or over
6.0–6.49	6																											
6.5–6.99	14	15																										
7.0–7.49	10	47	1																									
7.5–7.99	9	30																										
8.0–8.49	1	33		1	1																							
8.5–8.99	1	47	5																									
9.0–9.49		65	30	4																								
9.5–9.99		8	51	8																								
10.0–10.49		2	53	30	4																							
10.5–10.99			63	80	18	5																						
11.0–11.49			53	151	21	13	3	1	2																			
11.5–11.99				30	33	36	10	2	3	1																		
12.0–12.49			1	25	79	18	43	6	43	14																		
12.5–12.99			1		42	7	45	7	39	14	1																	
13.0–13.49					2	2	27	7	45	30	2		1			3												
13.5–13.99							39	3	43	65	7																	
14.0–14.49				1			1	3	18	43	16		1	2														
14.5–14.99								1	12	17	66		2	1														
15.0–15.49								2	1	2	10	5		2	3													
15.5–15.99										4	7	10	8	19	7													
16.0–16.49													17	9	17	2												
16.5–16.99								1				8	3	25	32	6												
17.0–17.49													23	8	16	11	1											
17.5–17.99												5	6	4	9	11	1											
18.0–18.49																20	1											
18.5–18.99																10	2											
19.0–19.49																1	4											
19.5–19.99																			2	1								
20.0–20.49																		1	11	12								
20.5–20.99																		2	12	30	1							
21.0–21.49																			12	34	14							
21.5–21.99																			2	17	29							
22.0–22.49																		1	8	34	18	2						
22.5–22.99																				21	31	7	1			1		
23.0–23.49																				2	26	42	4	11		2		
23.5–23.99																				1		26	1	1	3	10		
24.0–24.49																						1		2		2		1
24.5–24.99																												1
25.0–25.49																											4	2
26 or over																												2
Totals	41	247	258	330	200	81	168	33	206	190	109	28	61	70	84	64	9	4	47	152	119	78	6	14	3	15	4	6

Source: Lloyd's Register of Shipping as at 31 December 1999.

Table 5.2 DWT and dimensions of ore and bulk carriers

Dry bulk carriers – number of vessels in divisions of DWT

Overall length (Division in metres)	6000–9999	10 000–19 999	20 000–29 999	30 000–39 999	40 000–49 999	50 000–59 999	60 000–69 999	70 000–79 999	80 000–89 999	90 000–99 999	100 000–109 999	110 000–119 999	120 000–129 999	130 000–139 999	140 000–149 999	150 000–174 999	175 000–199 999	200 000–224 999	225 000–249 999	250 000–274 999	300 000–324 999	350 000–374 999
120–124	9	17																				
125–129	7	26																				
130–134	11	23																				
135–139	10	45																				
140–144		81	1																			
145–149	1	171	16																			
150–154		76	93	1																		
155–159		47	127	1																		
160–164		71	189	5																		
165–169		3	195	22	2																	
170–174		1	168	175	17																	
175–179		2	155	136	129	1																
180–184		5	203	232	354	1																
185–189		4	61	61	117	4																
190–194		3	16	132	103	5																
195–199		1	21	37	14	9	1															
200–204		1	1	3	15	4																
205–209			3	5	4	49	3															
210–214			3	1	6	23	2															
215–219			3	41	7	6	11															
220–224			34		1	5	243	17	1													
225–229			3				204	88	5													
230–234			4	2		6	11	216	4	2												
235–239					1	1	11	21	1	4	1	1		1								
240–244								4	5	4	3	2	4	6								
245–249			1				2	30	2	1	1	4	14	6	3	1						
250–254							20	8	7	2	1	7	13	16	31	6						
255–259								2				1	3		48	48	1					
260–264								5		1						8						
265–269						1					1						6					
270–274														5	8	40	20					
275–279						1								2	1	66	11					
280–284														2		25	7					
285–289																4	2	4				
290–294															1	6		2				
295–299																						
300–304							6	1							1			6	1			
305–309							1	1	2	1								13	2			
310–314																		2				
315–319																		1				
320–324																			5	1		
325–329																				5	2	
330–334																					1	
335–339																						
340–344																						1
Totals	38	577	1297	854	770	115	515	393	27	15	7	15	34	38	93	204	47	28	8	6	3	1

Table 5.2 (continued)

Dry bulk carriers – number of vessels in divisions of DWT

Extreme breadth Division in metres	6000–9999	10 000–19 999	20 000–29 999	30 000–39 999	40 000–49 999	50 000–59 999	60 000–69 999	70 000–79 999	80 000–89 999	90 000–99 999	100 000–109 999	110 000–119 999	120 000–129 999	130 000–139 999	140 000–149 999	150 000–174 999	175 000–199 999	200 000–224 999	225 000–249 999	250 000–274 999	300 000–324 999	350 000–374 999	Totals
15–16	5	–	–	–	–	–	–	–	–	–	–	–	–	–	–	–	–	–	–	–	–	–	5
17–18	25	14	–	–	–	–	–	–	–	–	–	–	–	–	–	–	–	–	–	–	–	–	39
19–20	8	149	6	–	–	–	–	–	–	–	–	–	–	–	–	–	–	–	–	–	–	–	
21–22	–	367	402	29	–	–	–	–	–	–	–	–	–	–	–	–	–	–	–	–	–	–	
23–24	–	44	322	114	–	–	–	–	–	–	–	–	–	–	–	–	–	–	–	–	–	–	
25–26	–	2	444	157	–	–	–	–	–	–	–	–	–	–	–	–	–	–	–	–	–	–	
27–28	–	1	121	424	53	5	–	–	–	–	–	–	–	–	–	–	–	–	–	–	–	–	
29–30	–	–	1	93	415	15	–	–	–	–	–	–	–	–	–	–	–	–	–	–	–	–	
31–32	–	–	–	37	300	88	510	374	10	–	–	–	–	–	–	–	–	–	–	–	–	–	
33–34	–	–	1	–	2	1	–	–	–	2	–	–	–	–	–	–	–	–	–	–	–	–	
35–36	–	–	–	–	–	3	5	19	2	1	–	–	–	–	–	–	–	–	–	–	–	–	
37–38	–	–	–	–	–	1	–	–	9	3	1	–	–	–	–	–	–	–	–	–	–	–	
39–40	–	–	–	–	–	–	–	–	–	–	–	3	–	–	–	–	–	–	–	–	–	–	
41–42	–	–	–	–	–	2	–	–	1	2	4	10	24	–	–	–	–	–	–	–	–	–	
43–44	–	–	–	–	–	–	–	–	5	7	2	1	9	10	3	–	–	–	–	–	–	–	
45–46	–	–	–	–	–	–	–	–	–	–	–	1	1	27	81	75	13	–	–	–	–	–	
47–48	–	–	–	–	–	–	–	–	–	–	–	–	–	1	3	121	24	–	–	–	–	–	
49–50	–	–	–	–	–	–	–	–	–	–	–	–	–	–	6	8	10	26	–	–	–	–	
51–52	–	–	–	–	–	–	–	–	–	–	–	–	–	–	–	–	–	–	1	1	–	–	
53–54	–	–	–	–	–	–	–	–	–	–	–	–	–	–	–	–	–	2	4	4	–	–	
55–56	–	–	–	–	–	–	–	–	–	–	–	–	–	–	–	–	–	–	2	1	1	–	
57–58	–	–	–	–	–	–	–	–	–	–	–	–	–	–	–	–	–	–	1	–	–	–	
61 or over	–	–	–	–	–	–	–	–	–	–	–	–	–	–	–	–	–	–	–	–	2	1	
Totals	38	577	1297	854	770	115	515	393	27	15	7	15	34	38	93	204	47	28	8	6	3	1	

Table 5.2 (continued)

Dry bulk carriers – number of vessels in divisions of DWT

Draught (Division in metres)	6000–9999	10 000–19 999	20 000–29 999	30 000–39 999	40 000–49 999	50 000–59 999	60 000–69 999	70 000–79 999	80 000–89 999	90 000–99 999	100 000–109 999	110 000–119 999	120 000–129 999	130 000–139 999	140 000–149 999	150 000–174 999	175 000–199 999	200 000–224 999	225 000–249 999	250 000–274 999	300 000–324 999	350 000–374 999
6.0–6.49	10	2	–	–	–	–	–	–	–	–	–	–	–	–	–	–	–	–	–	–	–	–
6.5–6.99	8	5	2	–	–	–	–	–	–	–	–	–	–	–	–	–	–	–	–	–	–	–
7.0–7.49	6	16	1	–	–	–	–	–	–	–	–	–	–	–	–	–	–	–	–	–	–	–
7.5–7.99	11	49	6	–	–	–	–	–	–	–	–	–	–	–	–	–	–	–	–	–	–	–
8.0–8.49	3	97	49	3	–	–	4	–	–	–	–	–	–	–	–	–	–	–	–	–	–	–
8.5–8.99	–	90	13	12	1	–	2	2	–	–	–	–	–	–	–	–	–	–	–	–	–	–
9.0–9.49	–	213	87	15	1	–	–	–	2	–	–	–	–	–	–	–	–	–	–	–	–	–
9.5–9.99	–	99	609	48	3	–	2	–	–	1	–	–	–	–	–	–	–	–	–	–	–	–
10.0–10.49	–	4	335	65	17	2	–	–	–	–	–	–	–	–	–	–	–	–	–	–	–	–
10.5–10.99	–	–	186	404	117	5	–	–	–	–	–	–	–	–	–	–	–	–	–	–	–	–
11.0–11.49	–	2	5	270	319	7	2	–	–	–	–	–	–	–	–	–	–	–	–	–	–	–
11.5–11.99	–	–	3	28	257	76	1	1	2	–	–	–	–	–	–	–	–	–	–	–	–	–
12.0–12.49	–	–	1	9	52	17	87	14	6	1	–	–	–	–	–	–	–	–	–	–	–	–
12.5–12.99	–	–	–	–	2	8	106	15	1	6	–	–	–	–	–	–	–	–	–	–	–	–
13.0–13.49	–	–	–	–	1	–	280	95	7	2	1	1	–	–	–	–	–	–	–	–	–	–
13.5–13.99	–	–	–	–	–	–	30	203	9	2	1	–	–	4	–	–	–	–	–	–	–	–
14.0–14.49	–	–	–	–	–	–	1	59	–	1	3	–	10	–	2	–	–	–	–	–	–	–
14.5–14.99	–	–	–	–	–	–	–	4	–	2	1	–	1	–	4	–	–	–	–	–	–	–
15.0–15.49	–	–	–	–	–	–	–	–	–	–	1	5	11	7	5	–	–	–	–	–	–	–
15.5–15.99	–	–	–	–	–	–	–	–	–	–	–	6	6	21	10	2	–	–	–	–	–	–
16.0–16.49	–	–	–	–	–	–	–	–	–	–	–	3	2	6	60	26	1	–	–	–	–	–
16.5–16.99	–	–	–	–	–	–	–	–	–	–	–	–	4	–	12	54	15	1	–	–	–	–
17.0–17.49	–	–	–	–	–	–	–	–	–	–	–	–	–	–	–	104	29	2	–	–	–	–
17.5–17.99	–	–	–	–	–	–	–	–	–	–	–	–	–	–	–	16	2	18	5	–	–	–
18.0–18.49	–	–	–	–	–	–	–	–	–	–	–	–	–	–	–	2	–	2	1	–	–	–
18.5–18.99	–	–	–	–	–	–	–	–	–	–	–	–	–	–	–	–	–	5	2	–	–	–
19.0–19.49	–	–	–	–	–	–	–	–	–	–	–	–	–	–	–	–	–	–	–	2	–	–
19.5–19.99	–	–	–	–	–	–	–	–	–	–	–	–	–	–	–	–	–	–	–	1	–	–
20.0–20.49	–	–	–	–	–	–	–	–	–	–	–	–	–	–	–	–	–	–	–	3	3	–
20.5–20.99	–	–	–	–	–	–	–	–	–	–	–	–	–	–	–	–	–	–	–	–	–	–
23.0–23.49	–	–	–	–	–	–	–	–	–	–	–	–	–	–	–	–	–	–	–	–	–	1
Totals	38	577	1297	854	770	115	515	393	27	15	7	15	34	38	93	204	47	28	8	6	3	1

Source: Lloyd's Register of Shipping as at 31 December 1999.

57

Table 5.3 DWT and dimensions of general cargo ships

Division in metres Overall length	General cargo – number of vessels in divisions of DWT										
	6000–6999	7000–7999	8000–8999	9000–9999	10 000–14 999	15 000–19 999	20 000–24 999	25 000–29 999	30 000–39 999	40 000–49 999	50 000–59 999
100–104	88	38	57	1	–	–	–	–	–	–	–
105–109	290	100	22	2	4	–	–	–	–	–	–
110–114	56	51	25	41	6	–	–	–	–	–	–
115–119	35	89	85	13	11	–	–	–	–	–	–
120–124	57	64	25	15	23	–	–	–	–	–	–
125–129	4	31	61	39	76	–	–	–	–	–	–
130–134	32	32	89	32	61	–	–	–	–	–	–
135–139	–	39	9	5	71	26	–	–	–	–	–
140–144	–	2	2	1	93	176	1	–	–	–	–
145–149	–	–	2	1	92	167	6	–	–	–	–
150–154	–	–	–	8	143	97	28	–	–	–	–
155–159	1	1	–	3	73	127	57	–	–	–	–
160–164	–	–	–	1	73	79	75	2	–	–	–
165–169	–	–	1	–	5	50	52	1	1	–	–
170–174	–	–	–	–	1	40	33	28	4	–	–
175–179	–	–	–	–	1	17	78	30	9	–	–
180–184	–	–	–	–	–	3	24	11	8	–	–
185–189	–	–	–	–	–	3	3	12	38	19	–
190–194	–	–	–	–	–	–	6	–	11	33	5
195–199	–	–	–	–	–	–	–	–	3	–	10
200–204	–	–	–	–	–	2	–	1	3	39	–
205–209	–	–	–	–	–	–	2	–	–	10	–
210–214	–	–	–	–	–	–	–	–	–	2	2
215–219	–	–	–	–	–	–	–	–	–	1	–
Totals	563	447	378	162	733	787	365	85	77	105	17

Table 5.3 (continued)

General cargo – number of vessels in divisions of DWT

Division in metres	6000–6999	7000–7999	8000–8999	9000–9999	10 000–14 999	15 000–19 999	20 000–24 999	25 000–29 999	30 000–39 999	40 000–49 999	50 000–59 999
Extreme breadth											
13–14	9	–	–	1	–	–	–	–	–	–	–
15–16	375	56	17	11	–	–	–	–	–	–	–
17–18	169	309	271	53	101	149	–	–	–	–	–
19–20	9	80	84	93	337	431	1	–	–	–	–
21–22	1	2	6	4	276	182	107	–	–	–	–
23–24	–	–	–	–	19	23	104	1	13	–	–
25–26	–	–	–	–	–	2	111	51	19	–	–
27–28	–	–	–	–	–	–	41	33	44	67	2
29–30	–	–	–	–	–	–	1	–	1	38	15
31–32	–	–	–	–	–	–	–	–	–	–	–
Totals	563	447	378	162	733	787	365	85	77	105	17
Draught											
6.0–6.49	39	24	3	–	–	1	–	–	–	–	–
6.5–6.99	400	60	11	1	4	–	–	–	–	–	–
7.0–7.49	91	186	119	28	74	3	–	–	–	–	–
7.5–7.99	29	160	95	54	163	–	–	–	–	–	–
8.0–8.49	3	16	144	47	178	10	4	–	–	–	–
8.5–8.99	1	–	6	29	229	123	14	–	–	–	–
9.0–9.49	–	1	–	3	75	298	112	43	1	–	–
9.5–9.99	–	–	–	–	9	199	196	21	4	2	–
10.0–10.49	–	–	–	–	1	122	31	14	17	2	–
10.5–10.99	–	–	–	–	–	28	5	2	13	4	–
11.0–11.49	–	–	–	–	–	1	3	5	38	52	8
11.5–11.99	–	–	–	–	–	2	–	–	3	43	9
12.0–12.49	–	–	–	–	–	–	–	–	1	2	–
12.5 or over	–	–	–	–	–	–	–	–	–	–	–
Total	563	447	378	162	733	787	365	85	77	105	17

Source: Lloyd's Register of Shipping as at 31 December 1999.

59

Table 5.4 Carrying capacity in TEUs and dimensions of cellular container ships

Container ships – number of vessels in divisions of TEU capacity

Overall length (Division in metres)	0–99	100–199	200–299	300–399	400–499	500–599	600–699	700–799	800–899	900–999	1000–1099	1100–1199	1200–1299	1300–1399	1400–1499	1500–1599	1600–1699	1700–1799	1800–1899	1900–1999	2000–2099	2100–2199	2200–2299	2300–2399	2400–2499	2500–2599	2600–2699	2700–2799	2800–2899	2900–2999	3000–3099	3100–3199	3200–3399	3400–3599	3600–3799	3800–3999	4000–4199	4200–4399	4400–4599	4600–4799	4800–4999	5000–5499	5500–5999	6000–6499	6500–6999
60–64	1	1																																											
65–69	3	2																																											
70–74	3	1																																											
75–79	3	12																																											
80–84	1	16																																											
85–89	1	18	1	4																																									
90–94	1	8	7	5																																									
95–99	1	8	1	16																																									
100–104		4	6	13	8	6																																							
105–109		9	34	17	10	2	1																																					
110–114		4	17	14	15	3																																							
115–119		7	25	14	9																																								
120–124	2		6	24	8	2	14																																						
125–129		2	10	26	9	14	19																																						
130–134			6	12	10	11	10	2																																					
135–139	4		3	6	13	16	13	13																																					
140–144					2	9	2	9																																					
145–149		2					7	16																																					
150–154				9		2	8	9	2																																				
155–159						9	7	7	13																																				
160–164						2	2	2	9	11																																			
165–169								3	10	7	7																																		
170–174							3	2	2	5	4																																		
175–179								10	4	6	2																																		
180–184								4	1	3	9	2																																	
185–189									1		4	17	4																																
190–194									1			13	6																																
195–199										1		30	2	2																															
200–204											2	24	7	12	13	8	8	5		5																									
205–209									4	6		30	22	12	20	24	9	9	1	16	13																								
210–214									7	8	9	16	7	3	7	6	2	7	3	2	2	2	6		4			6																	
215–219									5	5	2	4	1	6	2	12	6		3	10	10	8	4	8	9			5	2	2															
220–224								2	6	5	9	1	1	8	2	2	12		2	2	2	4	2		27	3	2	4	4	1			4												
225–229								10	5	6		1		3	5	9	6	6	8	7	8	6	6				16	21	8		2		11	1											
230–234								4	3	1		2	3	3	5	12	12	2	2	10	2	4	2	9		6		6	7	1	13			17	16										
235–239								1	1					2	4	2	1	6	8	2	8	8				2	16	4	3	1	5		8	2	5										
240–244								1		6				3	2	5	9	3	1	7	4	4	3		4	1		3	11	2	2		4	8	2	1									
245–249														1			4	1	2	3	2	2		9	9				1	7	4	2		1		6	3								
250–254								1	4	2		1		9		2	1	1	2	1	1	1	1	3	1		1		1	1	4	2	1		2	1	9	1							
255–259																		3	3							1				2	2			1	8	3	2		1						
260–264																		1							1		1				3			13	16	6	14	5	12	2					
265–269																				2		1							3		5					1	2	9	13	6		3	7		
270–274																								1					1	3	2		8							1		12			
275–279																									1				1		4			17	16		2	5			12	12	4		
280–284																									1	1	1				2		2	1	2		2	9			6	6	6		
285 or over																									1		1			2				1	6	37	28			2	5	5	3	21	4
Totals	14	79	44	139	91	116	83	101	71	80	116	159	75	50	47	83	87	77	37	39	59	49	19	25	62	23	33	44	41	28	51	9	27	59	29	51	43	37	28	33	22	13	26	14	21	4

Table 5.4 (continued)

Container ships – number of vessels in divisions of TEU capacity

Extreme breadth — Division in metres	0–99	100–199	200–299	300–399	400–499	500–599	600–699	700–799	800–899	900–999	1000–1099	1100–1199	1200–1299	1300–1399	1400–1499	1500–1599	1600–1699	1700–1799	1800–1899	1900–1999	2000–2099	2100–2199	2200–2299	2300–2399	2400–2499	2500–2599	2600–2699	2700–2799	2800–2899	2900–2999	3000–3099	3100–3199	3200–3299	3400–3599	3600–3799	3800–3999	4000–4199	4200–4399	4400–4599	4600–4799	4800–4999	5000–5499	5500–5999	6000–6499	6500–6999
10–10.9	–	1	–	–	–	–	–	–	–	–	–	–	–	–	–	–	–	–	–	–	–	–	–	–	–	–	–	–	–	–	–	–	–	–	–	–	–	–	–	–	–	–	–	–	–
11–11.9	5	2	–	–	–	–	–	–	–	–	–	–	–	–	–	–	–	–	–	–	–	–	–	–	–	–	–	–	–	–	–	–	–	–	–	–	–	–	–	–	–	–	–	–	–
12–12.9	2	5	–	–	–	–	–	–	–	–	–	–	–	–	–	–	–	–	–	–	–	–	–	–	–	–	–	–	–	–	–	–	–	–	–	–	–	–	–	–	–	–	–	–	–
13–13.9	1	26	1	1	–	–	–	–	–	–	–	–	–	–	–	–	–	–	–	–	–	–	–	–	–	–	–	–	–	–	–	–	–	–	–	–	–	–	–	–	–	–	–	–	–
14–14.9	1	14	–	–	–	–	–	–	–	–	–	–	–	–	–	–	–	–	–	–	–	–	–	–	–	–	–	–	–	–	–	–	–	–	–	–	–	–	–	–	–	–	–	–	–
15–15.9	1	16	10	14	2	3	–	–	–	–	–	–	–	–	–	–	–	–	–	–	–	–	–	–	–	–	–	–	–	–	–	–	–	–	–	–	–	–	–	–	–	–	–	–	–
16–16.9	–	7	8	25	3	1	–	–	–	–	–	–	–	–	–	–	–	–	–	–	–	–	–	–	–	–	–	–	–	–	–	–	–	–	–	–	–	–	–	–	–	–	–	–	–
17–17.9	1	3	10	35	5	3	7	1	–	–	–	–	–	–	–	–	–	–	–	–	–	–	–	–	–	–	–	–	–	–	–	–	–	–	–	–	–	–	–	–	–	–	–	–	–
18–18.9	3	5	6	32	24	27	21	15	2	–	–	–	–	–	–	–	–	–	–	–	–	–	–	–	–	–	–	–	–	–	–	–	–	–	–	–	–	–	–	–	–	–	–	–	–
19–19.9	–	–	3	13	10	21	26	20	4	3	–	–	–	–	–	–	–	–	–	–	–	–	–	–	–	–	–	–	–	–	–	–	–	–	–	–	–	–	–	–	–	–	–	–	–
20–20.9	–	–	3	14	23	37	4	32	21	19	44	21	20	–	–	–	–	–	–	–	–	–	–	–	–	–	–	–	–	–	–	–	–	–	–	–	–	–	–	–	–	–	–	–	–
21–21.9	–	–	–	2	22	20	12	5	7	4	13	31	29	6	–	–	–	–	–	–	–	–	–	–	–	–	–	–	–	–	–	–	–	–	–	–	–	–	–	–	–	–	–	–	–
22–22.9	–	–	–	–	2	4	6	7	5	19	31	20	3	14	12	4	3	–	–	–	–	–	–	–	–	–	–	–	–	–	–	–	–	–	–	–	–	–	–	–	–	–	–	–	–
23–23.9	–	–	–	–	2	–	2	7	7	4	13	21	6	1	17	15	36	1	6	3	–	–	–	–	–	–	–	–	–	–	–	–	–	–	–	–	–	–	–	–	–	–	–	–	–
24–24.9	–	–	2	–	–	2	5	10	3	17	18	24	3	–	12	24	18	18	21	4	4	6	4	–	24	3	–	2	2	–	–	–	–	–	–	–	–	–	–	–	–	–	–	–	–
25–25.9	–	–	1	–	–	1	–	1	–	–	4	2	3	8	10	10	28	9	6	5	5	7	15	17	20	1	1	7	2	–	–	–	–	–	–	–	–	–	–	–	–	–	–	–	–
26–26.9	–	–	–	–	–	–	–	–	–	–	–	–	8	6	8	6	10	28	21	22	19	13	17	7	16	19	32	13	6	–	–	–	–	–	–	–	–	–	–	–	–	–	–	–	–
27–27.9	–	–	–	–	–	–	–	–	–	1	–	–	10	3	6	2	2	9	9	5	22	22	4	7	24	3	1	1	2	28	51	9	27	59	29	51	43	33	22	32	14	26	14	21	4
28–28.9	–	–	–	–	–	–	–	–	–	–	–	–	–	–	–	–	–	–	–	–	5	6	–	20	20	1	1	–	10	–	–	–	–	–	–	–	–	–	13	24	–	6	20	–	–
29–29.9	–	–	–	–	–	–	–	–	–	–	–	1	2	–	–	–	–	3	6	9	19	3	–	1	16	19	32	44	31	–	–	–	–	–	–	–	41	28	9	8	–	20	–	–	–
30–30.9	–	–	–	–	–	–	–	–	–	–	–	2	3	1	6	6	6	7	3	3	22	5	4	7	–	–	–	–	–	–	–	–	–	–	–	–	2	5	–	–	–	–	–	–	–
31–31.9	–	–	–	–	–	–	–	–	–	–	–	–	–	1	3	4	2	6	8	18	27	9	–	–	–	–	–	–	–	–	–	–	–	–	–	–	–	–	–	–	–	–	–	–	–
32–32.9	–	–	–	–	–	–	–	–	–	–	–	–	–	–	–	–	–	–	–	–	–	–	–	–	–	–	–	–	–	–	–	–	–	–	–	–	–	–	–	–	–	–	–	–	–
37–37.9	–	–	–	–	–	–	–	–	–	–	–	–	–	–	–	–	–	–	–	–	–	–	–	–	–	–	–	–	–	–	–	–	–	–	–	–	–	–	–	–	–	6	–	–	–
40 or over	–	–	–	–	–	–	–	–	–	–	–	–	–	–	–	–	–	–	–	–	–	–	–	–	–	–	–	–	–	–	–	–	–	–	–	–	–	–	–	–	14	20	14	21	4
Totals	14	79	44	139	91	116	83	101	71	80	116	159	75	50	47	83	87	77	37	39	59	49	19	25	62	23	33	44	41	28	51	9	27	59	29	51	43	33	22	32	14	26	14	21	4

Table 5.4 (continued)

Container ships – number of vessels in divisions of TEU capacity

Division in metres (Draught)	0–99	100–199	200–299	300–399	400–499	500–599	600–699	700–799	800–899	900–999	1000–1099	1100–1199	1200–1299	1300–1399	1400–1499	1500–1599	1600–1699	1700–1799	1800–1899	1900–1999	2000–2099	2100–2199	2200–2299	2300–2399	2400–2499	2500–2599	2600–2699	2700–2799	2800–2899	2900–2999	3000–3099	3100–3199	3200–3399	3400–3599	3600–3799	3800–3999	4000–4199	4200–4399	4400–4599	4600–4799	4800–4999	5000–5499	5500–5999	6000–6499	6500–6999
3.0–3.49	3	2	1	–	–	–	–	–	–	–	–	–	–	–	–	–	–	–	–	–	–	–	–	–	–	–	–	–	–	–	–	–	–	–	–	–	–	–	–	–	–	–	–	–	–
3.5–3.99	2	4	–	2	–	–	–	–	–	–	–	–	–	–	–	–	–	–	–	–	–	–	–	–	–	–	–	–	–	–	–	–	–	–	–	–	–	–	–	–	–	–	–	–	–
4.0–4.49	4	8	–	6	–	–	–	–	–	–	–	–	–	–	–	–	–	–	–	–	–	–	–	–	–	–	–	–	–	–	–	–	–	–	–	–	–	–	–	–	–	–	–	–	–
4.5–4.99	1	29	–	5	–	–	–	–	–	–	–	–	–	–	–	–	–	–	–	–	–	–	–	–	–	–	–	–	–	–	–	–	–	–	–	–	–	–	–	–	–	–	–	–	–
5.0–5.49	1	9	3	22	1	–	–	–	–	–	–	–	–	–	–	–	–	–	–	–	–	–	–	–	–	–	–	–	–	–	–	–	–	–	–	–	–	–	–	–	–	–	–	–	–
5.5–5.99	1	20	11	1	20	–	–	–	–	–	–	–	–	–	–	–	–	–	–	–	–	–	–	–	–	–	–	–	–	–	–	–	–	–	–	–	–	–	–	–	–	–	–	–	–
6.0–6.49	2	6	19	57	30	9	2	4	–	–	–	–	–	–	–	–	–	–	–	–	–	–	–	–	–	–	–	–	–	–	–	–	–	–	–	–	–	–	–	–	–	–	–	–	–
6.5–6.99	–	1	1	30	7	24	15	15	2	–	–	–	–	–	–	–	–	–	–	–	–	–	–	–	–	–	–	–	–	–	–	–	–	–	–	–	–	–	–	–	–	–	–	–	–
7.0–7.49	–	–	1	10	23	42	19	7	6	–	–	–	–	–	–	–	–	–	–	–	–	–	–	–	–	–	–	–	–	–	–	–	–	–	–	–	–	–	–	–	–	–	–	–	–
7.5–7.99	–	–	–	7	9	18	18	31	21	1	–	–	–	–	–	–	–	–	–	–	–	–	–	–	–	–	–	–	–	–	–	–	–	–	–	–	–	–	–	–	–	–	–	–	–
8.0–8.49	–	–	–	3	–	22	16	20	22	9	2	3	–	–	–	–	–	–	–	–	–	–	–	–	–	–	–	–	–	–	–	–	–	–	–	–	–	–	–	–	–	–	–	–	–
8.5–8.99	–	–	–	2	1	4	7	15	9	8	1	6	–	–	–	–	–	–	–	–	–	–	–	–	–	–	–	–	–	–	–	–	–	–	–	–	–	–	–	–	–	–	–	–	–
9.0–9.49	–	–	–	–	–	1	4	4	8	18	52	6	–	–	–	–	–	–	–	–	–	–	–	–	–	–	–	–	–	–	–	–	–	–	–	–	–	–	–	–	–	–	–	–	–
9.5–9.99	–	–	–	–	–	–	2	5	3	18	17	35	–	–	–	–	2	–	–	–	–	–	–	–	–	–	–	–	–	–	–	–	–	–	–	–	–	–	–	–	–	–	–	–	–
10.0–10.49	–	–	–	–	–	–	–	–	–	8	7	38	10	–	–	6	11	–	–	3	6	–	–	–	–	–	–	–	–	–	–	–	–	–	–	–	–	–	–	–	–	–	–	–	–
10.5–10.99	–	–	–	–	–	–	–	–	–	8	24	14	6	–	2	21	27	40	2	5	9	10	–	–	5	–	–	–	–	–	–	–	–	–	–	–	–	–	–	–	–	–	–	–	–
11.0–11.49	–	–	–	–	–	–	–	–	–	5	5	24	8	1	18	14	10	8	2	8	2	9	–	2	6	–	–	–	–	–	–	–	–	–	–	–	–	–	–	–	–	–	–	–	–
11.5–11.99	–	–	–	–	–	–	–	–	–	5	2	34	24	5	11	27	27	12	3	5	16	9	7	1	23	–	–	3	10	–	–	–	–	–	–	–	–	–	–	5	–	–	3	–	–
12.0–12.49	–	–	–	–	–	–	–	–	–	–	6	5	11	20	7	7	5	3	12	18	18	16	7	6	17	3	4	26	3	19	5	–	1	–	6	5	1	–	–	5	2	8	–	–	–
12.5–12.99	–	–	–	–	–	–	–	–	–	–	–	–	13	11	7	7	5	13	12	8	2	–	4	7	6	11	16	9	5	5	2	4	1	28	12	11	2	1	1	8	–	–	–	–	–
13.0–13.49	–	–	–	–	–	–	–	–	–	–	–	–	2	9	–	1	–	–	–	–	16	3	–	10	5	5	4	3	5	1	16	5	17	17	4	16	2	25	1	13	–	2	–	–	–
13.5–13.99	–	–	–	–	–	–	–	–	–	–	–	–	1	3	2	1	–	1	2	–	–	2	1	–	–	3	9	3	17	3	8	–	5	12	6	7	37	–	10	1	–	4	1	–	–
14.0–14.49	–	–	–	–	–	–	–	–	–	–	–	–	–	1	–	–	–	–	–	–	–	–	–	–	–	1	–	–	1	–	13	–	1	2	–	12	–	7	10	–	12	12	10	10	4
14.5–14.99	–	–	–	–	–	–	–	–	–	–	–	–	–	–	–	–	–	–	–	–	–	–	–	–	–	–	–	–	–	–	7	–	2	–	1	–	1	–	–	–	–	–	–	11	–
Totals	14	79	44	139	91	116	83	101	71	80	116	159	75	50	47	83	87	77	37	39	59	49	19	25	62	23	33	44	41	28	51	9	27	59	29	51	43	33	22	32	14	26	14	21	4

Source: Lloyd's Register of Shipping as at 31 December 1999.

it has to be done as percentages of the total fleet. The number of ships of 2000 TEU and more increased from 8 per cent to 34 per cent. In 1980 less than one-half per cent of container ships carried 3000 TEU and more. This had changed to 18 per cent by 1999. Of these 9 per cent were of 4000 TEU and more and 3 per cent of 5000 TEU and more.

The pronounced cut-off value of 32.0 to 32.9 m for the breadth of container ships corresponds to the maximum value allowed for the Panama Canal which is 32.24 m. Only 103 ships were, so-called, post-Panamax with a larger breadth. This limitation is also clearly noticeable for dry-bulk carriers.

For general cargo ships carrying capacity data would normally be more important for charter traffic, which would load and/or discharge full ship-loads at the port in question. This would generally not be the case for ships in liner traffic, which call at several ports and only partly load and/or unload at any one port. Both tankers and dry-bulk carriers normally load or discharge full shiploads. For bulk terminals and some container terminals the carrying capacities as well as the dimensions are often known in advance, fully or partly, for all of the relatively few different ships which call.

The dimension distributions for ships calling at the port in question or for a comparable port may or may not be close to the world-fleet distributions. The present distributions have to be determined and future distributions forecast, considering the likely feedback from new or modified port facilities on the ships which are going to call in the future. The statistical distributions for ships' dimensions have to be used in different ways depending upon the purpose and dimension in question. This is discussed below for each of the dimensions.

Length overall

For one or two berths an overall ship length close to the maximum value would have to be used, adding 10 to 20 m spaces for moorings between ships and for stretches of quay between ends of ships and ends of the quay. This does not mean that the berth or quay length should be equivalent to the sum of ship length and mooring spaces. In principle, a quay is only required along the hatches for general cargo and containers. For liquid bulk, and certain types of dry bulk, only a loading/unloading platform for cargo-handling equipment is required, supplemented of course by berthing and mooring platforms. For a long stretch of quay an average overall ship length should be used, as the probability of having to accommodate only large ships simultaneously is low. For computer simulation of ship arrivals, along a stretch of quay, an actual or forecast ship length distribution is generated as a subroutine. An arriving ship is accommodated using the assumption that two or more non-occupied quay stretches may be aggregated and considered as one if required. In addition to berth lengths, access channel configurations (widths and bends) and basin sizes are influenced by the need to accommodate ships of certain lengths, although here limitations of an existing access channel may prevent the arrival of too-long ships.

Access channel configurations and basin sizes are also influenced by the beam and by the ship's manoeuvrability at low speed with or without tug assistance as the case may be. They should be determined by using available navigational simulation models, except for simple cases. Such simulation models are discussed in 'Manoeuvrability at low speed', later in this chapter.

Extreme breadth

The breadth governs the reach of cargo-handling equipment and influences the width of access channels and basins. Access channels that are not straight will also have their widths influenced by ships' length and low-speed manoeuvrability.

Shore-based cargo-handling equipment should be able to reach the far side of the ships' holds and, where direct transfer of cargo to or from lighters takes place and ships' gear cannot be used, the shore equipment has to reach the far side of the holds of the lighters.

Draught

The fully loaded draught is obviously important for the depth of access channels and basins as well as for the depths along berths. Variations of the water depth during the tidal cycle, as well as the draught of partly loaded ships, should be considered.

Cargo-carrying capacity and cargo units carried

The size of ships is normally expressed as Gross Register Tonnage (GRT), Net Register Tonnage (NRT) or Dead-Weight Tonnage (DWT), which are defined as follows:

(a) the GRT is the total volume of the ship including its superstructure and hatches measured in units of $100\,\text{ft}^3$ ($2.83\,\text{m}^3$). Thus GRT is clearly not a measure of cargo-carrying capacity. As a rough approximation GRT is 0.68 to 0.82 of DWT for smaller general-cargo ships, and decreases to 0.6 for large general-cargo ships. For tankers and dry-bulk carriers GRT decreases from 0.6 to 0.5 of DWT with increasing vessel sizes;

(b) NRT is measured in the same units as GRT and is arrived at by subtracting from GRT the total volume of the engine room, compartments used for navigation and other operations, crews quarters, and ballast tanks. NRT thus measures the volume available for cargo, fuel, water and other supplies. Roughly, NRT is 0.4 of DWT for cargo ships larger than 2000 DWT; and

(c)   DWT is the displacement tonnage (the mass of water displaced by the ship when fully loaded to its Plimsoll mark) minus the ship's own mass, all measured in long tonnes of 1016 kg. DWT is thus the total carrying capacity for cargo, fuel, water and other supplies.

While there is no direct relationship between GRT and DWT, there is to some extent a quantitative relationship between NRT and DWT. If the NRT volume was filled with a liquid (sea water) which has a density of 1 long tonne per cubic metre and assuming loading to the Plimsoll mark, NRT would be 0.35 of DWT (DWT divided by 2.83). This is not far from the average value, which is 0.4 of DWT.

For relatively heavy cargo, DWT less fuel, water, etc. will be the carrying capacity of the ship, which is normally 0.85 to 0.9 of DWT. However, for light cargoes (high stowage factors measured in cubic metres per tonne) the volume of the cargo holds of the ship will limit the cargo carrying capacity. This volume is generally not NRT, less the volumes available for fuel, water, etc., except for tankers. It is close, however, to this value for modern ships with large open holds and wide hatch covers. For cellular container ships the volume available for containers (including those on deck) measured in TEUs will normally be the cargo carrying capacity, as the average container stowage factor exceeds 2.5.

General-cargo ships in liner traffic typically load and discharge only part of their cargo in any one port, whereas such ships in charter traffic as well as bulk carriers normally load and discharge full loads. Container vessels, which are in liner traffic, are found in both of the above categories. The minimum storage requirement for full shiploads of bulk or containers should correspond to the carrying capacity of the largest ship that calls, and the total handling rate for equipment should be roughly proportional to the carrying capacity. Thus, for large oil tankers the loading/unloading installations are often geared to a 24-hour service time independent of the size of the tanker.

Tables 5.1, 5.2 and 5.3 show the end of year 1999 world fleet DWT distributions for oil tankers, dry-bulk carriers and general-cargo ships respectively. Table 5.4 shows the world fleet TEU carrying capacity distribution for cellular container ships, at the same time.

The cargo units carried by ships are important for handling equipment and handling rates as well as for stowage of cargo both on board and in port. General cargo, in cartons, crates, bales, bags, drums, etc., of widely varying sizes and shapes, is obviously slow and difficult to handle and stow. Larger and more uniform units, such as palletised cartons, pre-slung bags and, in particular, containers are easier to handle and store.

Cargo-handling gear

Ship's gear is normally limited to cranes or derricks and winches for general cargo, and pumps for liquid cargo. For general cargo handling, great strides have been made during recent decades in regard to improvement of ships' gear. The old derrick-and-winch system has been replaced by cranes, which cover the area of the hatches and often the area of the holds, if the hatch covers are close to the same size as the holds. Modern ships' gear makes it feasible for ports to avoid investing in costly quay cranes and to eliminate existing cranes. Where quay cranes are not available, old ships with inefficient or poorly maintained gear are less likely to call. The use of ship's cranes and winches do however restrict the ship motion that may be tolerated without interruption of cargo handling. This is the case both for rapid motion induced by waves and for very slow tidal motion.

Some of the early container ships and some modern smaller feeder ships have their own container-handling gear. This is not the case for container ships in service on major routes.

Larger oil tankers are normally equipped with pumps for unloading in less than 24 hours, provided the storage tanks are relatively close both in regard to distance and elevation. Loading is usually by gravity flow. If the storage tanks are not close to the terminal, booster pumps on shore may be required both for loading and unloading.

Dry bulk carriers do not normally have any cargo-handling gear. For obvious reasons, ramps on roll-on/roll-off (ro/ro) vessels greatly reduce the need for port facilities to load and unload such ships. LASH carriers have their own gear for the handling of barges.

Size and shape of hull and superstructure

The length of the straight and vertical sides of the hull is important for the layout of berths, cargo-handling platforms, berthing and mooring platforms and fender systems. The choice between relatively many small fenders and relatively few large fenders should be influenced by the size distribution for ships, which call at the berth or quay in question. The horizontal curvature of the bow determines the spacing of fenders and their protrusion from the quay structure. The fenders should be placed in such a manner that, for a normal tidal range, the stern will not touch the quay structure, bollards or other fixed facilities, when the fenders are fully compressed. Further, the lower part of fenders should not be exposed to the propellers of light ships. Bulbous bows and their shape are also of importance for the design of fender systems. Below the part of the system that is supposed to be in contact with the hull there should not be any exposed support piles or other members, which may be damaged by, or do damage to, the bulbous bow.

The size and shape of the submerged hull for fully loaded ships are needed to determine maximum current forces. Similarly the size and shape of the non-submerged hull and superstructure for light ships are needed to determine maximum wind forces. Resulting maximum mooring and fender forces may correspond to a combination of simultaneous wind and current conditions at an intermediate loading condition for the ship.

The following equations are used to determine current forces and moments:

Longitudinal force $\quad F_L = 0.5 C_L \rho_w V_C^2 TL$

Transverse force $\quad F_T = 0.5 C_T \rho_w V_C^2 TL$

Yaw moment $\quad M_Y = 0.5 C_Y \rho_w V_C^2 TL^2$

where C_L, C_T and C_Y are non-dimensional shape coefficients, which vary with current direction, draught and draught to water-depth ratio; ρ_w is the density of water; V_C is the average current velocity over the draught; T is the average draught; and L is the length overall or between perpendiculars.

The values of shape coefficients, which are determined by hydro-dynamic model tests, depend upon whether length overall or length between perpendiculars is used in the equations.

The longitudinal and transverse wind forces, as well as the wind yaw moment, are determined in a similar fashion. However, the projected front area of the ship is used in the equation for the longitudinal wind force, and the projected side area is used in the equations for the transverse wind force and the yaw moment. Further, the density is the one for air and the velocity is the one at a certain height above the water surface.

Ship-testing laboratories have more or less comprehensive libraries of wind and current coefficients.

As far as we have been able to determine, generalised wind and current coefficients have only been developed for tankers ranging in size from 150 000 to 500 000 DWT (Oil Companies' International Marine Forum, 1977). It is believed that their C_L values for currents are too large by a factor of more than 2. However, this is fortunately of limited importance for our applications, as longitudinal current forces are generally not governing for mooring forces.

Hull strength

The hull structure and the ultimate hull stress govern the ultimate fender-reaction-force distribution. From the point of view of fender-system design, the problem is to determine the lengths and heights of the fender contact surface, which are required to avoid stresses that will cause plastic deformations of the weakest parts of the side of the hull, when exposed to the design impact force. For vertical and parallel sides of tankers and dry bulk carriers an approach was developed (Svendsen and Vinther Jensen, 1970). For tankers the solutions are shown in Figs 5.1 and 5.2, and for dry-bulk carriers in Figs 5.3 and 5.4. Figure 5.1 shows the plastic moment of resistance for the side longitudinals (horizontal members) divided by the product of the spacing of the side transverses (vertical members) and the yield stress ($M_{sl}/s\partial_y$), the spacing of side transverses (s) and of side longitudinals (t), all as functions of the draught of the tanker. Figure 5.2 shows the relationship between the ratio of fender length to side transverse spacing (l/s) and fender height to side longitudinal spacing (h/t) for constant values of the maximum fender force multiplied by the side transverse spacing and divided by the plastic moment of resistance (Ps/M_{sl}). The yield stress for hull construction steel is usually $260\,\text{N/m}^2$ ($36\,980\,\text{lb/in}^2$). The solution is only valid below the limit for the fender force, which would also cause plastic

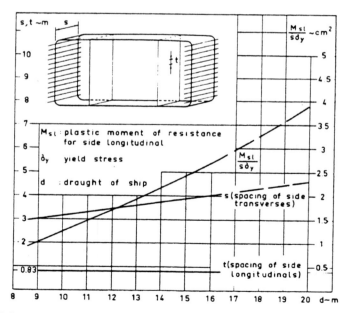

Fig. 5.1 Input data for Fig. 5.2. Reproduced by permission of the Dock & Harbour Authority

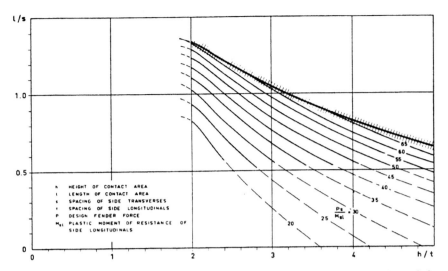

Fig. 5.2 Fender dimensions for tankers. Reproduced by permission of the Dock & Harbour Authority

deformation of the side transverses. For 10 000 to 100 000 DWT tankers this limit is indicated by the hatched curve. The acceptable fender surface dimensions are, of course, those above the moment curve in question. The ratio between fender length and side transverse spacing should not be less than 0.5 to 0.65 and the ratio between fender height and side longitudinal spacing not less than about 2.

For dry-bulk carriers, plastic deformations of the vertical side frames between upper wing tanks and lower hopper side tanks are normally governing. Figures 5.3 and 5.4 are used for determination of fender surface areas in the same manner as for tankers.

The solutions relate only to bending of critical hull members as described above. They do not consider membrane stresses which, if added, would result in somewhat smaller fender areas.

Design requirements for avoidance of hull damage caused by minor ship impacts with fenders and floating objects are dealt with by Lützen (1998) as well as by Lützen and Terndrup Pedersen (1998).

Superstructure configuration

The superstructure height for light vessels at high tide, as well as its width relative to the width of the hull, is important for the positioning and design of quay cranes, container cranes and bulk-handling installations. Care should be taken to eliminate the possibility of collisions between cargo-handling equipment and the superstructure of ships.

69

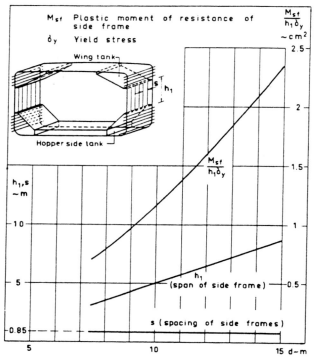

Fig. 5.3 Input data for Fig. 5.4. Reproduced by permission of the Dock & Harbour Authority

Motion and moorings

The six translatory and rotation ship motions are shown in Fig. 5.5. Natural roll, pitch and heave periods are governed mainly by the ship itself, whereas natural surge, sway and yaw periods, to a large extent, are governed by the berth structure (solid wall or open pile supported deck structure) and in particular by the mooring and fender systems. Of course, surge, sway and yaw occur only for restrained ships.

To give a simplified illustration of the problem, the following basic equation for translation motion of a restrained floating body may be used:

$$M'\ddot{X} = C_m M_f \ddot{U} + 0.5 C_d \rho A (U - X)|U - X| - R;$$

where

$M' =$ virtual mass of the body;
$M_f =$ mass of displaced fluid (equal to the mass of the floating body);
$C_m =$ coefficient of virtual mass ($M' = C_m M_f$);
$C_d =$ drag coefficient;

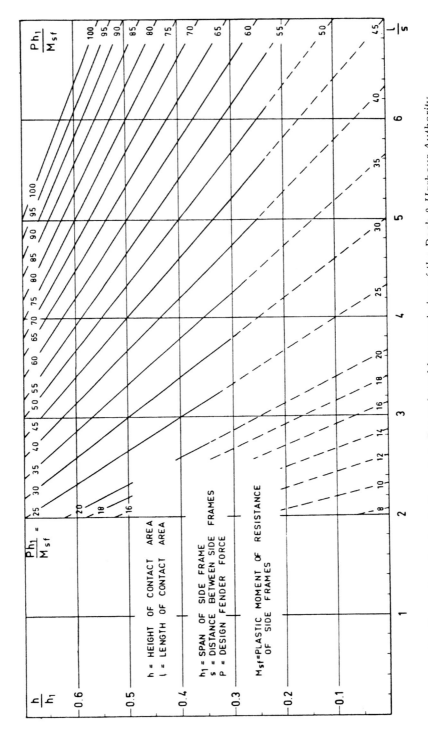

Fig. 5.4 Fender dimensions for dry-bulk carriers. Reproduced by permission of the Dock & Harbour Authority

$\rho =$ density of fluid;

$A =$ area of body perpendicular to direction of motion;

$R =$ restraining force;

$U =$ velocity of fluid;

$\ddot{U} =$ acceleration of fluid;

$X =$ velocity of body; and

$\ddot{X} =$ acceleration of body.

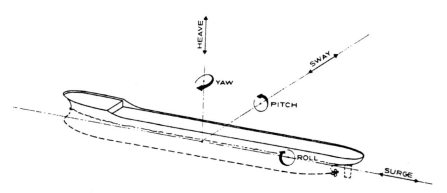

Fig. 5.5 Ship motions

The general problem is the vastly more complex one of coupled motions in 6 degrees of freedom of a ship exposed to irregular wave action and under non-linear restraint by mooring and fenders. Attempts at arriving at analytical solutions have been made (Wilson, 1973(a), Seidl, 1973 and others) and are continuing. Finite difference techniques used in a computer program have been suggested for solving the six complex and simultaneous differential equations.

One of the greatest problems associated with analytical determination of ship motions and associated mooring and fender forces is the calculation of the hydrodynamic forces and pressure distributions, in particular for small under-keel clearances which are common, and for ships alongside solid quay walls. The required hydrodynamic coefficients, have to be determined by hydraulic model tests, as full-scale measurements under desirable and well-defined conditions would be very expensive and time consuming, or by advanced computational fluid dynamics tools. Some leading hydraulic laboratories claim to have developed computer programs yielding results which compare favourably with those from hydraulic-model tests. Other leading laboratories do not agree with this. Until further theoretical and analytical advances have been made it may be prudent to continue to employ hydraulic-model tests. Such model tests have to accurately reproduce irregular wave conditions at the berth, as well as the elastic and other, often non-linear, properties of moorings and fenders.

The natural roll, pitch and heave periods cannot be influenced appreciably by moorings or fenders. If wind wave or swell periods, which are properties that cannot be changed by breakwaters or other protective structures, are close to natural roll, pitch or heave periods or harmonics thereof, there may be problems. Considerable roll, pitch or heave would adversely affect cargo handling for all types of ships except liquid bulk–carriers, because the liquid-handling system consists of flexible hoses or of loading arms with swivel joints which are able to follow the motion of the mid-ships manifold, where the pitch is negligible anyway.

Surge is influenced mainly by the moorings, sway and yaw by both the moorings and fenders. Moorings, which are usually part of the ship's equipment, consist of manila or nylon ropes, steel wires with or without nylon tails and, for some newer ships, constant-tension winches for each of the moorings. However, available constant-tension winches are often locked by the crew and thus not used as intended. Elastic properties of moorings have been presented (Wilson, 1973(b)). The reason for using nylon tails with steel wires is, of course, the need to increase the elasticity, that is, to soften the moorings.

General guidelines for tanker moorings have been developed (Oil Companies' International Maritime Forum, 1978). They are equally valid for other types of ships, and those of particular importance for berth designers are summarised below:

(a) the mooring system should be symmetrical;
(b) the spring line direction should be as close to the longitudinal direction of the ship as possible;
(c) the breast moorings should be roughly perpendicular to the ship;
(d) bow and stern moorings are not necessary, provided mooring points are suitably designed and arranged;
(e) the vertical mooring angle should be as small as possible, and not exceed 30°;
(f) the lines should be of the same type, diameter and, if possible, also length, in particular for lines in the same direction; and
(g) same tail for same type of line.

Minimum requirements for the number, length and breaking strength of mooring lines have been established (Lloyd's Register of Shipping, 1989), but are rather difficult to summarise because of the complicated equation employed for characterising different ships.

A number of ships and moorings exposed to various wave conditions are discussed in various publications (Van den Bunt, 1973, van Oorschot, 1976, Khanna and Birt, 1977 and Stammers and Wennink, 1977).

Ship motion, which would damage the moorings or the fenders, is determined as discussed above by hydraulic model tests using scale models of ships, moorings and fenders exposed to careful reproductions of natural, irregular waves. However, considerably less motion (amplitude and/or velocity and/or acceleration) make certain cargo-handling operations impossible or may reduce cargo-handling rates appreciably. The motion, which may be tolerated and the relationship between motion and cargo-handling rates vary very much with the following types of cargo-handling operations:

(a) loading of dry bulk into large, open holds is quite insensitive to any kind of motion. A surge amplitude of the order of magnitude of the length of the holds would be acceptable. A combined amplitude for sway and yaw of the order of magnitude of the width of the holds could be tolerated. Roll would generally not matter within a wide range. The acceptable combined pitch and heave amplitude would be influenced by dust control requirements;

(b) discharge of dry bulk carriers with large holds tolerate much less motion than their loading, as the discharge devices have to extend inside the holds, be in contact with the cargo (perhaps dig into it) and come close to the bottom of the hold when the discharge is nearing completion. At this stage of the operation, earth-moving equipment (bulldozers and front-end shovels) is normally used in the hold to bring the remaining cargo to the discharge device. It would be difficult to determine the motion, which may be tolerated as well as reduced cargo-handling rates caused by motion;

(c) loading and discharge of liquid bulk is done through flexible devices (loading arms or hoses) which can accommodate considerable motion of the ships' manifold relative to the fixed installations on the loading platform. The manifold is normally located mid-ships and thus pitch and yaw would be the least troublesome motions. Certain liquefied natural gas (LNG) carriers are not designed for the sloshing of LNG in partly full tanks. Sloshing would result in damage to tank insulation with potential catastrophic consequences. These vessels therefore tolerate very little motion, both during loading and during discharge;

(d) loading and discharge of general cargo require that stevedores work in the hold close to the hook of a shore crane or of ship's own gear. For discharge, forklifts are commonly used in the holds to bring cargo units to the hook. This obviously limits ship motion or reduces cargo-handling rates;

(e) loading and discharge of cellular container ships are extremely sensitive to all types of ship motion because of the small tolerances (a few inches) between horizontal dimensions of containers, which are being lowered or raised, and horizontal dimensions of cell guides and of cells; and

(f) ro/ro operations are adversely affected both by ship motion and by resulting ramp motion. Only very limited motion may be tolerated.

Very little work has been done concerning these important relationships between ship motion and cargo-handling rates for different ship types, cargoes and handling equipment. For cellular container ships a beginning has been made. Observed loading rates for varying vessel surge have been presented (Hwang and Divoky, 1977). Only surge motion was measured. The effects of surge and sway on container-loading rates have been investigated experimentally (Slinn, 1979). A container crane was used with many different operators, each working one or two hours. A horizontal frame equipped with cell guides was installed on the quay and used to simulate the cell. The frame motion reproduced actual recordings of a ship's surge and sway motion. Each test started with the container about 10 m above the quay. Generally the operator lowered the container rapidly to about 20 cm above the cell guides and waited until the cell passed under the container; rarely, and with little success, did the operator attempt to follow the moving cell with the swinging container. Frequently, the operator was unable to place the container inside the cell guides in his first attempt, in which case the container was raised about 20 cm and the exercise was repeated. This continued until the container was placed correctly or 2 minutes had elapsed. Jamming of containers in cell guides (or in cells) because of angular motion with or without vertical motion caused by roll, pitch, yaw and perhaps heave could not be reproduced. Correlation between many relevant parameters were sought. The best correlation was found between the number of attempts to place the container (it was assumed that two attempts were made per cell-motion cycle) and the maximum cell velocity. The relationship was linear. The number of attempts is converted to time and added to a known or assumed container-loading cycle time, which corresponds to no appreciable ship motion. Thus a reduced loading rate is arrived at.

General criteria for allowable ship motion are presented in PIANC (1995), which discusses many aspects in great detail and contains many references. However, it does not seem quite clear exactly how the criteria have been arrived at. For certain categories of ships the criteria are apparently based solely on educated guesses by ships' crew members and/or port operators.

Manoeuvrability at low speed

General considerations

Low-speed manoeuvres are characterised by very low engine settings from slow ahead to slow astern, supplemented by higher engine settings during short periods of time, either to stop the ship or to change its turning rate. Bow and/or stern thrusters may be used as well as tug assistance.

Low-speed manoeuvres may conveniently be divided into:

(a)   bringing the ship into the harbour;
(b)   turning;
(c)   coming alongside berth; and
(d)   departing from berth.

Ships normally have to sail through access channels, port entrances and basins, with or without tug assistance, before they come alongside a berth. Traditionally, rules of thumb have been used for determining ships' trajectories but size increases for various ship types have made this approach unsatisfactory. One of the most important rules of thumb was that the protected turning basin should have a diameter of four times the length of the longest ship expected to call. This rule of thumb does not relate to how ships actually manoeuvre and it is not being used today.

The first attempts at determining the influence of ships manoeuvrability on port layout took place in the early sixties (Svendsen, 1973). Experienced pilots and port engineers co-operated in an effort which resulted in the so-called graphic method for determining ship manoeuvres. Since then manoeuvring tests, using scale models of ships in models of the actual harbour configuration, have been used. This method suffers from a number of shortcomings as described below. The large ships, and a need to take these ships into existing harbours originally designed for smaller ships, has led to the need for more advanced tools. The primary tool has become computer simulation using advanced mathematical models of ships and the physical environment in which they operate.

Model tests

The ship models are either radio controlled from shore or operated by a navigator on-board the model. Testing of the models in basins with correct bottom configurations and port structures (breakwaters, berths, etc.) is expensive and time consuming. There are several other problems associated with model tests:

(a)   Reynolds number, VL/γ where V is the ship's speed, L is the ship's length, and γ is the viscosity of water, will be wrong for the model. In some cases this will influence the hydrodynamic forces;
(b)   wind forces, which are of overriding importance to manoeuvres at dead-slow speed, are virtually impossible to generate and control;
(c)   everything happens much faster in the model than in the prototype. The time scale is the square root of the length scale. It may be difficult to perform required manoeuvring fast enough; and
(d)   tug assistance is difficult to reproduce.

Computer simulation models

Calculation of ship manoeuvres on a computer was developed during the late sixties and seventies. At first, only simple manoeuvres such as turning circles and zigzag manoeuvres could be accurately reproduced. During the seventies the mathematical models became more advanced resulting in realistic simulation of complicated approach channel and harbour manoeuvres. At present the simulation technology is evolving further, more or less with the same speed as the increase in available computer power.

Simulation requires first of all a realistic mathematical model of the ship and the environment to which it is exposed. Simulation models must be able to include various hydrodynamic and aerodynamic effects as well as other effects:

(a)   forces acting on the underwater part of the hull, including shallow-water effects;
(b)   forces from rudder and propeller;
(c)   forces from thrusters;
(d)   tug forces with realistic time lag;
(e)   wind forces acting on the hull;
(f)   current forces;
(g)   hydrodynamic effects from vertical or sloping walls (bank effects);
(h)   wave effects;
(i)   passing ships in channels; and
(j)   use of anchors and moorings and contact with fenders.

A mathematical model, which contains all of the above effects, is described in Jensen *et al.* (1993).

Realistic modelling of the physical environment is important. This includes bathymetry, currents outside and inside the harbour, banks and dredged slopes, wind effects, including gusts and lee effects, and wave effects in certain cases. The current pattern may be complicated and calculations using advanced hydrodynamic tools such as the DHI MIKE 21 may be necessary.

All of these effects require accurate input of aero- and hydrodynamic forces. Such inputs are determined from various sources. Hydrodynamic forces are determined from both forced oscillation model tests in deep and shallow water as well as advanced computational fluid dynamic tools. Aerodynamic forces can be determined from either wind tunnel experiments or computational aerodynamic simulation models.

Experienced navigators, such as captains or local pilots, command the ship during the simulation. This is necessary to make the manoeuvres realistic and be able to draw valid conclusions from the results. The navigator observes the ship status from the visual view, the compass, the log, the rate of turn indicator and other instruments, and corrects the rudder angle, handle settings, thruster settings, etc., and even gives commands to assisting tugs.

The ship simulators can be divided into three types with increasing realism and complexity:

(a)   the desktop simulator;
(b)   the part-task simulator; and
(c)   the full-mission simulator.

The desktop simulator usually runs on a high-end PC, which calculates the ship motions, environmental conditions and graphic user interface, while it controls inputs at the same time. Such a simulator cannot only show a birds-eye view of the ship and surroundings but sometimes also a three-dimensional, out-of-the-window look. An example of such a desktop simulator is shown in Fig. 5.6. The radar, birds-eye view, the control handles and the three-dimensional representation are shown on one screen.

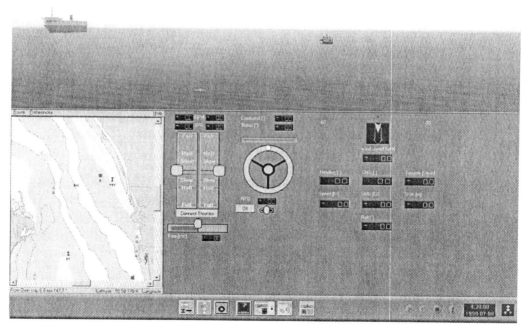

Fig. 5.6 Example of a desktop simulator

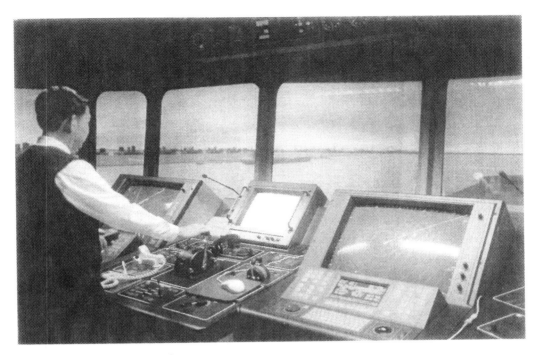

Fig. 5.7 Full-mission simulator

The next step is a so-called part-task simulator, which has a variety of combinations of CPUs and visual channels. The major improvement over the desktop simulator is the addition of real hands-on controls, which enable the navigator to concentrate on screens emulating real instruments and/or visual channels.

The final step is the full-mission simulator, which includes handles and instruments as found on-board real ships and a full three-dimensional, out-of-the-window view. The ship's bridge is a mock-up, physical representation of a real bridge layout. In the full-mission simulator the navigator experiences the feeling of being onboard a ship. An example of such a mock-up is shown in Fig. 5.7.

Each approach or departure run can be evaluated afterwards in a replay system. During each run all relevant parameters such as ship position, speed, engine RPM, rudder angle, etc. are stored for use later on. A very important part of each run is the debriefing, where the navigator, experienced simulator instructors, port planners and naval architects discuss the results. A computer program logs all relevant parameters as function of time and works as a sort of tape recorder, where the timing and use of manoeuvring devices (rudder, thrusters, etc.) can be investigated in detail on a graph during the debriefing session. A typical approach to a harbour with subsequent turning and berthing is shown in Fig. 5.8.

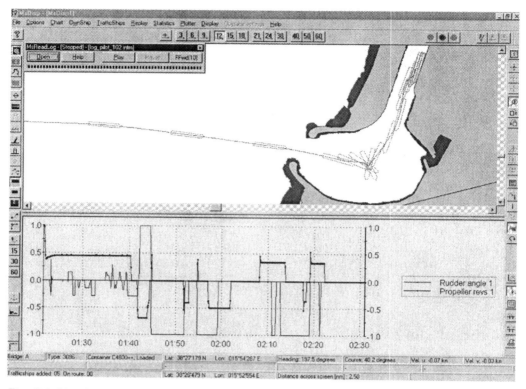

Fig. 5.8 Simulated harbour entry for large container vessel shown in replay program (reproduced by permission of the Gioia Tauro Harbour Authority, Italy)

From such simulations the approach and berthing of various ships can be investigated. It is also possible to determine:

(a) manoeuvring approach strategy;
(b) turning strategy;
(c) weather limits with respect to wind, current and waves;
(d) required tug assistance, number and sizes of tugs;
(e) required bow or stern thruster capacity; and
(f) training requirements for navigators/pilots.

When a new harbour or a harbour modification is being designed, perhaps the most important result of real-time simulation is the identification of high-risk areas of grounding or collision (Sand *et al.* (1994)). This may influence the required dredging volume. To be able to determine high-risk areas of grounding it is essential to perform many runs in various environmental conditions. This can sometimes be both time consuming and expensive. A so-called fast-time simulator can be employed (Arndal and Yañez,1999; Sand and Petersen, 1998). An artificial navigator, in

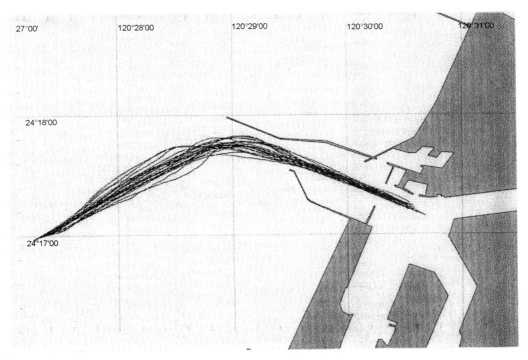

Fig. 5.9 Results of runs produced by a numerical navigator

the form of a computer program, controls the ship and many runs can be carried out rapidly so that a track envelope is obtained. Figure 5.9 shows the result of many runs produced by such a numerical navigator.

References

Arndal, N. and Yañez, M. A. (1999) 'Use of fast-time and real-time simulation in harbour design: The Dos Bocas case', *MARINA 99, Lemnos, Greece*, May.

Bunt, J. D. van den (1973) 'Model study on mooring forces due to regular and irregular waves', in *Analytical Treatment of Problems in the Berthing and Mooring of Ships*, Wallingford: NATO Advanced Study Institute.

DHI MIKE 21: http://www.dhisoftware.com

Hwang, L. S. and Divoky, D. (1977) 'The design factors affecting container-ship loading rates', *Ports '77. ASCE*.

Jensen, P. S. *et al.* (1993) 'DEN-Mark1 – An innovative and flexible mathematical model for simulation of ship manoeuvrability', *MARSIM, St Johns, Newfoundland*.

Khanna, J. and Birt, C. (1977) 'Two mooring dolphin concept for exposed tanker terminals', *Ports '77. ASCE*.

Lloyd's Register of Shipping (1980) *Statistical Tables,*. London: Lloyd's Register.

Lloyd's Register of Shipping (1989) *Rules and Regulations for Classification of Ships*, London: Lloyd's Register,

Lützen, M. (1998) *Dimensionering for mindre kontaktskader*, (in Danish) PhD Dissertation (unpublished).

Lützen, M. and Terndrup Pedersen, P. (1998) 'Design Against Minor Impacts', *International Symposium: Advances in Bridge Aerodynamics, Ship Collision Analysis, Operation and Maintenance*, Technical University of Denmark, Lyngby, May.

Oil Companies' International Marine Forum (1977) *Prediction of Wind and Current Loads on VLCCs*, London: OCIMF.

Oil Companies' International Marine Forum (1978) *Guidelines and Recommendations for Safe Mooring of Large Ships at Piers and Sea Islands*, London: OCIMF.

Oorschot, J. H. van. (1976) 'Subharmonic components in hawser and fender forces', *Proceedings of the 15th Coastal Engineering Conference, ASCE.* New York.

PIANC (1995) *Criteria for Movements of Moored Ships in Harbours – A Practical Guide*, Supplement to Bulletin 88.

Sand, S. E. *et al.* (1994) 'Risk Analysis of Simulated Ship Approaches to Ports', *PIANC CONGRESS*, Seville, Spain.

Sand, S. E. and Buus Petersen, J. (1998) 'Use of Ship Simulators in Bridge Design', *International Symposium: Advances in Bridge Aerodynamics, Ship Collision Analysis, Operation and Maintenance*, Technical University of Denmark, Lyngby, May.

Seidl, L.H. (1973) 'Predictions of motions of ships moored in irregular seas', in *Analytical Treatment of Problems in the Berthing and Mooring of Ships*, Wallingford: NATO Advanced Study Institute.

Slinn, P. J. B. (1979) 'Effect of ship movement on container handling rates', *The Dock & Harbour Authority*, London, August.

Stammers, A. J. and Wennink, C. J. (1977) 'Investigation of vessel mooring subject to wave action', *The Dock & Harbour Authority*, London, April.

Svendsen, I. A. (1973) 'Ship Manoeuvres in Harbours', in *Analytical Treatment of Problems in the Berthing and Mooring of Ships*, Wallingford: NATO Advanced Study Institute.

Svendsen, I. A and Vinther Jensen, J. (1970) 'The form and dimensions of fender front structures', *The Dock & Harbour Authority*, London, June.

Wilson, B. W. (1973a) 'Progress in the study of ships moored in waves', in *Analytical Treatment of Problems in the Berthing and Mooring of Ships*, Wallingford: NATO Advanced Study Institute.

Wilson, B. W. (1973b) 'Elastic characteristics of mooring ropes', in *Analytical Treatment of Problems in the Berthing and Mooring of Ships*, Wallingford: NATO Advanced Study Institute.

CHAPTER 6

Access channels and basins

Dr Ian Dand

Introduction

The access channel and basin features, discussed below, comprise the following:

(a)   location;
(b)   alignment;
(c)   length;
(d)   width;
(e)   side slopes;
(f)   water depth;
(g)   dredging; and
(h)   aids to navigation.

It is assumed that a number of alternative port locations have been chosen and cargo throughput forecasts have been made for them. It is also assumed that the numbers of ship arrivals in each major category per annum (or per season) have been estimated.

Some of the properties listed above clearly influence each other. For example, location, width and alignment affect both capital and maintenance dredging requirements, while width and depth are related through their effects on the navigability of ships moving along the channel.

Access channels

General considerations

An access channel is defined as any stretch of waterway linking the basins of a port to the open sea.

The access channel is characterised by its width, depth and alignment, and is delineated for its users by aids to navigation such as marker buoys and beacons. Its location is important because:

(a)   it determines the wave, current and wind conditions met by the ships in the channel; and

(b)   it determines the amount of dredging required and hence both capital and maintenance costs of dredging.

Existing channels may be improved by deepening, widening and straightening. Deepening means that deeper draught ships can use the channel and those already using it can do so with a greater margin of safety. It also means that a greater part of the tidal cycle is available for the larger ships, thus helping to reduce congestion around high water.

Widening and straightening is generally carried out with the ultimate aim of improving the safety margins for navigation and reducing the time it takes to sail through the channel. Widening a channel still further allows safe two-way passage, thereby increasing its ship traffic capacity.

The design of both safe and efficient access channels is important. Their value to the port is clear and, by improving their efficiency, a greater number of ships and/or larger ships of greater carrying capacity may be enabled to use the port safely.

The access channel, therefore, is an important part in the transportation system of which the port is a key component. Provided the channel is properly designed, the system can work at its proper pace and capacity, thereby enabling the port to act as an efficient interface between sea and land transport.

There is a need for a straightforward and logical approach to access-channel design.

The design of access channels
The passage along the access channel usually takes only a comparatively short period of time. The channel may, therefore, be designed with this transient occupation in mind; it may be stepped so that vessels can use it on a falling tide, or have a fixed depth so that certain vessels will have a limited window of opportunity for its use during a given tidal cycle.

The channel design may be contrasted with that for berth and anchorage areas, where ships remain for one or a number of tidal cycles. Their depth must be adequate to allow a sufficient under-keel clearance at all states of the tide, and the available bottom area must be adequate to allow the ship to respond to wind, waves and currents if at anchor. At a berth, a suitably deep pocket must be provided. In certain cases, provision must be made for vessels to wait as a temporary measure near the channel, until the designated berth or anchorage becomes available.

All these facets must be considered when access channels are being designed or modified.

PIANC (1997) produced by a joint working group comprising members from, among others, the Permanent International Association of Navigation Congresses (PIANC) and the International Association of Ports and

Harbours (IAPH), provides methods and guidelines for design. It covers both concept design and detailed design, the former providing a rapid estimate of width and depth, while the latter recommends guidelines for more detailed analysis of critical features.

Concept design

For a new channel (or a new leg of an existing channel) an important early consideration is the capital cost of its dredging. As this is directly related to the channel's location, length, width, depth and side slopes, rapid estimates of these parameters are required for trade-off studies as the concept is developed. Simple design methodology is therefore required, which will define the channel dimensions accurately enough for meaningful trade-off studies to be carried out. This is achieved in PIANC (1997) by the following steps:

(a)   choose a design ship;

(b)   select alignment, using rules derived from best practice. This is going to determine the channel length;

(c)   derive the width, using rules relating elements of width to the beam of the design ship according to best practice; and

(d)   derive the depth from consideration of design ship draught, tidal variation, vertical ship motions and squat, bearing in mind certain vessel speed limitations imposed by the water depth.

The design ship

The purpose of the design ship is to provide a basis for channel design, which ensures safe passage of not only the design ship, but all the other ships which are forecast to use the channel. It may be the largest ship likely to visit the port, that which is the most difficult to steer (due to a large windage, perhaps) or that which carries a particularly hazardous cargo.

It may be that more than one design ship is necessary if the channel is to cater for a wide range of vessel types. Analysis of the actual and forecast traffic for the port can provide information on the probable design ship type(s) and size(s).

Alignment and length

Alignment and length considerations include:

(a)   the environmental conditions along alternative alignments;

(b)   the need to avoid obstacles (rocky bottom) or areas of extreme sediment accretion, which are difficult to dredge or require extensive maintenance dredging respectively, both of which are costly;

(c)   prevailing wind, waves and currents. Ideally prevailing wind, wave and current directions should not lie across the channel;

(d)   avoidance of bends near port entrances; and

(e)   the edge of the channel should be such that ships passing along it do not cause disturbance or damage. This is especially relevant for craft which make excessive wash. A revised alignment may direct wash away from sensitive areas.

Channel length should be the smallest possible, compatible with safety and the discounted stream of annual dredging costs. This minimises time taken to transit the channel, bearing in mind hydrodynamic limits on speed mentioned in 'Depth and speed' in this chapter.

Width

Width is discussed in PIANC (1997) for straight channel legs and for the more complex case of channel bends. The basic method is developed for straight legs as described below.

Straight legs   It is assumed that the width of a straight channel may be built up from a basic width to which are added a number of increments (see Fig. 6.1). The basic width is that required by the design ship to move in calm water with no wind. It is somewhat greater than its beam to allow for the inherent manoeuvrability possessed by the vessel, which governs its ability to maintain a straight course; a ship with poor manoeuvrability will generally require a greater width to maintain its course than one with good directional stability. Lest it should be thought that any ship should be able to maintain a straight course in calm water, in the absence of external effects such as wind and current, it may be mentioned that a single screw ship experiences a phenomenon known

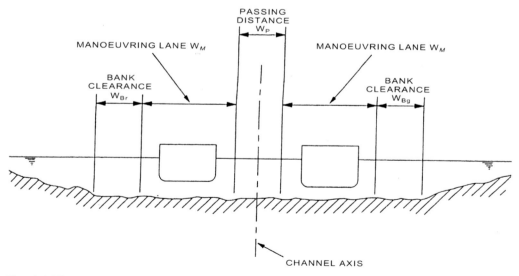

Fig. 6.1 Elements of channel width

as screw bias. This is the tendency of a rotating screw to move laterally (as well as forward) when behind a ship, and is caused by, among other things, the flow into the propeller being modified by the shape of the ship's aft body and the presence of the rudder. If, in addition, the ship is directionally unstable, it may have some trouble in maintaining a straight course, even in benign conditions.

Tables are given in PIANC (1997) for basic channel widths (for manoeuvrability), in relation to the beam B of the design ship; they range from 1.3B for a ship with good manoeuvrability to 1.8B for one whose manoeuvrability is poor.

In Table 6.1 width increments are given, which must be added to the basic width. These are in terms of the beam of the design ship B, and allow for the following:

(a)   vessel speed;
(b)   prevailing cross winds;
(c)   prevailing cross and longitudinal currents;
(d)   waves;
(e)   aids to navigation;
(f)   bottom surface;
(g)   water depth in the channel; and
(h)   level of cargo hazard.

Additional width increments are recommended for two-way traffic (to allow sufficient width between the opposing traffic streams for safety) and bank clearance. The latter allows for hydrodynamic effects which can cause the ship to sheer uncontrollably due to the proximity of an underwater or a protruding bank (see Tables 6.2 and 6.3).

The width increments in the above tables were derived empirically from best practice in a number of countries worldwide. Values from many existing access channels were collected and considered by PIANC (1980) before refinement for PIANC (1997). Some, such as the bank clearance increments, were derived from the results of physical model tests, while others were derived from multi-disciplinary discussion and the practical experience of local mariners and harbour masters. They were not developed using computer simulations of low-speed ship manoeuvring, because this important tool did simply not exist at the time the channels in question were designed. The term 'best practice' implies general satisfaction, which translates into very infrequent collisions and/or groundings. Thus it is likely that the recommended width increments err on the side of excessive safety, rather than the opposite.

Channel bends   While it is preferable to choose an alignment that minimises the number of bends, such features are not uncommon and must

Table 6.1 Additional widths for straight channel sections

Width w_i	Vessel speed	Outer channel exposed to open water	Inner channel protected water
Vessel speed (knots)			
Fast >12		0.1B	0.1B
Moderate >8–12		0.0	0.0
Slow 5–8		0.0	0.0
Prevailing cross wind (knots)			
Mild ≤15 (≤ Beaufort 4)	all	0.0	0.0
Moderate >15–33	fast	0.3B	–
(> Beaufort 4–Beaufort 7)	mod	0.4B	0.4B
	slow	0.5B	0.5B
Severe >33–48	fast	0.6B	–
(> Beaufort 7–Beaufort 9)	mod	0.8B	0.8B
	slow	1.0B	1.0B
Prevailing cross current (knots)	all	0.0	0.0
Negligible <0.2	fast	0.1B	–
Low 0.2–0.5	mod	0.2B	0.1B
	slow	0.3B	0.2B
	fast	0.5B	–
Moderate >0.5–1.5	mod	0.7B	0.5B
	slow	1.0B	0.8B
	fast	0.7B	–
Strong >1.5–2.0	mod	1.0B	–
	slow	1.3B	–
Prevailing longitudinal current (knots)			
Low ≤1.5	all	0.0	0.0
Moderate >1.5–3	fast	0.0	–
	mod	0.1B	0.1B
	slow	0.2B	0.2B
Strong >3	fast	0.1B	–
	mod	0.2B	0.2B
	slow	0.4B	0.4B
Significant wave height H_s and length λ (m)			
$H_s \leq 1$ and $\lambda \leq L$	all	0.0	0.0
$3 > H_s > 1$ and $\lambda = L$	fast	~2.0B	–
	mod	~1.0B	–
	slow	~0.5B	–
$H_s > 3$ and $\lambda > L$	fast	~3.0B	–
	mod	~2.2B	–
	slow	~1.5B	–
Aids to navigation			
Excellent with shore traffic control		0.0	0.0
Good		0.1B	0.1B
Moderate with infrequent poor visibility		0.2B	0.2B
Moderate with frequent poor visibility		>0.5B	>0.5B

Table 6.1 (continued)

Width w_i	Vessel speed	Outer channel exposed to open water	Inner channel protected water
Bottom surface			
If depth $\geq 1.5T$		0.0	0.0
If depth $< 1.5T$ then			
Smooth and soft		0.1B	0.1B
Smooth or sloping and hard		0.1B	0.1B
Rough and hard		0.2B	0.2B
Depth of waterway			
$\geq 1.5T$		0.0	$\geq 1.5T$
$1.5T–1.25T$		0.1B	0.0
$1.25T$		0.2B	$<1.5T–1.15T$
			0.2B
			$<1.15T$
			0.4B
Cargo hazard level			
Low		0.0	0.0
Medium		~0.5B	0.4B
High		~1.0B	0.8B

Notes: L = Ship length, B = Ship beam and T = Ship draught.
Source: PIANC (1997). Reproduced with permission from PIANC–IAPH.

Table 6.2 Additional width increments for two-way traffic

Width for passing distance, W_p	Outer channel exposed to open water	Inner channel protected water
Vessel speed (knots)		
Fast >12	2.0B	–
Moderate >8–12	1.6B	1.4B
Slow 5–8	1.2B	1.0B
Encounter traffic density		
Light	0.0	0.0
Moderate	0.2B	0.2B
Heavy	0.5B	0.4B

Source: PIANC (1997). Reproduced with permission of PIANC–IAPH.

be designed properly. Although the detailed design of a bend is best accomplished with computer simulation, some guidelines allow an early estimate of a bend design to be made. PIANC (1997) makes the following recommendations:

(a)   use as few bends as possible and attempt to avoid concatenating multiple bends, especially if they turn in opposite directions. Keep

89

the length of straight connecting legs between bends greater than five ship lengths if possible;

(b)   choose a bend radius compatible with the turning radius of the design ship. Turning radii vary with rudder angle and water depth/draught ratio, and Fig. 6.2 from PIANC (1997) gives typical values. They vary from 2.5 to 16 times the ship length;

(c)   choose a bend width large enough to accommodate the design ship when turning. When a ship turns, it does so with an angle of drift, which results in an increase in its basic width for manoeuvrability; in effect, the ship 'skids' around a corner. This width increase also depends on rudder angle and water depth and typical values are given in Fig. 6.3 from PIANC (1997). They vary from 1.05 to 2.0 times the beam. In very shallow water (relative to the draught of the ship) it should be noted that ships do not develop large drift angles in a turn and, as a consequence, do not need as much width increase; and

(d)   if possible keep the channel banks as low as possible to avoid the effects of ship-bank interaction. Generally this is not feasible, so width allowances may be made if bank effects are severe. Ship handlers sometimes use bank rejection to aid turning and this should enter into any considerations regarding the shape and placement of the outer bank on a channel bend. Such matters can only be resolved with the use of computer simulation at the detailed design stage.

Table 6.3 Additional width increments for bank effects

Width for bank clearance (W_{Br} or W_{Bg})	Vessel speed	Outer channel exposed to open water	Inner channel protected water
Sloping channel edges and shoals	Fast	0.7B	–
	Moderate	0.5B	0.5B
	Slow	0.3B	0.3B
Steep and hard embankments, structures	Fast	1.3B	–
	Moderate	1.0B	1.0B
	Slow	0.5B	0.5B

Source: PIANC (1997). Reproduced with permission from PIANC–IAPH.

Depth and speed

In channels, water depth and ship speed are linked. This is due to the large resistance experienced by ships when moving in shallow water at, or near, the so-called critical speed V_{crit} for which the Froude Depth Number F_{nh} equals unity:

$$F_{nh} = V_{crit}/\sqrt{gh} = 1.0$$

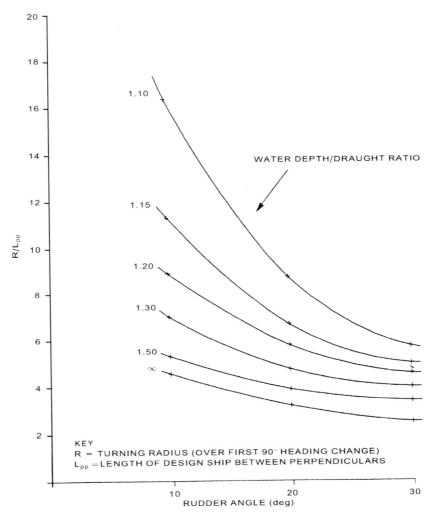

Fig. 6.2 Turning radius as a function of rudder angle and water depth (based on a single screw/single rudder container ship)

where h is water depth and 'shallow water' is defined as that having a water depth/ship draught ratio of less that 3.5.

For merchant ships this increase in resistance is a limiting factor on speed and many ocean-going vessels are unable to proceed at a value of F_{nh} greater than 0.60 to 0.70. Therefore, when determining channel depth, care should be taken to check the effect it may have on speed to ensure that the desired transit speed can in fact be achieved.

The minimum depth must allow for:

(a)   maximum draught of the design vessel;

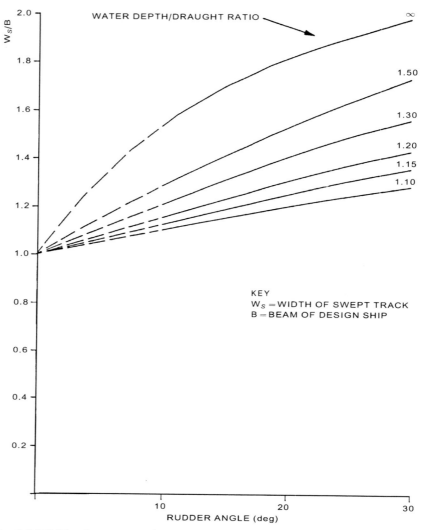

Fig. 6.3 Width of swept track in a turn as a function of rudder angle and water depth (based on a single screw/single rudder container ship)

(b) squat;

(c) vertical motions due to waves and swell (heave, pitch and roll);

(d) vertical motions induced by passing vessels;

(e) seabed type (mud, silt or hard bottom);

(f) dredging tolerance (usually 0.1 to 0.6 m). Sounding accuracy is normally about 0.2 m;

(g) heel induced by wind or other factors;

(h) water density; and

(i) tidal variation.

The draught of the design ship will have been derived from an analysis of the fleet of ships likely to visit the port (see Chapter 5).

Squat is the tendency for a vessel to change its under-keel clearance as it moves ahead or astern, or is passed by another vessel close by. It derives from the changes induced in the hydrodynamic pressure forces over the hull by the presence of the seabed and/or by the presence of the other vessel. Squat may be estimated in a number of ways and many are discussed and compared in PIANC (1997). The following simple expression from its 'Reference 2' is suggested in the *Guidelines for Concept Design*:

$$\text{Squat(m)} = 2.4 \frac{\nabla}{L_{pp}^2} \frac{F_{nh}^2}{\sqrt{(1 - F_{nh}^2)}}$$

where

∇ = volume of displacement $(\text{m}^3) = C_B \cdot L_{pp} \cdot B \cdot T$
L_{pp} = length between perpendiculars (m);
B = beam (m);
T = draught (m);
C_B = block coefficient; and
F_{nh} = Froude Depth Number.

A simple graphical method for large tankers is shown in Fig. 6.4, which is valid for full-bodied (bluff) tanker forms and will give conservative estimates for finer-formed ships.

Vertical motions due to waves are usually only considered at the detailed design stage by using mathematical models for sea keeping. For concept design a depth/draught ratio, based on experience, is used to allow for wave-induced vertical motions. In ports subject to wave action, depth/draught values of 1.3 or more are used.

In other ports, where wave or swell action is a dominant feature, and it is commercially important for ships to arrive or depart with as deep a draught as is compatible with safe operation, specialist computer programs are used. These use complex mathematical models to estimate ship motions (notably those in pitch, heave and roll) for a ship at a given loading in the wave and swell conditions prevailing at the time, or predicted for the future. Such models give the probability of some part of the ship striking the channel bed; knowledge of such information allows the shipping and harbour authorities to make a decision on the advisability of a ship using the channel in the prevailing conditions with her existing draught (if arriving) or her proposed draught (if departing). These computer models for under-keel clearance have to take account of the present and predicted states of the tide and variations in channel soundings below datum. They are used, for example in the iron-ore exporting ports of north-west Australia, where loading bulk carriers to

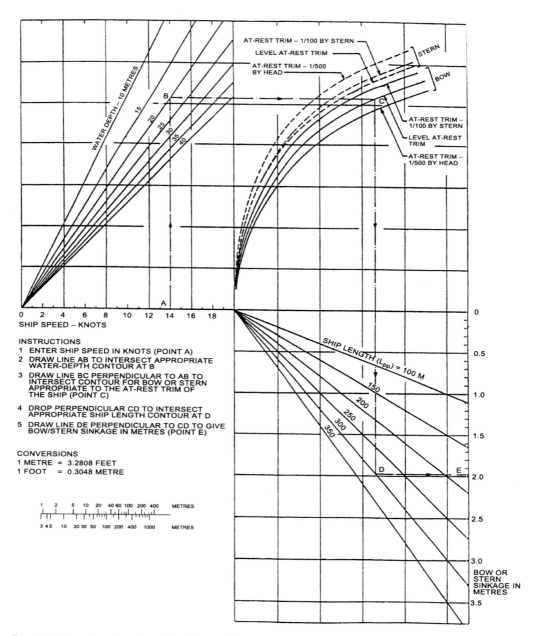

Fig. 6.4 Squat estimation chart for full-bodied ships

draughts, which are the deepest compatible with safety, is commercially very attractive.

When vessels pass each other, the balance of hydrodynamic pressure forces over their hulls may be affected. The effect is magnified in shallow or restricted water and, whereas it can cause one or both vessels to sheer, it can also cause significant increases in squat. As shown in PIANC (1997), up to ten times the squat of the same ship in isolation can occur. A further feature of this type of squat is that it can occur to a moored or berthed ship, when passed by another ship.

A further source of vessel-induced motion can arise if the channel is shared with a high-speed vessel. While such a vessel (a high-speed ferry for example) may be much smaller than the design ship, its wash could be substantial. This is because it may be operating in the trans-critical speed range, forming large, long period, diverging waves. These may be sufficient to cause a nearby large vessel to move vertically. If the under-keel clearance is small, a transient grounding could occur.

The type of seabed can affect the determination of depth. If the seabed is hard, then soundings are well defined and depth is known with some certainty. If the seabed is composed of mud or silt, then the bottom is ill defined. In this case, the concept of a 'nautical bottom' is introduced, defined as that level below datum for which the density of the seabed material is at least $1200 \, kg/m^3$.

The tidal cycle for the access channel will be an important consideration in the determination of depth. When the range is large, depth must be determined by comparing the costs of providing and maintaining a depth, which allows access at all states of the tide and the benefits of:

(a) reduced waiting times for ships entering or leaving the port. Limitations on access may force ships to wait elsewhere (or even use another port) or reduce their speed of approach below its optimum value; and

(b) lower freight rates resulting from the use of larger ships.

If restrictions to certain tidal windows are inevitable, the windows or tidal 'slots', should be as wide as possible to allow the required density and mix of vessels to have access to the port. For long channels a stepped depth profile may be appropriate, so that adequate depth is available as the tidal elevation changes along the channel.

Detailed design

While concept design will yield the broad parameters of an access channel, it is often necessary to carry out more detailed design for all, or part, of the waterway. In addition to reducing capital and/or maintenance costs, this may serve the purpose of making operators and users confident of the design. However, often the need arises because it has proved impossible, for various reasons, to stay within the boundaries and recommendations of the concept design. For example, vessels may have to negotiate a bend and then reduce speed to berth. If the distance in which the vessel must slow down is limited, a dilemma may arise because speed must be high enough to maintain control in the bend, but not too high to compromise slowing. Detailed assessment of the bend width and radius, together with stopping distances is therefore essential. Other examples, where more detailed considerations are needed, include bends which are tighter than normally recommended or of unusual shape, the effects of strong and variable currents, the effects of waves on manoeuvring in exposed channels and accurate estimation of vertical motions in channels through materials which are costly to dredge.

Further considerations at the detail design stage relate to the degree of risk involved in operating in the proposed channel and whether this is acceptable. In addition, it will be almost certain that knowledge of channel capacity will be required.

All of these require detailed consideration and this is almost universally accomplished by means of computational or physical models.

Computational models

For the detailed design of approach channels the following computational models are used:

(a) models of the environment including current/tidal stream, and, if appropriate, wind;

(b) tidal models, especially if the local tidal elevations are complex or likely to be modified by proposed developments;

(c) squat models;

(d) ship-motion models, used to predict the effect of the wave climate on under-keel clearance, with special emphasis on the probability of striking the bed of the channel;

(e) ship manoeuvring models to assess width, alignment and stopping zones. These models are also invaluable in allowing mariners (especially local pilots) to gain a 'feel' for the channel at the design stage and to provide empirical input to its design. They may also be used for assessing the number, type and location of the aids to navigation required to mark the channel; and

(f) if it is a busy channel, its effect on marine risk can be assessed by a variety of risk models, the most valuable being the marine traffic simulator. Such models determine the number of 'encounters' which occurs between ships over a given time period and, from this information, it is possible to deduce a collision frequency probability, from which collision risk can be deduced.

Some of the above models are discussed briefly below. A more detailed description of their use is given in PIANC (1997).

Squat models Methods for the estimation of squat are commonplace and in PIANC (1997) several are discussed. They range from simple empirical methods to complex computer models. A comparison shows a wide scatter between estimates made by these means, although there is some 'common ground' in the middle of the scatter (see Fig. 6.5). Almost all squat estimation methods are for sub-critical speeds.

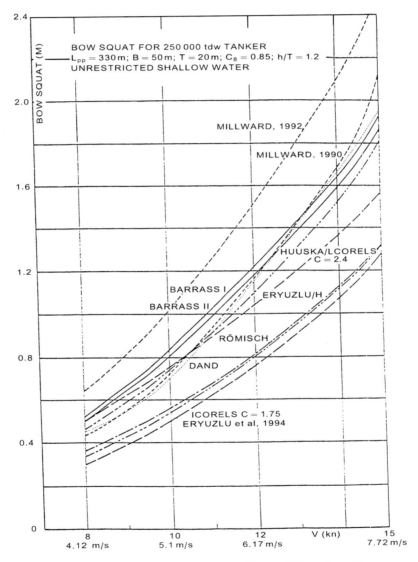

Fig. 6.5 Bow squat for 250 000 tdw tanker ($L_{pp} = 330\,m$; $B = 50\,m$; $T = 20\,m$; $C_B = 0.85$) *in unrestricted shallow water* ($h/T = 1.2$)

Ship motion models   Models to compute all six degrees of freedom of motion of an arbitrary floating body are common and well established. Many operate in the frequency domain and usually assume the system to be linear so that extreme non-linear motions are not accounted for. Others operate in the time domain, using a time-stepping technique, which allows non-linearity to be incorporated.

Frequency domain models derive Response Amplitude Operators (RAO), similar to transfer functions, from the actual shape of the ship in deep or shallow water. The RAO values are then used with the actual or assumed wave spectrum to provide motion, velocity and acceleration spectra. These are valuable because they allow various probabilities to be derived, not least of which is the combined probability of pitch, heave and roll causing some part of the vessel to strike the seabed.

Ship manoeuvring simulation models   Ship manoeuvring models are common in many maritime countries and universally operate in the time domain. Their use ranges from the full mission bridge simulator to the PC-based design tool. The former is the marine equivalent of the aircraft simulator and is generally used for a similar, training role. The latter are used largely for design of ships, ports and approach channels. As they can operate at speeds faster than real time, they are useful in generating large amounts of data in comparatively short time scales.

It is normal to use the 'fast-time' simulator to assess and, if necessary, modify the proposed channel design. Alignment and width are the parameters for which such a process is best suited, although some indirect information on depth (especially its effect on speed and manoeuvring) is also obtained. The fast-time simulator may be manually or automatically controlled. Care should be taken in assessing the results, when the simulation is automatically controlled, as accuracy depends on the type of control used. Some controllers have a built-in randomness which mimics human variability and can give a reasonable estimate of the spread of tracks and hence the suitability of the proposed channel width. Automatic control, using prescribed waypoints, provides a rapid means of obtaining multiple tracks for this purpose. Figure 6.6 shows an example. In addition to such measures, the fast-time simulator also shows the ease, or otherwise, with which the design ship can be manoeuvred along the proposed channel.

The ways in which the ease of manoeuvring can be measured are mentioned below, but one advantage of the fast-time simulator is that many repeat runs can be carried out and overlaid. The resulting envelope (Fig. 6.6) soon reveals areas where more or less dredging is required, thereby allowing dredging costs to be minimised.

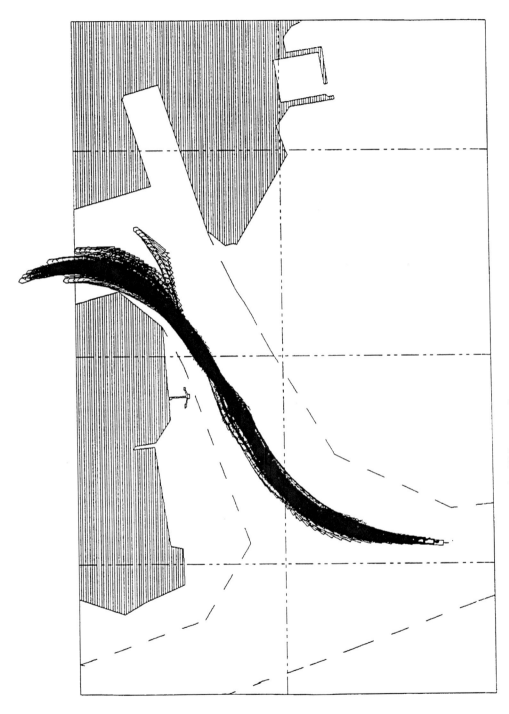

Fig. 6.6 Large inbound container vessel – Ebb current runs

Once the channel design has been refined in this way, or earlier if possible, it should be subjected to the scrutiny of the professional mariner. This is best done in a full-mission simulator, where the mariner can experience control and manoeuvrability in the channel. This usually results in constructive comments on width and alignment, together with the important assessment of the necessary aids to navigation. These are the mariner's main visual cues, which, together with his radar, define for him the channel's width and alignment.

Manoeuvrability at low speed, its computer simulation and other evaluation tools are further discussed in Chapter 5.

Marine traffic models   The values of a marine traffic simulation model are many, especially if it includes a berthing/dwell/unberthing model for the vessels visiting the port. As far as the approach to channel design is concerned, the traffic simulator can explore the feasibility of one- or two-way operation, together with congestion, waiting and queuing if certain ships are given exclusive use.

Berth operations may be affected by such a channel, if its capacity is too low to satisfy the expected traffic and ships must queue for access. Alternatively, if the potential capacity of the channel far exceeds the present and anticipated traffic, it will be unnecessarily expensive. Both of these aspects can also be studied using marine-traffic simulation.

Finally, if the channel's width, depth and alignment is such that a number of ships can pass only at certain states of the tide (rather than at all times) then berth operations (at a ro/ro or ferry terminal for example) would be severely compromised. With its ability to model queuing and the number of encounters in a given channel, the marine traffic simulator is able to assess the channel's suitability for a given system of berths. Moreover, it is able to investigate the effects of changes to the channel width, depth and alignment so that suitable modifications may be made. Special berthing procedures, such as 'Berth A must be sterilised while operations are underway at Berth B' can also be studied, as can the effect of changing channel or basin dimensions to eliminate the need for such procedures.

The essence of the marine traffic simulator is that it deals with a multiplicity of ships, their behaviour and operation, rather than concentrating solely on one vessel, the design ship, as in manoeuvring simulation. The traffic model is usually calibrated against the existing traffic flowing through the port, an exercise that itself needs further computer models to read and analyse field measurements, generally obtained from the port radar. The model then has the approach channel inserted and its effect on the following are measured:

(a)   berth operations due to delays in an arrival at available berth, loading, and in departure from the berth (see above);

(b)   congestion (measured by the surface density of vessels);

(c)   vessel speed profiles and delays caused by congestion or exclusive use channel operation; and

(d)   encounters (defined when one vessel needs to take avoidance action because of the proximity of another vessel or vessels). Encounters are triggered when one or both vessels have to make a change of course to avoid the other.

On completion of the simulator runs, the results are used to determine the level of marine (or, more specifically, collision) risk. Collision risk comprises two components:

(a)   frequency; and

(b)   consequence.

Traffic simulations provide quantitative information for the former. It is tempting to assume the encounter frequency to be a measure of the risk of collision (generally the main focus of marine-traffic risk), but a direct connection between the two has so far proved elusive. Use of encounter frequency alone is not without its problems, because very congested harbours generally allow a much higher encounter frequency with no significant increase in collisions.

The consequences of collision or grounding frequencies, deduced from marine traffic simulations, are assessed in a number of ways:

(a)   the potential for loss of human life not only on the ship itself, but also on shore. This can be determined from a knowledge of the number of passengers or type of cargo the vessel is carrying as well as the population density on shore;

(b)   the potential for damage to the environment, in terms of both flora and fauna. This would overlap with any Environmental Impact Assessment (EIA) being undertaken in connection with a particular marine-traffic development; and

(c)   if neither of the above are relevant then commercial consequences in terms of loss of time, business and cargo must be assessed.

Evaluation of all of these aspects is in human, environmental and, generally, in monetary terms.

In an attempt to provide more realistic results, recent traffic models allow each vessel to manoeuvre and avoid other vessels by adhering to the International Regulations for the Prevention of Collisions at Sea (the ColRegs) or any local variant of these, which may be in force. If such a model is properly calibrated, the way is open to collision frequency modelling, so that a better assessment of marine risk and its relation to channel design, may be made.

Further information on marine traffic simulation can be obtained from PIANC (1997), Dand and Colwill (2001) and Lyon and Koh (2001).

Physical models

The navigational aspects of channel design can, if necessary, be studied using physical models. Although the advent of sophisticated computer models has diminished the need for physical models, there are some occasions when their use is preferable. These are:

(a) if the design vessel is of a new or unusual type for which no suitable computer model exists;
(b) if the waterway passes through complex structures such as a flood barrier or a narrow bridge;
(c) if ships in an approach channel pass berthed vessels, all subject to complex wave and current effects;
(d) if complex, multiple ship passing situations are likely;
(e) if the current regime is particularly complex;
(f) if wave action is complex and will have a marked effect on ship handling; and
(g) if there is propeller scour in the channel and associated basins.

Added to this list are the model tests necessary to develop a manoeuvring simulation model. These may be commissioned if the design vessel is particularly innovative and a complex simulation is needed. Finally, there are, of course, 'one-to-one' full-scale model tests (i.e. trials) whose results are used to calibrate a manoeuvring simulation model of a particular ship.

Physical model tests cannot be carried out at distorted scales and have the disadvantage that, as the scale should be as large as possible, large areas are needed for the channel model. Occasionally it may be possible to run the manoeuvring/ship-handling experiments in concert with a suitable hydraulic model, but often the scale of a conventional hydraulic model is too small for adequate ship-handling studies.

Ship models for this work are generally free-running and radio-controlled. While valuable for this purpose, they suffer from some problems deriving from the effects of scale:

(a) for a given speed, the propeller thrust will be too high for scale and may make the rudder more effective than it should be;
(b) all motions take place in an accelerated time scale so that care must be taken in assessing ship-handling motions in which human judgement and reaction times are important; and

(c)   in many models the human operator has a 'bird's-eye' view from above, rather than from the wheelhouse. The advent of miniature cameras and large manned models minimises problems from this source, but, should these not be available, care must again be taken in assessing the results.

In spite of these and other caveats, physical model tests can provide valuable information and, in some cases, may give the only means by which a solution can be obtained.

Analysis of the results of physical model tests, and the criteria by which they are judged, are the same as those used with computational models.

Safety criteria

When the models described above are being used, some means of judging the results is necessary. Sometimes expert judgement will suffice, but often an appeal to more quantitative safety criteria is required.

International agreements for such marine safety criteria are lacking, although some exist which enable the designer to determine whether his channel design is likely to be both safe and cost-effective. PIANC (1997) list a number of these, which are both deterministic and probabilistic. Examples are:

(a)   rudder activity criteria;
(b)   width criteria;
(c)   depth criteria; and
(d)   alignment criteria.

Rudder activity criteria

Rudder activity can be measured easily when ship manoeuvring simulations or physical models are used and it seems intuitively obvious that it must provide a measure of the adequacy and safety of a channel design. Although there is some truth in this, it should be treated with caution because:

(a)   rudder activity is also a measure of the ship's inherent manoeuvrability;
(b)   rudder activity is as much a measure of the ship handler's technique as of the channel design; and
(c)   rudder activity is not very sensitive to changes in channel design.

Some criteria are:

(a)   the mean rudder angle required to counter different phenomena should vary from less than $5°$ for bank effects to $25°$ for cross wind and currents;

(b)   the standard deviation of the mean rudder angle should not be greater than 17°;

(c)   the root-mean-square (rms) rudder angle should not be greater than 10° for large tankers in straight channels; in this context, the rms is the root-mean-square of the rudder-angle time history obtained over the channel transit;

(d)   the maximum rudder angle should be less than 15° for 75 per cent of the channel transit;

(e)   the maximum rudder angle should not exceed 20° for a ship with good manoeuvrability; and

(f)   the number of crossings of the mean rudder-angle value should not be more than 0.1 per second.

Width criteria

Width criteria are both deterministic and probabilistic. An example of the former is: 'the maximum swept lane width should not be greater than 70% of the available lane width' (Ammar, 1969).

Probabilistic criteria are related to groundings and the result of a detailed survey of grounding incidents in north European ports was a remarkably consistent grounding rate of 0.03 incidents per 1000 ship movements or one grounding per 33 333 ship movements.

Techniques are available with multiple runs on manoeuvring simulations, whereby the probability of passing beyond the channel boundaries can be estimated. This can then be compared with an 'acceptable' historical level, such as the above, to determine the adequacy of channel width.

Depth criteria

Deterministic depth criteria usually relate to the under-keel clearance, which is to be allowed once all other components of depth (see 'Concept design' earlier in this chapter) have been taken into account. The following values are normally used:

(a)   0.3 m for a muddy seabed;

(b)   0.5 m for a sandy seabed; and

(c)   at least 1.0 m for a hard or rocky seabed, measured from the top of the rocks.

Probabilistic criteria are usually of the form: 'The probability of one seabed contact in x years use shall not exceed y.'

Alignment criteria

Criteria for the channel alignment are an amalgam of those related to width and depth above as well as to channel bends. Track envelopes, derived from multiple simulation runs, have to satisfy the width criteria.

However, it is in regard to this criterion that expert opinion is perhaps most valuable in assessing the adequacy of the design.

Basins

The design of basins should give due consideration to the operational requirements of the ships which use them. Basins are classified as open basins or closed basins. An open basin is an area (usually sheltered), which allows stopping and turning manoeuvres to be made. A closed basin is an enclosed area with an entrance within which ships berth and may or may not turn.

The open basin may well be formed as the termination of an access channel, while the closed basin can be part of a dock system, where ships may be locked in and out. In both areas, ships will be manoeuvring at very low speeds, and are likely to be turning, breasting (or crabbing i.e. moving sideways), berthing and unberthing.

Open basins

These require special design considerations with respect to size and depth, the former being well suited to solution by simulation using multiple runs. The vessels may be able to manoeuvre using their own control devices (such as propellers, rudders and thrusters) or they may be under the control of tugs. Information is needed on:

(a)   thruster power available and hence the time to swing;
(b)   the tidal streams and currents in the area, which may cause the vessel to drift while swinging;
(c)   waves;
(d)   wind;
(e)   depths;
(f)   the number and size of tugs available;
(g)   tidal range, if the vessel has to be swung in a given time for a given tidal window; and
(h)   area swept out by the vessel when manoeuvring.

Although it is usual to allow some multiple of the ship length to define a 'swinging circle' it is seldom that a vessel, when swung by tugs or when using its own thrusters, will in fact sweep out a true circle. Local currents, slight differences in tug control and centripetal effects all combine to cause the vessel to move ahead or astern while swinging. Similarly, it may be appropriate to swing a vessel off the berth, in which case the vessel may need to rotate through 180° or less before beginning to move ahead, long before any approximation to a circle has been swept out.

These issues are best resolved using manoeuvring simulation. Multiple tracks can be superimposed to show the probable area swept out by the ship(s) and the dredging requirements determined accordingly. The required number and size of tugs can be determined and their disposition around the ship assessed by this means, not forgetting the space needed for the tugs to operate. If significant wind or current regimes occur in the region, safe operating limits for berthing and unberthing can also be determined.

Closed basins

In the open basin, tides, wind and currents will be present from natural causes. They are measurable and predictable. Currents may also exist in closed basins, but they are not all likely to be of a natural origin, some being generated by the ship itself through the slipstream of its propellers and thrusters while turning or swinging. This is not predictable and, if the manoeuvre is prolonged, may have a marked effect, not only on the swinging ship, but also on any other ship in the basin. The more powerful the tugs or thrusters are, the more pronounced will such currents become.

Not only do the currents affect ships within the basin, they also stir up the basin bed material. This in turn can result in bed material being re-deposited along the periphery of the basin and in the berth pockets. It is difficult to alleviate this effect, other than to keep the power used on propellers, thrusters and tugs to a practicable minimum, consistent with control.

These effects, while important, are not easy to model mathematically and it is usual to resort to physical models should they be perceived as a potential problem. An example is the modelling used to investigate the effects of propeller scour. This is of particular interest in berths used by high-powered ships (such as container vessels), where local scour at the berth and swinging basin can be severe. Further information on this topic can be obtained in Oebius (1984).

A further effect arises from the current generated by a tug's slipstream impinging on the hull of the vessel it is attending. This can have the following effects:

(a)   the slipstream can 'push' the ship the tug is trying to 'pull', thereby reducing its effective bollard pull. Model tests have shown that such reductions can be very large; and

(b)   under the action of the slipstream near the ends of the vessel, it may actually turn in a manner which is the opposite of that intended by the tug.

These effects can be alleviated by ensuring that:

(a)   the tugs operate on long lines; and/or

(b)   the tug slipstream is directed away from the vessel it is attending by pushing instead of pulling.

The design of a closed basin should therefore allow tugs as much room as possible in which to operate.

A final point to consider in closed, unlocked basin design is the necessity to give a vessel adequate room to reduce speed and, if necessary, stop when entering the basin. If a vessel has to lose speed too rapidly, its wake can overtake it and, in extreme cases, move it forward and turn it in an uncontrolled manner (see Corlett, 1979). This can be to the detriment of the ship itself and to other ships and/or damage berths or jetties within the basin. Care should therefore be taken to ensure that the basin approaches are of adequate length to ensure a controlled and gradual loss of speed. Manoeuvring simulation can be used to determine an appropriate length.

References

Ammar, A. A. *et al.* (1969) 'Design of navigation canal cross section and alignment', *PIANC 22nd International Congress, Section 2, Subject 3*, Paris.

Corlett, E. C. B. (1979) 'Studies on interaction at sea', *Journal of Navigation*, 32: 2.

Dand, I. W. and Colwill, R. D. (2001) 'The development and application of a dynamic marine traffic simulator for the assessment of marine risk', *Inaugural International Conference on Port and Maritime R and D and Technology*, Singapore, October.

Lyon, P. R and Koh, K. S. H. (2001) 'Marine traffic simulation for port planning', *Inaugural International Conference on Port and Maritime R and D and Technology*, Singapore, October.

Oebius, H. U. (1984) 'Loads on beds and banks caused by ship propulsion systems', *Conference on Flexible Armoured Revetments Incorporating Geotextiles*, Institution of Civil Engineers, London, March.

PIANC (1997) Report of the Joint PIANC/IAPH Working Group, in collaboration with IMPA and IALA, *Approach Channels. A Guide for Design*, supplement to Bulletin 95, June.

PIANC (1980) International Commission for the Reception of Large Ships (ICORELS). *Report of Working Group IV*, supplement to PIANC Bulletin 35.

CHAPTER 7

Breakwaters

Ole Juul Jensen and Professor, Dr Hocine Oumeraci

Part I by Ole Juul Jensen

Introduction

Influence of breakwaters on site selection

In cases where there are two or more potential sites for a port on an open coast, the breakwaters may have considerable influence on the site selection because of their relatively high costs.

In order to minimise breakwater costs, the following factors should be considered:

(a)   *Total length* to be as small as possible.

   On a coast where the waves come from a narrow sector only (Fig. 7.1), minimum length is obtained if the port is placed at a headland with an offset of the coastline sufficiently large to allow for the necessary port area, at the same time as one relatively short breakwater can reach sufficient depth, by taking advantage of the curvature of the contour lines.

   On the coast in Fig. 7.2, where the wave sector is wide, the most favourable site would be between two neighbouring headlands particularly if the necessary dredging inside the port area is only of loose sediments.

(b)   *Depths along breakwater* to be as small as possible.

   The rocky shoal in Fig. 7.3 would present a most favourable location for a breakwater to be placed along its ridge and thus may dictate the site selection.

(c)   *Rocky material* available within a short distance.

   The coastal rocky ridge shown in Fig. 7.3 will allow a short haul by dump truck. However, environmental constraints may prevent this in many cases.

(d)   *Rock quarry* should yield large blocks.

   It is often a problem to get a sufficiently high percentage of larger blocks resulting from rock blasting.

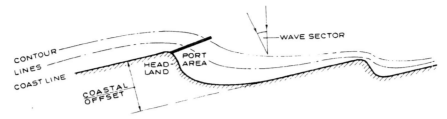

Fig. 7.1 Port site selection and breakwater alignment for narrow wave sector

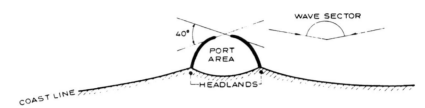

Fig. 7.2 Port site selection and breakwater alignment for wide wave sector

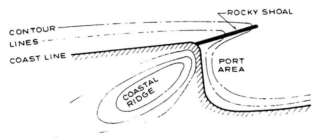

Fig. 7.3 Port site selection and breakwater alignment for shoal and coastal ridge

(e)   *Foundation conditions.*

    Along the breakwater alignment foundation conditions should not be so poor as to cause stability or settlement problems.

(f)   Production and transport of caissons.

    If floating caissons have to be used the feasibility of fabricating and transporting them to the site must be studied.

Relations between breakwater and layout of port

Figures 7.1 to 7.3 also illustrate the influence of port layout on breakwater costs. The *locations and alignments* of the breakwaters are determined by:

(a)   size of port area to be protected;

(b)   degree of shelter required at berths;

(c)   ship manoeuvre requirements in entrance and basins;

(d)   influence of breakwaters on currents;

(e)   influence of breakwaters on harbour resonance (long waves and tsunamis); and

(f)   influence of breakwaters on sediment transport, as well as on accumulation and erosion.

Figure 7.2 shows as one example that an angle of 40° between the breakwaters gives a minimum of sedimentation inside the port. This was determined by hydraulic model tests with waves combined with a strong tidal current along the coast.

Breakwater costs may be reduced considerably if such location and alignment are possible, which will make the higher storm waves break (because of depth reduction) before they reach the breakwater.

While a breakwater is constructed on an open coast there will normally be a tendency to *erosion around the end* of the structure. Such erosion may be avoided by placing a scour protection blanket on the seabed prior to the construction of the rest of the breakwater.

In some cases the breakwater will cause heavy *sedimentation along its outside*. This is particularly true of breakwaters on coasts exposed to swells during a major part of the year, as illustrated by Fig. 7.4. With the shown average wave direction and the breakwater alignment A_1 a reservoir R_1 is formed, which has to be filled with sand before sand accumulates along the outside of the breakwater. With the other alignment A_2, the

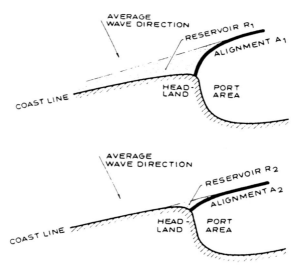

Fig. 7.4 Breakwater alignment for coast with large littoral drift

reservoir R_2 is much smaller in volume and thus accumulation along the breakwater starts sooner.

With proper planning of the location, alignment and construction programme a large littoral drift may be utilised for a considerable cost reduction. The depth at which breakwater construction takes place at any stage may be much smaller for alignment A_2 than for A_1. There are examples of breakwaters, with alignment such as A_1, which have largely been built in, say, a depth of 10 m, whereas A_2 would have led to such sedimentation along the outside of the breakwater that construction could have taken place in a depth varying from, say, 5 m on its outside to 3 m on its inside. This would reduce not only the volume of rock but also the weight of armour units.

Types of breakwaters

There are three main types of breakwaters, characterised by their *seaward faces*:

(a) The *sloping breakwater* (S), which has a seaward face with a slope of 1 on 1 or flatter. It usually consists of a rubble mound, which is often supplemented by concrete armour blocks and/or a concrete superstructure;

(b) The *vertical breakwater* (V), which has a vertical or near vertical seaward face and usually consists of concrete caissons; and

(c) The *composite breakwater* (C) consisting of a vertical breakwater, placed on a foundation, usually a rubble mound.

Functional requirements

Functional requirements are defined as all factors, which influence the design of the breakwater.

These factors are summarised in Table 7.1.

Climatic, oceanographic and hydrographic conditions

It should be noted that design of a breakwater cross section is in particular a function of:

(a) wave heights;
(b) wave directions; and
(c) water levels.

Each of these factors has its statistical distribution. Therefore the results of model tests have to be considered with a *risk analysis*, which combines these three distributions. The water level is often a combination of the deterministic astronomical tide and a stochastic storm surge. Since

Table 7.1 Master list for breakwater design

Item no.	Function	Design requirement for breakwater (BW)
1	Adequate construction materials available	Choose proper type S, V or C
2	Adequate contractors' experience and equipment available	Choose proper type and construction method
3	Adequate labour available	Choose proper type and construction method
4	Acceptable wave disturbance in harbour	Height of BW sufficient to limit generation of waves inside harbour by overtopping. Permeability of BW sufficiently small to limit transmission of waves
5	Reduce wave action in front of BW	Choose type S if waves reflected from type V adversely affect small craft navigation
6	Increase wave action in front of BW	Choose type V or C if reflected waves are desirable for increasing sediment transport in front of BW
7	Withstand wave action on seaward face	S and C: stability and strength of blocks in armour layer. V and C: static and dynamic stability of caissons. Strength of walls
8	Withstand wave action from beneath	Wave pressures on bottom of superstructure or caissons to be reduced by ventilation, drains, etc. and to be included in stability analysis (7 above)
9	Withstand wave action over the top	S: stability of superstructure or crest blocks. V and C: strength of deck slab
10	Withstand wave action on harbour side	S: stability of blocks on harbour side. V and C: strength of harbour-side wall
11	Avoid erosion on sea side	Protect seabed against wave action
12	Avoid erosion on harbour side and/or underneath BW	Place filter on harbour side and/or underneath BW (or seal at sea side)
13	Secure foundation	Stability and settlement for combinations of own weight, wave load and other live load, if any
14	Avoid sliding	Let superstructure elements and caissons interlock and make bottom sufficiently rough
15	Resist earthquakes	Secure foundation from liquefaction
16	Resist pressures from sand fill	V and C: design walls of caissons for inside pressures plus wave suction
17	Resist differential water pressures due to tides	V and C: design for excess inside water pressures or secure efficient drainage of fill
18	Reduce sedimentation	Permeability of BW sufficiently small to limit sediments from being washed through by wave action. Avoid excessive overtopping of water with suspended material
19	Allow for ice conditions	S: consider piling up of ice floes. V and C: consider impact from ice floes
20	Allow for cold weather	Avoid excessive overtopping that might lead to heavy icing of quay and ships berthed along harbourside

Table 7.1 (continued)

Item no.	Function	Design requirement for breakwater (BW)
21	Retain fill for reclaimed area behind BW	Design for earth pressure. Reduce overtopping and provide drainage
22	Allow passage of maintenance vehicles	Avoid overtopping in periods during which maintenance takes place
23	Utilise BW for ship manoeuvres inside harbour	Design for mooring and/or berthing forces. Reduce overtopping sufficiently for mooring crew to work on BW
24	Accommodate waiting ships at berths along inside of BW	Design for berthing and mooring forces. Reduce overtopping sufficiently for ships' crew to pass
25	Load or unload ships along inside of BW	Overtopping must be rare and wave spray reduced. Provide drainage for water on quay
26	Use of cranes travelling on rails	Check differential settlements and horizontal displacement of top of BW, due to foundation or to compaction of rock and sand fill by vibrations from wave action
27	Accommodate pipelines and other services along BW	Reduce overtopping and spray adequately
28	Survive collisions from ships loosing steering control	Check vulnerability of the part of the BW remaining after damage
29	Resist salt water, high temperatures and frost	S: use of durable materials. V and C: consider corrosion of concrete and steel
30	Resist abrasion from sand suspended in breaking waves	Minimise use of steel structures. Use high quality concrete
31	Resist abrasion from ice	Use high quality concrete. Minimise use of asphalt
32	Resist marine borers	Use durable timber
33	Safeguard the environment	Use acceptable construction methods
34	Have acceptable visual appearance	Involve architect if required
35	Allow for public use	Consider possible recreational facilities for the public

storm surges and waves are statistically independent of the astronomical tide (albeit that there may be a minor second-order effect in their super-position), the tide may normally also be treated as a stochastic factor. The storm surge level and the waves, however, will normally not be statistically independent. While the risk analysis is based on rare events, it will also be necessary to know the typical seasonal oceanographic conditions, so that contractors may evaluate related construction risks.

Besides the normal hydrographic work, a detailed survey is required along the chosen alignment for the determination of quantities. This is particularly true for rocky beds with local steep slopes.

Construction

It is of paramount importance to the costs of a breakwater that construction is simple and fast, whereas it is much less significant to attempt to save a few per cent of the materials required (rock, concrete and steel). Therefore, a breakwater cannot be properly designed without a thorough study of the construction conditions. Construction is dealt with in 'Construction of rubble-mound breakwaters' and 'Construction of caisson breakwaters' later in this chapter.

Design of rubble-mound breakwaters

General

The rubble-mound breakwater is probably the oldest type of breakwater structure, having been used for protection of man-made harbours and landing places on coasts at least since Roman times.

The more recent history of rubble-mound breakwaters dates back to the second half of the nineteenth century, when breakwaters like the ones in Alderney (1860) and Cherbourg (1837), in the Channel Islands and France respectively, were built. These breakwaters were made by dumping stones into the sea and allowing the waves to reshape the prototype. By adding more and more stones a stable structure was eventually reached. This is a type of berm-breakwater.

At the beginning of the twentieth century, when vessels became larger, the breakwater structures moved into deeper water and heavier armour than could be provided by quarried rock became necessary. Some of these early structures are still in existence in the Mediterranean, e.g. on the coast of north Africa. One example inspected by the author is in Oran, Algeria. Here an armour layer consisting of rectangular massive box-shaped concrete blocks was used. This had the advantage, compared to the use of quarry rock, that the limit to the size of armour units was not governed by the natural sources of quarry rock, but by the capacity and reach of available cranes. Hence, instead of quarry rock up to say 5 t, and if larger in very limited quantities, the weight was increased to perhaps 100 t or more by using concrete blocks. These however, typically have a lower density of say $2.4\,t/m^3$ instead of about $2.7\,t/m^3$ for rock.

Cubes and blocks with rectangular cross sections were the typical armour unit for the first half of the last century until a remarkable change occurred in the 1950s by the invention of the Tetrapod in France. Tetrapods have been used for more than 600 breakwaters worldwide, and were the most used units for about 20 years until the Dolos unit was invented in South Africa. This unit gained a considerable share of the market and was used until the 1970s, when the dramatic failures of the very large breakwaters in Sines, Portugal and San Ciprian, Spain occurred. For these structures Dolos units of about 42 t and 50 t respectively had been used. The cause of the failures, as well as of damage to other breakwaters, was breakage of the Dolos units. These failures and the failure of a number of Tetrapod breakwaters, such as those in Arzew El Djedid, Algeria and in Tripoli, Libya spurred a lot of interest and research into the field of rubble-mound breakwaters. This research revealed that the units had become so large that they were too structurally weak for the wave loads and the loads from impacts of adjacent units. The problem arose because units made from mass concrete, without steel reinforcing, were being made larger and larger using the same material, with the same structural strength. Hence the relative breakage resistance of the units was reduced, for example in the Sines breakwater which is in a water depth exceeding 40 m and is exposed to the full impact of waves generated in the North Atlantic.

The trend, after these events and the subsequent research, was to make more robust units and the principal invention was again French, namely the Antifer Cube, first used in Port Antifer in Normandy. It is a cube with grooves, tapered to ease its release from the formwork. This unit was used for some time in parallel with Tetrapods until the invention of the Accropode in France in 1981. This unit was a complete revolution in breakwater technology, as it was designed to be used in one layer only, which is different from all previously mentioned units. The Accropode is cast in a form consisting of two pieces without bottom and top. Other one-layer units also exist, such as the Tribar and the Shed. However they have not become widely used.

In 1995 another one-layer unit, the CORE-LOC, was invented by the US Army Corps of Engineers. This unit is relatively similar to the Accropode in overall shape, but the structural members take their shape more after the Dolos. Figure 7.5 shows the armour units mentioned above.

Nowadays the tendency is to use Accropodes or CORE-LOCs, which have almost similar porosity (50–52 and 56–58 per cent respectively) and stability coefficients, according to the inventors. However with the larger porosity the CORE-LOC requires slightly less mass of concrete per unit area of armour layer. As was the case for Tetrapods and Dolos, the units tend to be used in more exposed locations and/or in deeper water. Although a factor of safety is used when determining the required size, it is clear that with the steep slopes often used, such as 1 on 1.33 or even 1 on 1.2, the reserve stability, if any, is quite limited if something happens to a limited area of the armour layer, because the filter layer underneath consists of stones with typically 10 per cent of the weight of

Fig. 7.5 Types of armour units: Cube, Tetrapod, Dolos, Antifer Cube, Shed, Accropode and CORE-LOC

the main armour units and is built with a slope approaching their natural angle of repose. However, no major failures have been experienced yet, but the above factors call for caution and some degree of conservative design of exposed structures in deep water.

With thousands of rubble-mound breakwater structures in existence, the extensive research and practical experience accumulated over hundreds of years, one might expect that design of a rubble-mound breakwater would today be a trivial matter. However, we are still far from this being the case. Lack of ability to understand and describe accurately the environment, in particular the wave conditions a breakwater has to withstand, has until very recently rendered it impossible to completely systematise the empirical knowledge and thereby transfer experience gained under a certain set of conditions to other sites with different conditions. Much knowledge exists allowing for preliminary design of structures based on empirical relations. However, for major structures hydraulic-model testing is a necessity for final verification of structural stability and overtopping conditions, etc. Another important fact is that a rubble-mound breakwater is by its very nature an ill-defined structure, highly three dimensional and often constructed under conditions, that make it very difficult to know exactly what the actual cross sections look like. It is virtually impossible to verify anything but surface characteristics after construction, and in particular, after a breakwater failure.

The introduction of hydraulic-model tests as the main design tool for rubble-mound breakwaters has made it clear that the proper (economically optimal) design philosophy for such structures is to accept a certain amount of damage during its lifetime. This is a very delicate balancing act, in which a significant error almost anywhere in the design process may well lead to disaster. Of the latter there are many, also recent, examples.

One fundamental principle, which is often not given sufficient attention, is that the design of a rubble-mound breakwater cannot be properly carried out without a clear and realistic concept of how it is going to be constructed. This applies to all stages of construction for which at least one feasible method must be an integral part of the design. Moreover, every construction stage must have a reasonable degree of safety against damage and erosion by waves and currents, which occur during construction.

Main elements

The simplest form of a rubble-mound breakwater is one consisting entirely of loose materials (quarry rock and/or concrete blocks) without concrete superstructure (see Fig. 7.6). Design of such involves determination of the following:

(a)   crest elevation;
(b)   side slopes;
(c)   sizes and layer thickness for crest and lee-side armour units;
(d)   top level for main armour layer;
(e)   bottom level for main armour layer;
(f)   type and size of main armour units;
(g)   dimensions and rock size for toe berm;
(h)   top level for filter layer;
(i)   filter stone sizes; and
(j)   core material requirements.

The values of the above parameters are determined by one or more of the following criteria:

(a)   functional requirements;
(b)   hydraulic stability;
(c)   availability of construction materials;
(d)   constructability; and
(e)   geotechnical stability of the subsoil and of the mound itself.

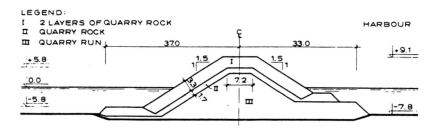

Fig. 7.6 Rubble-mound breakwater without superstructure

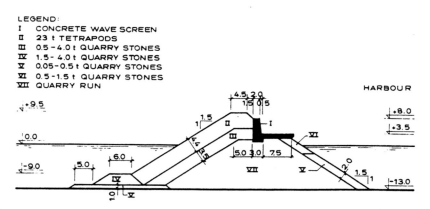

Fig. 7.7 Rubble-mound breakwater with superstructure

These criteria are, of course, inter-linked and the final choice of solution is always a matter of cost optimisation among various alternatives.

Since the total volume of the breakwater is approximately proportional to the total height of its cross section in the second power, the crest level is very important for the cost, especially in deep water. The crest level should be as low as functional requirements, and crest and lee-side armour stability will permit. The adverse effect of reducing the crest level is increased overtopping, when high waves and high water level occur simultaneously. For a breakwater without a concrete superstructure or wave wall, the critical factor in regard to crest level is normally hydraulic stability of crest and lee-side armour. If overtopping generated waves on the harbour side are acceptable a lower crest level may be selected.

In water depths of more than 10 to 15 m, or where vehicular traffic on the breakwater is a functional requirement, a concrete superstructure may be preferred (see Fig. 7.7). It will permit a much greater degree of over-topping without damage to the crest and the lee-side slope and thus a lower crest level from a stability point of view. Accordingly, the crest

Fig. 7.8 Sinking of ship caused by overtopping

level would normally be determined by acceptable wave disturbance generated by overtopping and by constructability considerations. The importance of waves generated by overtopping may be illustrated by the storm of December 1980 at Oran, Algeria, during which four moored ships sank in the port due to battering against the quay structure (see Fig. 7.8). Part of the reason for the extensive damage was that a significant portion of the breakwater wave wall (crest level approximately 5.5 m above mean sea level) due to heavy wave impacts was pushed across the breakwater into the port basin.

For a breakwater with a solid superstructure the crest-armour stability aspect is obviously replaced by considerations of strength and stability of the superstructure. The principles involved in determination of the various elements of the cross section are described in following subsections.

They are chiefly concerned with interaction between waves and the structure, but, where appropriate, constructability considerations are mentioned. General construction considerations are, however, presented in the section titled 'Construction of rubble-mound breakwaters', later in this chapter.

Rubble-mound hydraulics

Incident waves may be reflected and/or overtop the breakwater and/or be transmitted through the breakwater. In breakwater design both over-topping and transmission have to be limited to acceptable levels to prevent excessive wave disturbance behind the breakwater. Transmission is limited by having enough small stone sizes in the core and thus reducing permeability.

Waves will generate run-up over and in the main armour layer as well as in the filter layer below. The run-up water may exert forces on the super-structure and may overtop the breakwater and exert forces on the crest and the lee-side armour layer. However, it has been proved by model tests and practical experience that the run-down of water over and in the main armour layer, as well as in the filter layer below, governs the stability of armour units for normal front slopes of 1 on 1.5 to 1 on 3.0. Only for slopes flatter than about 1 on 3.5 is run-up governing. Run-down is also governing for the berm, which supports the main armour layer, and for toe protection.

Further, water flows in and out through filters and in and out of the core in varying directions. This means that filter criteria for breakwater layers have to be selected based on other considerations than those for uni-directional flow in soils.

An important exception to this is the case where a rubble-mound struc-ture is used to contain and protect a reclaimed area. In this case, the finer material resting upon coarse breakwater material requires extremely stringent filter criteria and the solution should normally include the use of a geotextile.

While run-up, and in particular run-down, is dealt with under the break-water elements which are exposed to such, primarily armour layers, over-topping as it affects port operations on and behind the breakwater is dealt with separately.

As mentioned above, theory can only be applied for preliminary design. Hydraulic model tests with reproduction of irregular waves are indispen-sable for projects of any significance.

Modes of damage due to waves and design of main elements

Wave induced damage may occur in various ways. The most common mode of damage occurs when the force generated by the run-down (or

run-up) on armour-layer units exceeds the stabilising effects of gravity
and interlocking between the units. When this occurs the units are dislocated from their original positions and roll or slide down the slope,
often all the way to the toe berm or to the seabed.

Main armour layer

Run-down is governing for rubble-mound breakwaters with slopes varying from 1 on 1.33 to 1 on 3.0. Only for slopes flatter than approximately
1 on 3.5 does run-up govern the stability. The velocity U in the run-
down is proportional to $\sqrt{gH}$, where H is the wave height and g is the
acceleration of gravity. The hydrodynamic force is proportional to
$\rho_w d^2 U^2$ (and thus a linear function of H), where d is a characteristic
linear dimension of the armour unit and ρ_w is the specific density of
seawater.

The submerged weight of the armour unit is proportional to $g(\rho_s - \rho_w)d^3$,
where ρ_s is the specific density of the armour unit material. Thus:

(a) the dislocating force is proportional to $\rho_w d^2 gH$; and
(b) the stabilising force is proportional to $g(\rho_s - \rho_w)d^3 \, F(\alpha, \varphi, I)$ where α
   is the slope angle, φ is the angle of friction of the armour units and I
   represents the interlocking effect, which especially for artificial units
   may be very significant, but hard to determine and describe in
   mathematical terms.

The limit of stability is described by equating the two above forces:

$$\frac{H}{(\rho_s - \rho_w)/\rho_w} = dF(\alpha, \varphi, I)$$

Normally it is the aim to determine the required weight above water (W)
of armour units, which is proportional to $\gamma_s d^3$ where $\gamma_s = g\rho_s$. The stability
criterion thus becomes:

$$\frac{\gamma_s H^3}{W(\rho_s/\rho_w - 1)^3} = F^3(\alpha, \varphi, I)$$

Various attempts have been made to describe the function $F(\alpha, \varphi, I)$. If
only lifting of units is considered, φ and I are eliminated and Svee (1962)
found $F(\alpha) = K_1 \cos \alpha$. If only sliding is considered I is eliminated and
Iribarren and Nogales (1954) found $F(\alpha, \varphi) = K_2(\tan \varphi \cos \alpha - \sin \alpha)$. In
the commonly used formula presented by Hudson (1959), which dis-
regards friction, $F(\alpha) = K_3 \sqrt[3]{\cot \alpha}$. Hudson's work was not concerned
with very steep slopes, for which friction would be an important parameter.
K_1, K_2 and K_3 above are assumed to be constants for a certain type of
armour unit. In general, the stability increases as α gets smaller and, in
principle, the hydrodynamic stability of the slope disappears, when α

becomes equal to the angle of friction φ. However, since we are dealing with the possibility of single-armour units being dislocated from their initial position and not with the stability of the slope as a whole, instability in this sense may occur at values of α substantially below φ, reflecting the fact that there is large scatter in the resistance to motion of single units. Hence the damage to an armour layer is expressed by the percentage of the total number of units pried loose from their position in one way or another under certain wave conditions over a specified length of time.

The influence of interlocking between armour units is particularly significant for concrete units, such as Tetrapods, Dolos or more recently Accropodes and CORE-LOCs. These were all designed specifically for this effect. Layers of such units are less sensitive to increased slope angles than layers of quarry rocks are. In the above formulae the effect of interlocking is expressed in the value of the constants K. The difference of the influence of the slope angle in the formulae is illustrated by Fig. 7.9, where $\tan \alpha$ is assumed to be 1.0 corresponding to a friction angle of $45°$. For slopes varying around 1 on 2 ($\alpha = 26.56°$) the values of $F(\alpha)$ resulting from the three formulae is shown. K has been adjusted for each formula to make $F(26.56°)$ equal to one.

While the Iribarren formula is more logical because it becomes invalid when the slope angle approaches the friction angle, experience indicates that it tends to exaggerate the influence of slope angles in their normal practical range of 1 on 1.33 to 1 on 2. In particular, this is the case for

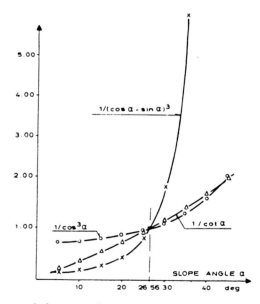

Fig. 7.9 Influence of slope angle in armour-unit stability formulae

highly interlocking units such as Dolos for which very little influence of the slope angle has actually been found within this range.

In addition, a very important parameter in the design of breakwaters is the wave period. Tests have shown (Gravesen and Sørensen, 1977; Jensen, 1984; Ahrens and McCartney, 1975; Van der Meer, 1993) that in general the stability decreases with increasing wave period (decreasing steepness). This may be explained by the longer and more voluminous run-up for longer waves, which results in larger run-down. It is therefore important for design of a breakwater structure that the wave period is evaluated along with the design wave height.

The K values to be used with Hudson's formula, as a first approximation in the design process (called K_D in the following) should be selected with great caution. The K_D value is a function of various factors of which the most important are:

(a)   type of armour unit;
(b)   type of structure;
(c)   slope and extent of the armour layer;
(d)   wave impact characteristics, i.e. wave height and period (length), and wave height to water–depth ratio; and
(e)   degree of acceptable damage to the armour layer.

The following is a brief summary of the literature on K_D values.

Rock armour, no or limited overtopping   Hudson formula: the mean weight of the armour units, W, is:

$$W = \frac{\rho_s \cdot H_s^3}{K_D(\rho_s/\rho_w - 1)^3 \cot \alpha}$$

Here ρ_s and ρ_w is the mass density of armour and water respectively, K_D the stability coefficient, H_s the significant wave height and α the slope of the armour layer.

This may be rewritten as:

$$N_s = \frac{H_s}{\Delta D_{n50}} = (K_D \cdot \cot \alpha)^{1/3}$$

Here N_s is another dimension less stability number, $\Delta = \rho_s/\rho_w - 1$, $D_{n50} = (M/\rho_s)^{1/3}$ is the nominal cube diameter of the armour and M is its mass.

Recommended K_D values are found in US Army Coastal Engineering Research Center (1973), which is the source for Table 7.2.

The damage to a breakwater armour layer is traditionally defined as the number of units displaced in reference area divided by the total number of units in the armour layer. Van der Meer (1993) introduced the now more commonly used, S, which is defined as A_e/D_{n50}^2, where A_e is the erosion

Table 7.2 K_D factors for two layer quarry rock armour, $1.5 < \cot \alpha \leq 3.0$

Stone shape	Placement	Damage, D			
		0–5% (initial)		5–10% (intermediate)	10–15% (severe)
		Breaking waves[1]	Non-breaking waves[2]	Non-breaking waves	Non-breaking waves
Smooth, rounded	Random	2.1	2.4	3.0	3.6
Rough angular	Random	3.5	4.0	4.9	6.6
Rough angular	Special[3]	4.8	5.5		

Notes:
1. Breaking waves means depth limited waves, i.e. wave breaking takes place in front of the slope.
2. No wave breaking, due to depth limitation, takes place in front of the slope.
3. Special placement with long axis of stone placed perpendicular to the slope face.
Source: Shore Protection Manual, 1973.

area around SWL. S has the values of 2, 4–6 and >8 for the three damage levels in Table 7.2. Further the so-called surf-similarity parameter or Irribarren Number, ξ, is used for describing the influence of wave period or steepness. It is defined as $\xi_m = \tan \alpha (S_{om})^{-0.5}$, where $S_{om} = H_s/L_m$. S_{om} is thus a measure of the wave steepness (m refers to the mean wave period T_z and derived wave length L_m). The Van der Meer Formulae for rock armour and deep-water conditions are:

Plunging waves, $\xi_m < \xi_{mc}$:

$$\frac{H_s}{\Delta D_{n50}} = 6.2 S^{0.2} P^{0.18} N_z^{-0.1} \xi_m^{-0.5}$$

Surging waves, $\xi_m > \xi_{mc}$:

$$\frac{H_s}{\Delta D_{n50}} = 1.0 (S^{0.2} P^{-0.13} N_z^{-0.1}) \cot \alpha^{0.5} \xi_m^P$$

where:

$$\xi_{mc} = (6.2 P^{0.31} \tan \alpha^{0.5})^{1/(P+0.5)}$$

For flat slopes with $\cot \alpha \geq 4.0$, the first equation should be used. Values for the notional permeability factor P range from 0.1 for a non-porous structure to 0.6 for a very porous structure. For a traditional breakwater with two armour layers, filter and permeable core, a value of 0.4 is recommended (Van der Meer, 1993). N_z is the number of waves.

The limits of the first two equations are:

$$0.1 \leq P \leq 0.6, \quad 0.005 \leq S_{om} \leq 0.06, \quad 2.0 \, \text{t/m}^3 \leq \rho_s \leq 3.1 \, \text{t/m}^3$$

$N_z \leq 7500$ after which number equilibrium damage is more or less reached. Similar formulae for shallow water conditions (breaking, depth limited waves and for overtopped breakwaters) were derived based on long series of model tests (Van der Meer, 1993).

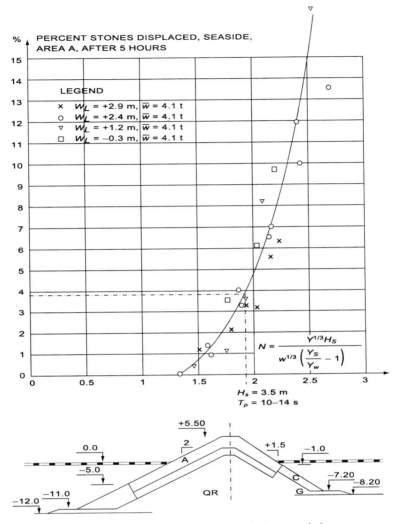

Fig. 7.10 Relationship between stability coefficient and damage
percentage found by hydraulic model tests

Figure 7.10 shows as an example the relationship between stability
coefficient and damage percentage found by hydraulic model tests.

Concrete Armour Units   Many results are reported in the literature; see
the detailed and informative summary by Burchart (1994).

For *Cubes* K_D is in the range of 2.5 to 8 for zero to moderate damage of 3
to 5 per cent. For *Dolos* the author found K_D values of about 8.0 for limited
damage, much lower than reported in the literature and for *Tetrapods*

$K_D = 7$ for breaking waves and 8 for non-breaking waves, at a damage level of 0 to 5 per cent.

Accropodes and *CORE-LOCs* have become the most commonly used armour units. Recent results for these units have been reported for one-layer systems on steep slopes, usually 1 on 1.33 or 1 on 1.50. Model tests show remarkably good stability performance, but it is important to realise that the margin between 'no damage' and failure by collapse, due to sliding of the entire armour layer, is not large. The Delft Hydraulics Laboratory reports $N_S = 3.7$ for 'no damage' and $N_S = 4.1$ for failure, corresponding to a difference in wave height of only 10 per cent. Hence it is very important to apply a larger safety factor than for the units which are used in two layers on gentle slopes. This is included in the recommendations by SOGREAH and the US Army Corps of Engineers for Accropodes and CORE-LOCs, which are shown in Table 7.3.

Table 7.3 K_D Values for Accropodes and CORE-LOCs

Unit	Breaking Waves	Non-Breaking Waves
Accropode Trunk	12	15
Accropode Roundhead[1]	9.2	11.5[2]
CORE-LOC, Trunk	16	
CORE-LOC Roundhead	13	

Notes:
[1] SOGREAH: Roundhead unit weight is 1.3 times unit weight for trunk section.
[2] Valid for slopes of 1 on 1.3 to 1 on 2.0.

Table 7.4 shows details with respect to layer thickness coefficients and porosity for quarry stones and the most commonly used concrete armour units. The following formulae apply:

Layer thickness $r = ncV^{1/3}$

where:

n = number of layers
c = coefficient
V = volume of unit

The number of blocks (or stones) per unit area is calculated by the formula: $N = nc(1 - p/100)V^{-2/3}$ where p = porosity.

It may be seen from the stability formula that the density of the armour relative to the density of the water plays a major role. An increase in armour unit density reduces the required volume of units, e.g. if the density of concrete units is increased from 2.3 to 2.65 t/m^3, the volume may be reduced to one-half (and the weight to less than 60 per cent). This may also be used to avoid having two or more concrete forms for armour units, when, for example, heavier units are required for the

Table 7.4 Data on layer thickness and porosity

Type of armour	Number of layers n	Placement	Layer thickness coefficient c	Porosity p	Reference
Quarry stones					
– narrow grading	2	Random	1.02	0.38	Shore Protection Manual (CERC, 1984)
– smooth	2	Random	1.00	0.37	Shore Protection Manual (CERC, 1984)
– rough	2	Random	1.00	0.40	Shore Protection Manual (CERC, 1984)
Cube	2	Random	1.00	0.33	Burchart (1994)
Modified Cube	2	Random	1.10	0.47	Shore Protection Manual (CERC, 1984)
Grooved (Antifer) Cube	2	Random	1.10	0.46	Burchart (1994)
Rectangular Block	2	Random	1.05	0.38	Burchart (1994)
Tetrapod	2	Random	1.02	0.50	SOGREAH
Dolos	2	Random	1.30	0.60	
Accropode	1	Special	0.90	0.50–0.55	SOGREAH
CORE-LOC	1	Special	0.93	0.60–0.65	US Army Corps of Engineers

round head than for the breakwater trunk. This has been accomplished in specific projects by using iron-ore aggregate in the concrete.

The stability coefficients (K_D values) in the literature all correspond to a certain degree of damage to the armour layers, although the degree of damage is often not defined precisely. However, Van der Meer's formula includes the damage in the form of the parameter S. There is not and cannot be a general rule for what an acceptable degree of damage should be. There is clearly a large difference between what would be considered acceptable damage for a shallow-water breakwater with natural rock armour, which can absorb considerable displacement and settlement, and for a deep-water structure with artificial armour units. For Accropodes or CORE-LOCs, the acceptable degree of damage (displaced units divided by total number of units) is for all practical purposes zero. For large units, even rocking may endanger the stability of the layer, if it causes breakage of individual units. Further, acceptable damage depends upon the residual stability of the damaged structure after a serious storm, and on how fast and at which cost it may be repaired.

In theory, the acceptable damage may be established by minimising the total discounted costs of capital investment, maintenance and repairs plus the value of losses of property protected by the breakwater and losses due to down time for the facilities, which the breakwater was built to protect. In reality, however, such considerations can only be order of magnitude evaluations owing to the uncertainty and the complexity of the parameters and processes involved. Nevertheless, such considerations can often provide a useful guide to major design decisions. This is especially the case where the value of facilities protected by the breakwater amounts to many times the value of the structure itself. In such cases it serves no purpose to under design the breakwater, as the consequences of damage or failure are very costly.

Movement or dislocation of concrete armour units is especially dangerous when the units are slender, such as Dolos and to some extent Tetrapods. However, these units are hardly used anymore after the development of the more robust one-layer systems of Accropodes and CORE-LOCs.

For all units it is true that when their sizes increase their strength decreases. This is because the strength against breakage increases with the linear dimension in the second power (the tensile strength of the concrete being constant), whereas the forces involved in rocking or displacement basically increase with the third power of the linear dimension. Practical experience from the 70s and 80s suggest that for this reason one should avoid Dolos units of more than approximately 10 t weight. This is based on the serious damage to deep-water breakwaters which took place at Sines, Portugal (Zwamborn, 1979) and at Arzew, Algeria.

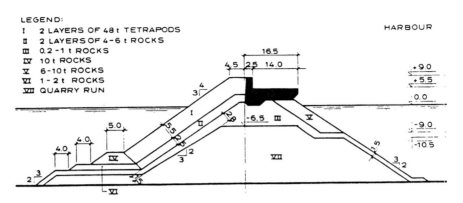

Fig. 7.11 Arzew breakwater cross section

At Sines, the armour layer consisted of 42 t Dolos units on a slope of 1 on 1.5. At Arzew 48 t Tetrapods and a 1 on 1.33 slope were used (see Fig. 7.11). In both cases a high percentage of the armour units broke during a severe storm. The profiles after the storm had many similarities. The superstructures became exposed when filter layer and core material slid down the slope after the armour layer had failed. The remaining slope is very flat from about 3 to 10–12 m below the still-water level. Below this level a mixture of broken and intact armour units and quarry stones from filter layers and the core stands very steep at almost its natural angle of repose. The material directly under the superstructure was eroded causing the superstructure to break and tilt forward as shown in Fig. 7.12). Once the armour layer failed complete destruction of the seaward face occurred because the eroded material was deposited below such levels, where it would have protected the remaining part of the structure. In other words, there was little or no residual stability.

A breakwater in shallow water with a slope of 1 on 2 and crest elevation adequate to avoid serious damage here and to the rear side, will normally have an S-shaped damage profile, which to a large extent has stabilised itself against further damage.

Rubble-mound breakwaters with a concrete superstructure (Fig. 7.7) are particularly sensitive to damage development. If the armour layer on the seaside begins to settle or disappear on the upper part in front of the wave wall its bared vertical face increases wave reflection. This results in increased run-down accelerating the damage into total erosion of the armour in front of the wave wall and erosion of the filter material and the base of the superstructure sets in. Breaking up and dislocation of the superstructure may then result. One might say that such breakwaters have an inherent instability that requires a stronger and more stable armour layer than that required for breakwaters without a superstructure.

129

Fig. 7.12 Superstructure breakage and tilt due to erosion

For design of main armour layers the following major aspects have to be taken into consideration:

(a)   in the case of a long breakwater, distribution of damage along the length of the breakwater is random. If design is based on average values from model tests, some stretches will suffer much greater damage. Therefore, model testing should ideally attempt to determine the scatter of damage to enable forecasting of maximum likely damage in addition to average damage;

(b)   wave statistics are often based on insufficient available data. Thus the true wave height distribution over the intended life of the structure may be higher than the design distribution developed on the basis of available data. Even if the available data are sufficient there is a risk for higher waves than expected due to the random nature of occurrence of storms, e.g. if the intended life is 50 years there is a 10 per cent probability that the structure may be exposed to a storm occurring once every 500 years. Table 7.5 shows the probability of exceeding design conditions within the life of the structure; and

(c)   the strength of model armour units is much higher than it should be according to the model laws. A model unit does not break and/or do damage to other units when dislocated, whereas this is the case for some prototype units. Furthermore, breakage of concrete armour units may be caused by contact forces between individual units arising when the armour layer is being compacted (settled) during initial storms. This may occur if a sufficiently high packing density has not been achieved during construction. This effect is obviously particularly dangerous in armour layers with very steep slopes, say steeper than 1 on 1.5. Armour layers consisting of such units should always be designed for a very low damage percentage, based on model test results, to reduce the risk of breakage of units. In most cases the allowable percentage should be zero.

Table 7.5 *Calculated risk (R per cent) for exposure to design conditions within desired life*

Desired life (L) of structure (years)	Recurrence interval for design conditions (T_d) in years					
	10	20	50	100	200	500
1	10	5	2	1	0.5	0.2
5	41	23	10	5	2	1
10	65	40	18	10	5	2
25	93	72	40	22	12	5
50	99.5	92	64	39	22	10
75	–	98	78	53	31	14
100	–	99.4	87	63	39	18
200	–	–	98	87	63	33

Note: $R = 1 - (1 - 1/T_d)^L$.

In addition to suffering certain limited damage during an extreme storm, an armour layer should be designed for only limited accumulated damage during the storms occurring during its useful life. Model tests performed by the author have shown that for deep-water breakwaters, where the displaced units are lost downwards, the average damage develops linearly with storm duration for constant *(stationary)* wave conditions (significant wave height and peak wave period), except for moderate wave conditions for which incremental damage decreases with test duration. If the latter circumstances are ignored it is thus possible to make a conservative forecast of accumulated damage over any period of years.

Figure 7.13 shows results of many series of model tests for the breakwater in Bilbao, Spain, which was seriously damaged in 1976 and 1977. In addition to demonstrating the linearity of damage with duration up to

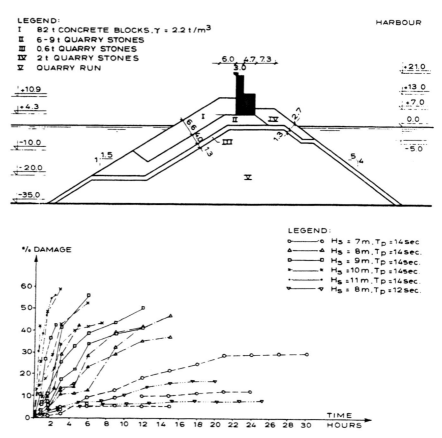

Fig. 7.13 Bilbao breakwater model test results (original profile)

rather high levels of damage, the tests show very significant scatter of results, for repeated exposure to the same wave train.

The major damage distribution has been shown by a number of model tests to conform approximately to a theoretical normal distribution. When its mean value and standard deviation are known, the maximum damage to one of a number of equally exposed breakwater sections may be determined.

Theoretically one can derive the distribution of damage to a breakwater (see Jensen, 1984). The distribution turns out to be binomial but approaches a normal distribution for major damage. The prototype scatter is likely to be considerably larger than the model scatter because of the length of the prototype breakwater relative to the limited length represented in the model and because of the more uniform construction of the model armour layer. Further, the tests showed that for the most severe wave train the armour layer once failed completely by sliding as a unit mass during the run-down of one single wave. Failure of the toe was probably the initiating factor.

The influence of the wave period may be taken into consideration by using the parameter $H_S^2 L_p$, instead of H_S^3 in the stability formula (Gravesen et al., 1979), where L_p is the wave length corresponding to the peak period of the wave spectrum. Van der Meer (1993) uses the surf-similarity parameter that is a function of the wave steepness. In site-specific model tests the correct relation between wave height and period should be introduced when selecting test parameters for model studies.

Cross-section model tests are normally carried out with wave fronts parallel to the structure. Tests have shown that oblique wave directions up to 45° do not cause less damage for quarry stones and Dolos. Further increase of the angle to 75° does not seem to reduce the damage for most units. However, attention should be paid to the fact that the damage from oblique attack is more localised around the SWL than from perpendicular attack. Another important aspect of oblique attack is reduced forces on superstructures and reduced overtopping. Thus the crest height of breakwaters, for which the highest waves will always have a large angle of incidence, may be considerably reduced compared to the case of frontal attack.

Crest

The crest should be so high and wide that it protects the rear side of the breakwater against damage caused by overtopping. Its width should be well defined for construction, which means that it should correspond to a certain number of armour units, usually three or four. A certain minimum width may be required for operation of construction equipment, which in some cases increases the width.

For breakwaters without a concrete superstructure, i.e. with rock or concrete armour units on the crest and on the upper lee-side slope, the height of the breakwater (crest level) is usually determined by the criterion that the same armour units may be used all over the exposed surface. This usually leads to a crest elevation of 1.0 to 1.5 H_S above the SWL in the design condition. The highest crest elevations are required on breakwaters exposed to long wave periods (low steepness) since the wave run-up increases with the wave period (see Fig. 7.14). For lower crest levels the armour on the crest and the lee side is less stable than on the seaside. It should be mentioned that on the crest and on the lee side interlocking units such as Accropodes, Dolos or Tetrapods show little or no advantage over quarry rock of the same weight, which if available, will normally be much cheaper. Also such breakwaters will often be so high that it may be economically advantageous to use quarry rock armour on the uppermost part of the seaside, where the hydraulic forces on the

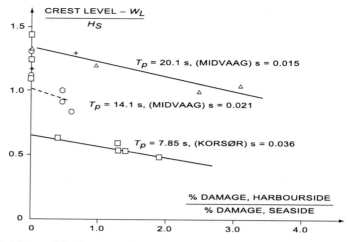

Fig. 7.14 Rear side damage (Jensen, 1984)

armour layer are smaller. In some cases, it has been advantageous to increase the crest elevation so almost no 'green water' overtopping occurs and much smaller armour units may be used on the crest and rear side. An example is the breakwater for Puerto de Carboneras in Spain, constructed in the 1980s, where the main armour layer on the seaside consists of 57 t rectangular cross-section concrete blocks up to +10 m. On the upper part of the seaward face, the crest between levels of +12 and +13.5 m and rear side quarry stones of 2 to 4 t were used as shown in Fig. 7.15.

Superstructure

Concrete superstructures are normally used to reduce the crest elevation, to reduce overtopping and/or to provide a roadway on top of the breakwater. When a superstructure is used to reduce overtopping it will often be exposed to large wave forces. No general guidelines exist for superstructure design, nor determination of wave forces. However some

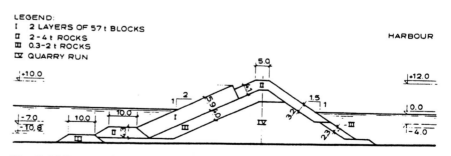

Fig. 7.15 Puerto de Carboneras breakwater cross section

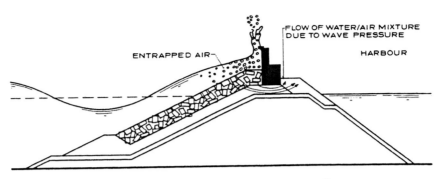

Fig. 7.16 Wave up-rush on breakwater with crown wall

results are available in the literature (Jensen, 1984). Model tests have to be carried out for the specific project. The wave forces are functions of the run-up volume and velocities, which in turn are functions of wave conditions and water level as well as breakwater slope, hydraulic roughness and porosity, rendering such wave forces very complicated. Further, the superstructure design is a combined hydraulic, geotechnical and construction material problem, because the foundation of the superstructure in the breakwater is important both for wave forces and for the structural stability. When the foundation material is permeable, as is the case for quarry stones, the base of the superstructure is exposed to wave-pressure lift forces. The combined effects of wave pressure on both the front side and on the base, which do not peak at the same time, are important. During extreme wave conditions armour in front of the superstructure may be removed and the superstructure exposed to larger wave forces. In principle, wave forces on superstructures are similar to forces on vertical face breakwaters (Lundgren, 1969). Figure 7.16 shows a breakwater with superstructure exposed to wave run-up. Generally durations of horizontal peak wave forces on superstructures are long compared with shock forces on vertical face breakwaters owing to the irregular shape of the water volume reaching the structure. In model tests, very brief wave pressures are often recorded, but they are confined to small areas and caused by compression of air bubbles or pockets. They cannot be reproduced according to a Froude model scale and should be interpreted by using the compression model described by Lundgren (1969). For oblique waves horizontal forces change more gradually and are of longer duration than for perpendicular waves.

Vertical uplift forces on the base are normally due to pressure in a mixture of water and air which is compressible. Thus pressure oscillations may occur. For oblique waves pressure changes occur more slowly. The run-up force is proportional to mv^2 where m is the mass of water, which hits the structure and v its velocity. v^2 is proportional to $g(H_s - H_{so})$, where H_{so}, is the minimum significant wave height required for run-up to reach the superstructure. Thus the force is proportional to $\gamma_w A(H_s - H_{so})$, where A is the area exposed to the wave force. The run-up pressure may be written $p = C_p \gamma_w, (H_s - H_{so})f(T, \alpha, \ldots)$ where C_p is a dimensionless

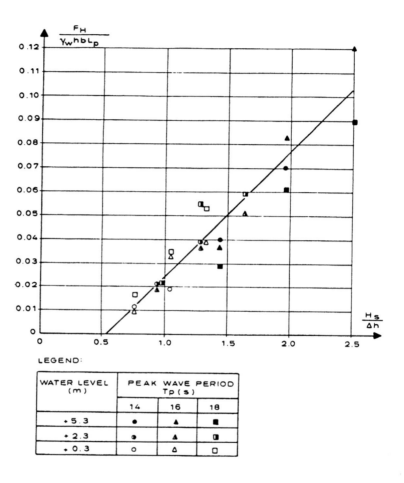

LEGEND:

WATER LEVEL (m)	PEAK WAVE PERIOD Tp (s)		
	14	16	18
+ 5.3	●	▲	■
+ 2.3	◑	▲	◪
+ 0.3	○	△	□

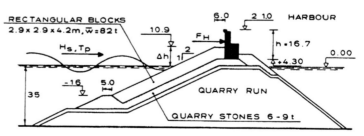

Fig. 7.17 Horizontal wave force on superstructure

coefficient and $f(T, \alpha, \ldots)$ is a dimensionless function of wave period, slope angle, type of armour units, permeability of filter layer, etc. Horizontal forces determined by model tests have been presented by Jensen (1983), which is the source for Fig. 7.17). The maximum horizontal force (F_h) per linear metre of breakwater for 1000 waves is made dimensionless by division with $\gamma_w h b L_p$, where h *is* the height of the front of the super-structure, b its width and L_p the wave length corresponding to the peak period of the wave spectrum. The results were that the maximum force is close to a linear function of $H_s / \Delta h$, where Δh is the vertical distance from the still-water level to the top of the armour layer and that wave forces are zero, when $H_s / \Delta h$ is less than 0.5, i.e. because the run-up does not reach the crown wall. Further, it was found that the force distribution over time is exponential, that the force increased with the wave period and that the force decreased linearly with increasingly oblique wave directions, in particular for the higher waves. Pressure distributions over the front of the superstructure, when the horizontal force is at a maximum, have been measured in model tests to be close to constant. Corresponding uplift pressure distributions on the base decrease linearly towards the rear.

The base of the superstructure should be at such a level that construction is not appreciably influenced by wave action. In case of large over-topping, it is necessary that the superstructure be cantilevered on its rear side to protect the breakwater slope. In general, a wave screen extending high above the top of the armour layer is not feasible, as it will be exposed to very large wave forces. Walls shaped with the aim of reversing the flow of water in the run-up generally do not function as intended for large waves and cause excessive wave pressures. A heel on the underside of the bottom plate, placed as far seaward as possible, will increase the stability of the superstructure against sliding.

Lee-side armour layer

The armour layer on the lee or rear side serves to protect the breakwater against damage from water, which passes over the crest or the superstructure. As overtopping is a very complex phenomenon (see 'Overtopping' later in this chapter) so is lee-side armour layer design, which often has to be based on model tests. For rubble structures without a superstructure, the required crest level decreases with decreasing wave period and with increasing armour unit size on the crest and rear side. For breakwaters with a superstructure, the geometry of the latter is very important for the armour layer, as it influences overtopping as illustrated in Fig. 7.18. The alternative B geometry is clearly to be preferred over alternative A as the wave energy is dissipated, when the overtopping water wedge hits the water behind the breakwater. Lee-side armour has been shown

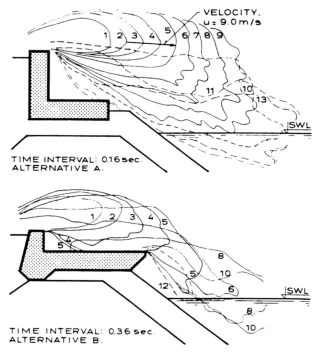

Fig. 7.18 Development of overtopping

to be considerably more sensitive to an increase in significant wave height than main armour, since the external hydrodynamic forces are a function of the amount of overtopping related to the parameter $H_s/\Delta H$, where ΔH is the freeboard, i.e. the height of the crest above sea level. This means that if the two armour layers are designed for the same degree of damage for the same design wave condition, the risk of serious damage to the lee-side, if the design condition is exceeded, is much larger than for the front side. Instead a probability approach should be employed, which results in the same risk of failure for the two armour layers during the useful life of the structure. It is important to realise that the over-topping water loses its force, when it hits the water surface on the lee-side. The requirements for protection of the lee-side slope from somewhat below the water level are therefore entirely determined by wave and current conditions behind the breakwater. For this reason, the core material may at times even be left unprotected from a distance below the water level.

Filter layers under armour

The purpose of filter layers is to prevent core material from being washed out through the armour layer. Published criteria for filter stability are

Table 7.6 Filter criteria

Reference	$\dfrac{d_{15,a}}{d_{85,f}}$	$\dfrac{d_{50,a}}{d_{50,f}}$	$\dfrac{d_{15,a}}{d_{15,f}}$
Terzaghi and Peck (1967)	≤ 4 to 5	–	≤ 20 to 25
US Army Coastal Engineering Research Center (1977)	~ 2.2	~ 2.3	~ 2.5
Thompson and Shuttler (1976)	≤ 4	≤ 7	≤ 7

shown in Table 7.6 where *a* refers to armour unit sizes and *f* refers to filter material sizes. The numbers correspond to the percentage of armour unit and stone sizes smaller than the size in question. The US Army Coastal Engineering Research Center (1977) criteria correspond to filter stone sizes (equivalent dimension) of 40 to 45 per cent of armour unit sizes. They are very strict compared with other criteria and with filter material actually used and will often mean that two filter layers are needed. Substantial experience suggests that they are excessively strict. Terzaghi and Peck's (1967) geotechnical criteria are meant for unidirectional flow perpendicular to boundaries between different layers and thus not directly applicable to breakwater filters. Due to the changes of flow directions in the latter, stricter criteria are required. Thompson and Shuttler's (1976) criteria are in between the two other ones and are considered to be a reasonable guideline.

Filter layers should be few and thick rather than many and thin for construction reasons. The minimum layer thickness should be equal to two times the dimension as recommended by US Army Coastal Engineering Research Center (1977), but often it is a good idea to make thicker filters. The stone dimension D is defined by $V = 0.75D^3$ where V is the volume (Stephenson, 1978). In determining layer thickness, due consideration should be given to inevitable construction inaccuracies and difficulties in accurate surveying. One should be reasonably sure that an intact filter exists over its entire area upon construction completion. Model tests (Gravesen *et al.*, 1979) have demonstrated that the armour layer stability increases with the filter layer thickness. The explanation is believed to be that more of the run-down takes place inside a thicker filter and thus less in the armour layer, where water velocities are reduced.

The Van der Meer formulae (1993) include a permeability factor P varying between 0.1 and 0.6. This is based on a long series of model tests and thus confirms the earlier findings that the permeability of layers below armour layers is an important aspect.

When large artificial armour units are used, quarry stones are used as filter layer or secondary armour layer, as it may be called, as it protects the core during construction, until the main armour layer is in place. Model tests with such different filter layers have shown less sensitivity to filter stone sizes than that indicated by the standard filter criteria. Model testing is the only reliable filter design methodology available at present.

Core

The core normally consists of quarry run, which should not be so coarse that wave action is transmitted through the breakwater. If considerable sediment transport takes place the core should be impermeable for suspended material. Sometimes a minimum stone size is specified to avoid washout of core material, which could cause settlement of the structure and perhaps accumulation of core material in dredged areas. Core material produced from normal sound rock, however, does not normally contain such large quantities of fines, that this becomes a problem. Experience has shown that the non-graded material tends to build up a natural filter losing only a small amount of its fines. In certain cases, the core may be made from hydraulic sand fill between banks of gravel or quarry run, usually built in several stages. Alternatively a sand fill core can be protected by a geotextile with quarry run filter layer(s) on top.

Berm

The seaward berm acts as the foundation and support for the main armour layer and may catch units dislocated from the armour layer. Hereby the slope of the latter becomes more gentle and its stability improves as some damage develops. It is normally recommended to construct the berm with a slope on its rear side as shown in Figs. 7.6 and 7.11, although model tests have not been able to demonstrate any beneficial effects. The reason for the recommendation is that if the berm surface is horizontal further back, filter stones may, during construction, fall down and make the berm surface and its transition to the filter layer surface concave, which would make placing of the first rows of armour units difficult. This would endanger the stability of the armour layer, especially for very steep slopes.

Besides meeting hydraulic and geotechnical requirements as a safe foundation for the armour layer, the berm should be sufficiently wide to compensate for construction inaccuracy and to allow for some damage during extreme wave action. For berms constructed with cranes working from the breakwater or other fixed structure, the minimum berm width should be three average berm stone dimensions. If a less precise construction method, such as dumping from a barge, is used the width should be five or more stone dimensions.

Berm stone sizes can be estimated from experience as presented in Fig. 7.19, from CIRIA (1991). The most critical conditions for a berm are combinations of low water and large waves. Higher breakwaters result in larger wave run-down, which may inflict larger velocity and inertia forces on the berm. For use with concrete armour units the largest

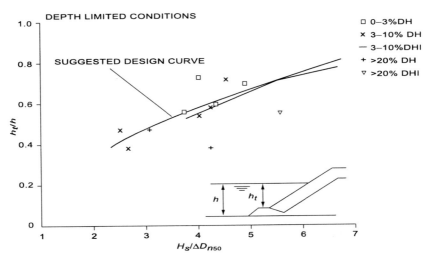

Fig. 7.19 *The stability as a function of h_t/t (from CIRIA, 1991)*

available quarry stones are normally used in the berm, which means that once this stone size is known it becomes the aim to determine the highest level at which it will be sufficiently stable. The longer the wave periods are for a certain significant wave height, the lower the berm surface level needs to be. For breakwaters in shallow water exposed to breaking waves it is often not possible to provide a stable berm and thus necessary to extend the armour layer all the way down to the toe protection. On a steep rocky seabed, it may be extremely difficult to design an armour layer for which the lower part is stable. In one case, 40 t concrete cubes, used as a toehold for the armour layer, were not stable. In such cases, it may become necessary to provide a toehold by blasting a trench in the rock bed or to use another fixed-wall-type structure.

Filter on seabed and toe protection

On a sandy seabed, a toe protection layer is traditionally used to protect the breakwater from undermining. The toe protection material is normally so coarse, compared with the sea-bottom material, that it does not conform to standard filter criteria. A filter layer is often used between the core and the bottom material. Very thin and/or geometrically complex multi-layer filters are sometimes designed for toe protection. This should be discouraged as such filters cannot possibly be constructed. Filters made from geotextiles may be used in certain cases.

Bends and corners

Armour layers on bends and corners generally have less stability than on straight stretches of breakwater, and thus may require heavier armour units. The required increase depends upon the breakwater profile, unit type, radius of curvature of bends, angle of corner and wave direction. In certain cases, especially at concave corners, it may also be necessary to increase the crest or superstructure top elevation, in particular if the wave direction is in the middle of the angle between the two alignments, which tends to increase run-up and overtopping.

Round heads

When a wave breaks over a cone-shaped breakwater head, large velocities occur. This is verified by Fig. 7.20 that shows as an example model test results for a Tetrapod round head having been exposed to waves from many different directions.

The highest velocities occur in an area just behind where the wave direction is tangential to the surface and slightly below SWL. Model tests with rubble-mound breakwater heads exposed to different wave directions have demonstrated the need for heavier armour units, which may need to be combined with a flatter slope to achieve the same stability as for the breakwater trunk. The armour unit weight may have to be increased by a factor of between 1.5 and 4. The higher values are for the more complicated armour units (the more their stability depends upon interlocking) and for the smaller radii of curvature of the round head.

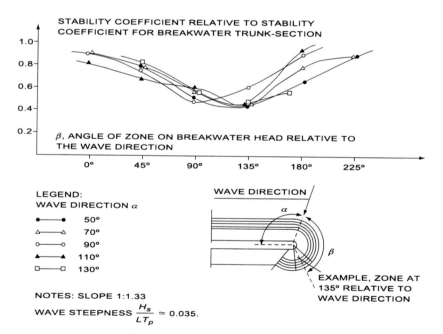

Fig. 7.20 *Stability coefficients for Tetrapod roundhead as function of wave direction and location*

Another alternative that has been used is to let one or more caissons constitute the breakwater end. Normally, for a certain increase in wave height the increase in damage is much higher for the round head than for the trunk. Therefore this should be taken into consideration in the design in order to obtain the same degree of safety for trunk and round head.

Overtopping

Where berths, reclaimed areas, roads, buildings, etc. are located behind and close to breakwaters, overtopping may cause inconvenience or problems.

Light overtopping may be characterised as 'spray-carry-over', while severe overtopping is called 'green-water'. The physical phenomenon of wave overtopping is to be considered as a two-phase flow with entrainment of the lower, dense media (water) into the upper media (air).

The horizontal distribution of the water falling behind the breakwater crest per width unit is on average found to be close to a theoretical exponential distribution. This implies that, on average, the vertical distribution of the water above the armour layer on the front side of the breakwater must also be close to exponential. The physical principles of the 'spray-carry-over' are shown schematically in Fig. 7.21. The parameters used later are defined here as well.

The overtopping intensity, q per second per linear metre of breakwater, is very irregularly distributed over time. The maximum values occur often for a few single waves and the instant intensity is many times larger than the average intensity. The overtopping quantity may be determined in hydraulic models by collecting the water in trays. Similar procedures have been used in the few prototype measurements.

The overtopping is dependent upon the geometry of the structure in question and the slope, porosity and roughness of the armour layer. It is

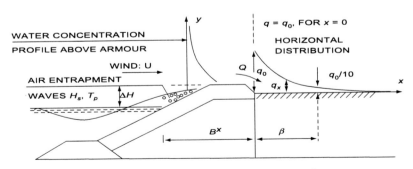

Fig. 7.21 Principles and definitions of overtopping and spray-carry-over

further dependent upon the impinging waves, their height H_S, period T_P (corresponding to peak of wave energy spectrum) and angle of incidence.

For a specific structure, the logarithm of the average overtopping q is found to be almost linearly dependent upon the parameter $H_S/\Delta H$, where ΔH is the freeboard of the breakwater (vertical distance from crest to SWL).

As shown schematically in Fig. 7.21, the average overtopping per area unit (m^2) decreases exponentially with distance, which may be expressed as $q_x = q_0 10^{-(x/\beta)}$, where β is a constant equal to the distance for which the overtopping intensity decreases by a factor of 10. The total over-topping Q, (per linear metre) may be determined by integrating q_x over x, which yields $Q = q_0\beta/\ln 10$. When Q and β are measured (or estimated from formulae or available test results) q_0 and thus q_x may be determined. Besides waves the overtopping depends upon the wind, which for low overtopping discharges (spray) has a significant influence on the discharge.

Prototype measurements

Prototype measurements of wave overtopping quantities are scarce, and to the authors' knowledge only four sets of measurements are available in the literature. The most important and extensive set of results is available from Fukuda *et al.* (1974), which present results of overtopping measurements carried out in Japan in 1972. Further, the Danish Hydraulic Institute made field measurements on a small Danish breakwater in the Port of Hundested in 1977. Supplementary model tests were made on the same breakwater. The results presented in Fig. 7.22 show excellent agreement between model and prototype. To the author's knowledge, the Japanese measurements have never been compared with model experiments with irregular waves. The model experiments reported by Jensen and Juhl (1987) were made with irregular waves and it was found that the over-topping discharge in the model was below that of the prototype.

The third and most recent result is by de Gerloni *et al.* (1991), who investigated critical overtopping discharges at vertical breakwaters in Italy. They found allowable overtopping discharges significantly larger than those in the Japanese study. Further results (Van der Meer, 1993) have been reported for a prototype, grass-covered dyke. The results are presented in Fig. 7.23.

The overtopping water has to be drained away in one way or another and unacceptable down-time for cargo handling or other operations must be avoided as well as damage to structures or equipment. For low breakwaters, with a harbour basin behind the structure, the overtopping may be so large that it generates dangerous waves, capable of sinking moored ships.

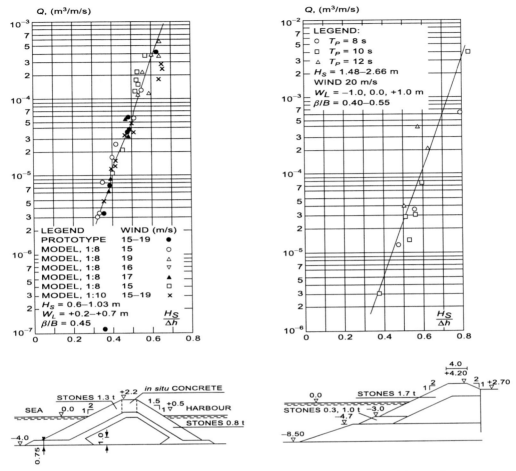

Fig. 7.22 *Examples of model test results of wave overtopping quantities and comparison with field measurements*

Recent changes in breakwater design and construction practice

In the last few years the international practice for port and breakwater construction has shifted towards 'design and build'. This means that contractors attempt to reduce as much as possible the construction costs of breakwaters.

It is important that consultants, who do the design work, realise that there is a very thin line between acceptable savings and savings which will cause prohibitively high maintenance costs and/or limited or total failure. An example from a major international breakwater project is shown in Fig. 7.24. The armour is $5\,m^3$ Accropodes, the slopes as steep as possible and the superstructure reduced in size and width to a bare

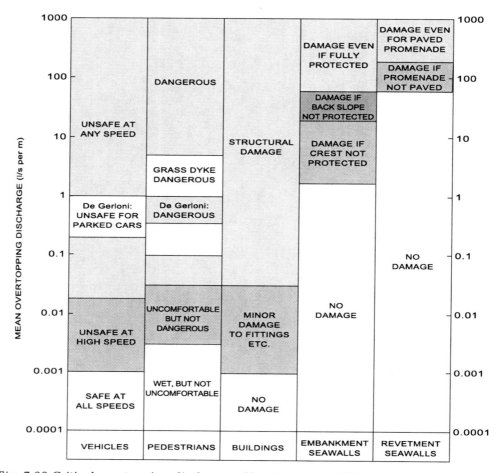

Fig. 7.23 Critical overtopping discharges (Van der Meer, 1993)

minimum. The hydraulic model tests, carried out for this design, played an even larger role than for more traditional projects.

Model tests

Hydraulic model tests remain the main design tool for determination of characteristics and dimensions of rubble-mound breakwaters. Standard practice is to test breakwater structures using model scales between 1 to 20 and 1 to 60, or as large a scale as the available facilities permit. The models are constructed and interpreted in accordance with the Froude model law. It is further standard practice to execute such tests with model waves, which are realistic reproductions of natural wave conditions, i.e. either based directly upon recorded natural wave trains, or most often generated from a well defined realistic wave spectrum such as for example

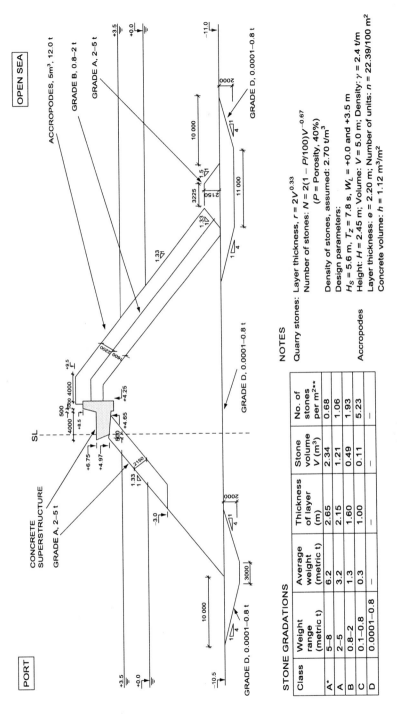

STONE GRADATIONS

Class	Weight range (metric t)	Average weight (metric t)	Thickness of layer (m)	Stone volume V (m³)	No. of stones per m²**
A*	5–8	6.2	2.65	2.34	0.68
A	2–5	3.2	2.15	1.21	1.06
B	0.8–2	1.3	1.60	0.49	1.93
C	0.1–0.8	0.3	1.00	0.11	5.23
D	0.0001–0.8	–	–	–	–

NOTES

Quarry stones: Layer thickness, $r = 2V^{0.33}$
Number of stones: $N = 2(1 - P/100)V^{-0.67}$
(P = Porosity, 40%)

Density of stones, assumed: 2.70 t/m³

Design parameters:
$H_S = 5.6$ m, $T_z = 7.8$ s, $W_L = +0.0$ and +3.5 m
Height: $H = 2.45$ m; Volume: $V = 5.0$ m; Density: $\gamma = 2.4$ t/m

Accropodes: Layer thickness: $e = 2.20$ m; Number of units: $n = 22.39/100$ m²
Concrete volume: $h = 1.12$ m³/m²

Fig. 7.24 Example of minimalistic breakwater design

the JONSWAP spectrum. The reason for this is that numerous aspects of breakwater performance (e.g. wave overtopping and stability) depend on the wave height in a complex non-linear fashion. Breakwater performance and stability depend on the following hydraulic parameters:

(a) wave height;
(b) wave period;
(c) wave direction relative to breakwater direction;
(d) duration of wave attack; and
(e) water level.

All of these parameters are more or less interrelated, and some of them may vary considerably along the length of a given breakwater, e.g. due to the seabed configuration. Programming of model studies and indeed the design philosophy of breakwaters is very different for breakwaters in deep water and breakwaters in shallow water. Deep water is here to be understood in terms of the wave height, meaning that waves do not break because of depth limitations before reaching the structure. For breakwaters in deep water, there is always a certain probability that any wave design conditions will be exceeded during the life of the structure.

On the other hand, for breakwaters in shallow water, where wave heights at the structure are limited by water depth, the extreme conditions are quite well defined but may occur many times and for a considerable accumulated period of time during the life of the breakwater. Furthermore, for shallow-water breakwaters, extreme high-water levels become a very important design and testing parameter since the maximum wave height reaching the structure is directly related to the water depth. Because of the many variables substantial simplification is required. This clearly implies risk of oversimplification, for which the penalty may be severe. Therefore, every breakwater should be treated as an individual problem and the test specifications should thoroughly reflect the local conditions as well as the cost of over design. It should be fully understood that breakwater-model tests are a far cry from standard construction materials testing or soil testing. A breakwater-model test is to be considered as a project by itself to be designed according to the specific conditions of the site and the performance requirements. Since interpretation of test results in terms of safety and adequacy of the structure is directly based upon the hydraulic climate (waves and water levels) of the site, it is evident that the testing programme should be planned accordingly. The testing programme should reflect realistic and comprehensive combinations of wave heights, periods and water levels. In most cases this requires that the specifications for the model study should include analysis of wave and water-level conditions on the basis of original data, often supplemented by hindcast and refraction analysis, nowadays often made by numerical models, simulating propagation of waves over complex seabed configurations in shallow water.

For breakwater stability models, Froude's law of similitude applies as gravity and inertia are the predominant forces. Thus the Froude number $F = v/(gl)^{0.5}$, where v is a characteristic velocity, g is the acceleration of gravity and l is a characteristic length, should be the same in prototype and model. This means that if the length scale is λ and knowing that the gravity acceleration scale is one (unity), it is clear that the time scale becomes $\lambda^{0.5}$ since the unit for acceleration is (m/s^2). The velocity scale is also $\lambda^{0.5}$, the volume and force scales are λ^3 and the stress scale is λ. For small-scale models also water surface tension and in particular non-turbulent flow in porous media play a role. Jensen and Klinting (1983) have demonstrated that the latter may be compensated for by using exaggerated stone sizes in certain parts of the model. The strength of concrete armour units cannot be reproduced to scale except in very special models. Thus, breaking of model units never takes place, whereas large prototype units do break as a result of dislocation, rocking or settlement (compaction). Different test procedures and strategies are used at different hydraulic laboratories. Traditionally, stability tests are carried out as a series of test runs in which wave height is increased step by step. Without reconstruction of the cross section the damage is recorded after each run and a relationship established between damage and wave height. The problem with this procedure is that it is difficult to relate test wave conditions to forecast wave conditions for the site in question. This procedure is often used with the concept of one and one only design wave height and wave period. It does not consider storm duration or the accumulated effects of less extreme wave action. The following test procedures should be considered in each case:

(a)   series of runs with increasing wave heights;
(b)   runs with design situation;
(c)   runs reproducing individual storms;
(d)   long duration runs with stationary wave conditions;
(e)   series of runs reproducing the accumulated effects of waves over a number of years, e.g. the useful life of the structure;
(f)   runs aimed at determining residual stability; and
(g)   runs with cross sections under construction.

For deep-water structures, as opposed to structures exposed to waves breaking in shallow water, there is a certain probability of occurrence of extreme wave conditions exceeding those assumed for the design. This should be considered in the testing strategy. Runs with increasing wave heights are suited for a first evaluation of stability. For a deep-water structure, they show approximately what significant wave height results in serious damage or complete failure. However, damage depends upon duration of exposure and the same damage may result for lower waves over a longer period of time. Runs with the design situation may be suitable if only a single element needs to be tested, while the others are considered adequate. This is typically the case for superstructure design or for crest elevation determination based on stability requirements for the rear side. Runs with individual storms may be advantageous if only a few storms are forecast to occur during a certain period of time, such as the construction period or during the life of structures exposed to hurricanes. Reproduction of water-level variations during a storm is often a problem. Long duration runs with stationary wave conditions are useful to determine whether damage stabilises or continues to develop until complete failure occurs. For steep armour layers damage will be a linear function of time, whereas for less steep layers, damage tends to stabilise. Series of runs reproducing accumulated damage over a number of years are used, together with knowledge of distributions of occurrence of wave conditions over time, to calculate long-term damage. However, such calculations are somewhat uncertain as they have to be based on various assumptions and linearisations. Instead long-term damage, that is the accumulated effects of many storms, may be determined by tests, usually of long duration, carried out to simulate sequences of storms.

Fig. 7.25 Model of breakwater for cooling water outfall

For deep-water structures, the residual stability should be determined for total-loss risk analysis, as the structure may not be economically feasible if a small increase in design conditions would result in complete failure. This is, in particular, important for structural elements for which one or a few waves may cause complete failure, such as armour layers on steep breakwaters, which may slide down in their entirety, for super-structures and for rear side armour layers. Tests of cross sections under construction should be planned carefully and would normally represent individual storms, since storm duration is a decisive parameter for the extent of damage to the core and filter layer(s).

Figure 7.25 shows as an example a model of the breakwaters for a cooling water outfall for Shoaiba Power Plant in Saudi Arabia.

Construction of rubble-mound breakwaters

General considerations

The required construction technology level is clearly higher for structures with artificial armour units and/or concrete superstructures than for pure rubble structures. Often armour units, which are just sufficient from a hydraulic point of view, are specified, although the use of larger units would not significantly increase the costs. The reason is that the total construction cost per armour layer surface area unit consists of the cost of fabrication, which increases with the unit size, and the cost of placing, which is reduced as the number of units to be placed is reduced. The volume in the layer is proportional to its thickness which is in turn proportional to $V^{1/3}$, where V is the armour unit volume, while the number of armour units per area unit is proportional to $1/V^{2/3}$. This means, for example, that a 50 per cent increase in armour unit weight, only increases the stone or concrete volume per area unit by 14 per cent, while the number of armour units is reduced by 24 per cent.

If larger armour units and/or a flatter slope are required for the round head, this will be governing for the lift capacity of the crane(s) to be used. Thus an increase of armour unit weight for the breakwater trunk may not influence the crane requirements. While the heavier lifts increase wear and tear of cranes as well as fuel costs, it may be assumed that the time it takes to place one unit does not vary appreciably within a certain weight range. Also a reduced number of armour units reduces the number of forms, casting operations and transports of units, all of which contribute to savings in time and costs. The conclusion is that one should not over-emphasise efforts to reduce the size of armour units.

Construction methods

During design, it is always necessary to consider how the structure is going to be built and it may be necessary to specify, at least to a certain extent, the construction method. Designs which are either impossible or very time consuming to construct, e.g. thin filter layers, should be avoided. In general, rubble-mound breakwaters are constructed by placing or dumping material from either floating equipment or equipment standing on the breakwater. The traditional method involves dumping of core material in the breakwater alignment from trucks. After the core has reached the required shape, filter stones are dumped and armour units placed on the slopes. After the final length of the breakwater has been reached the crest units are placed on the way back. A major problem with this method is that the slopes of stone material dumped from trucks become very steep and close to the natural angle of repose of the material, especially under water. The method may, however, be improved if the lower part of the front side is dumped from a barge.

As the top of the core serves as a roadway for construction equipment this may determine its width. The roadway should be at such a level that equipment is not endangered and the roadway destroyed by wave run-up under normal conditions. On the other hand, it may be necessary to accept that construction is interrupted and equipment removed from the incomplete breakwater under severe wave conditions. If a super-structure is constructed before armour units are placed it may be used to support a heavy crane on rails, if such is needed for placing armour units.

Large breakwaters are normally constructed by a combination of floating and breakwater-based equipment, as shown schematically in Fig. 7.26. The material for the seabed filter, toe protection and core is often most economically dumped from either a split barge or a side-unloading barge. Normally, the split barge deposits its load in a heap, while it is possible to place material in a more even layer using side unloading. The split barge may, however, be operated to place finer fractions in relatively thin, well-controlled layers. The maximum level of construction to which barges may be used depends, of course, on their draft, both when fully loaded and empty. Barges equipped with cranes may be used to place material at higher levels than otherwise possible or for placing of armour units, but the movements of the barge and the difficulty of maintaining it in a very accurate position may cause armour units to break or to be placed inaccurately. If filter material and armour units are delivered to the site by barge it is better to have breakwater-based equipment place it. For construction of berms, which serve as toe support for armour layers, placing by crane from the breakwater is recommended to achieve sufficient accuracy. If floating equipment is used, the berm stones should also be placed by crane, not dumped. If the latter, less accurate, method is used, it should be compensated for by widening of the berm.

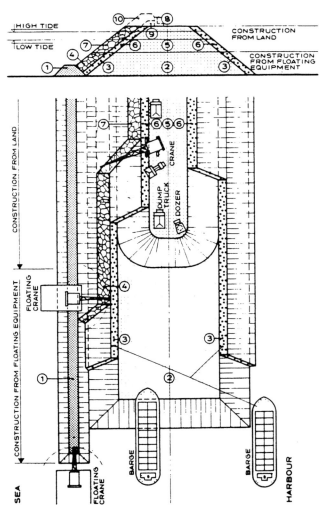

Fig. 7.26 Construction of rubble-mound breakwater (reproduced by permission of H. Vandervelden)

The filter layer under the armour is of major importance for the stability of the entire structure. It should be sufficiently thick to ensure that the required minimum thickness is present all over the surface to be covered and to make it possible to ascertain its presence by realistic survey methods. It should be kept in mind that determination of the exact location of a steeply sloping very rough surface is not a simple matter and requires considerable care. Unless special allowances have been made, the filter material should be placed by equipment working from the breakwater.

The main armour units are practically always placed one at a time by a crane normally working from the breakwater. However, for some deep-water breakwaters built during recent years, armour units have become very heavy and the armour layer width, measured horizontally, has become large, requiring a very considerable crane reach. This led to placing of armour units from floating equipment in one case or even to dumping of very large Cubes in another case. As the berm width was quite limited in both cases these construction methods seem rather unreliable or even too risky.

Armour units such as quarry stones, concrete Cubes and Dolos are normally specified to be placed at random. In order to assure uniform distribution of the units, a grid system indicating the position of each unit should be used. This is, in particular, important for Accropodes and CORE-LOCs, which are placed in a single layer only. Special positioning systems have been developed for placing. The structure should always be inspected by diver engineers before, during and after placing of the armour layer. The crest should be constructed in the same manner as the above-water part of an armour layer.

The concrete superstructure should have its base at such a high level that its construction is not affected by wave run-up, except under severe conditions. If it is at a lower level, a rather wide shoulder of armour and filter layers may provide some protection during construction. It is normally practical to cast superstructure elements in 5 to 10 m lengths.

Construction materials
The materials are required to be durable in seawater and solid enough to sustain impacts occurring during placing and throughout the life of the structure. Various kinds of rock are the most common materials. They are normally produced by blasting of solid quarry rock. When sufficiently large blocks cannot be produced from the quarry, artificial armour units made from concrete are used. In order to facilitate quarry and construction work the number of stone classes should be kept as low as possible. The difference in stone sizes between gradations should be so wide that it is easy to distinguish visually between different gradations. Further, gradations should normally not overlap but the various stone classes should be designed to utilise as much of the material produced by blasting as possible. Lack of knowledge of quarry output of different stone sizes has on occasions required modification of breakwater cross sections, after construction has started, to fit available stone sizes. Test blasting should be carried out in the designated quarry (quarries) during design to ensure that the output of different stone classes matches the design. The core usually consists of gravel or quarry run, which may be screened for removal of fines, if a large proportion of fines is expected.

About 40 different types of artificial armour units have been developed. There are basically two different kinds of units. One requires placing in a certain pattern to ensure interlocking with adjacent units. Due to the inherent inaccuracy of underwater construction such units cannot be recommended. The other type is designed for random placing and interlocking.

Construction cross sections and strategy

Rubble-mound structures are, in general, very vulnerable during construction. The core is easily eroded before the filter layer is placed and the filter layer may be endangered before the armour layer is placed. Similarly, the crest and rear side are exposed to overtopping during construction. The end of the incomplete breakwater poses a special problem during construction.

Construction strategy should aim at the earliest possible completion of the main armour layer and at keeping to a minimum the exposed length of core or filter layer, in particular the core at the temporary end of the incomplete breakwater. When core material is dumped from barges, partial core completion may be far ahead of the rest of the structure, as the top of the core is only brought up to a level, where it will not be disturbed except by extreme wave action. In this manner, the critical part of the construction, i.e. of the parts above and just below the water surface, may advance at a much greater speed than if the entire profile is constructed by placing material over the end of the breakwater.

Part 2 by Professor Dr Hocine Oumeraci

Caisson breakwater types and lessons learned from failures

Types of caisson breakwaters

Besides rubble-mound breakwaters, vertical-face breakwaters are among the oldest and most used breakwaters. Initially this type was developed to reflect all incident waves, which do not overtop the structure. In its initial form it was simply called vertical breakwater (Fig. 7.27). The monolithic structure has a plain vertical front wall over the entire water depth.

Another type is the composite breakwater (Fig. 7.27), which in particular is preferred in deeper water. The term composite is used to express the fact that the structure works as a vertical breakwater at high tide and as a rubble-mound breakwater at low tide (PIANC, 1976). If the low-water level (LWL) can reach the berm level, it is called a high-mound composite

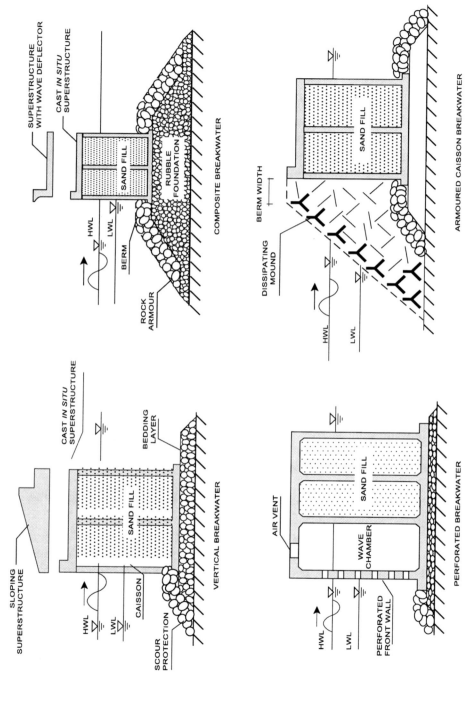

Fig. 7.27 Main types of caisson breakwaters

breakwater (HMCB). The rubble foundation of a HMCB (see Fig. 7.42) generally has a rather flat slope (cot $\alpha \geq 3$), because one of its key features is to cause breaking of the largest waves before they reach the monolithic structure, thus substantially reducing the wave loads (see 'Wave loads on armoured caisson breakwaters' later in this chapter).

To reduce wave reflection, which may constitute a hazard to navigation and/or to structural stability by causing scour, Jarlan introduced the per-forated caisson breakwater (Fig. 7.27). It is characterised by a perforated front wall and a wave dissipation chamber. To further reduce both wave forces and wave reflection the armoured caisson breakwater was devel-oped (Fig. 7.27). This type is particularly popular in Japan, where it is called the horizontally composite breakwater.

Besides the main types of caisson breakwaters shown, there are other vertical-face breakwaters such as blockworks, piled curtain walls and slotted wave screens (Fig. 7.28). A number of structural innovations in cross-section and plan view to reduce wave forces, reflection and over-topping have been made (see 'Innovative caisson breakwaters', later in this chapter).

Caissons may be given any shape and size, thus making them ideally adaptable for multi-purpose use such as incorporation of recreational facilities, turbines for wave-power generation, etc. Due to recent develop-ments, extensively discussed by Oumeraci et al. (2001), the competitive-ness of caissons for sea walls and breakwaters in terms of hydraulic performance, total costs, construction time, maintenance standardisation, quality control and environmental requirements is expected to increase substantially.

Lessons learned from vertical breakwater failures

To stress the limitations of the present tools for analysis of vertical break-waters and prediction of their behaviour as well as the need for using engineering judgement in design, construction and maintenance of caisson breakwaters, a brief summary is given below of lessons learned from past failures of vertical breakwaters (for details see Oumeraci, 1994a).

(a) *Inadequacy of the maximum design wave concept:* Generally the maximum wave height with a return period of 20 to 100 years is used for design. This is not justified in the sense that the largest waves may not reach the structures due to breaking and that lower waves may be even more critical. Further, incremental weak-ening of the foundation and other structural components sensitive to cyclic loads are caused by moderate wave conditions (see Oumeraci, 1994a). Therefore, and for reasons which will become more apparent (see 'wave loadings of caissons with plain front walls' later in this

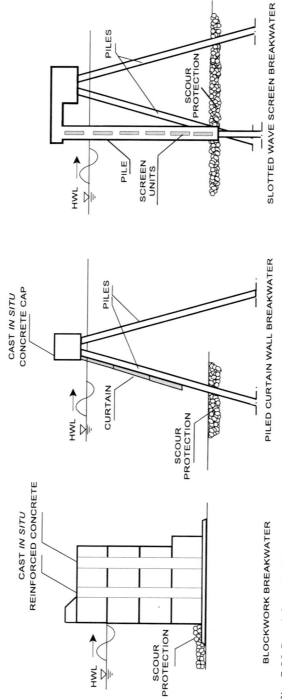

Fig. 7.28 Special vertical-face breakwaters

chapter) design wave load criteria should be adopted, which can account for both sudden failure caused by an extreme single wave and gradual failure under moderate but repetitive wave loads;

(b) *Outdated reflective breakwater concept:* Heavy storms are characterised by highly irregular and short-crested waves. Wave breaking at the wall may occur, even in deep water, due to wave–wave interaction. Therefore, vertical breakwaters do not always work as reflective structures. The occurrence of wave breaking must always be considered and does in fact represent the most important cause of damage (Oumeraci, 1994a). Moreover, the trend is rather towards increasing use of dissipating low reflection caisson alternatives (Takahashi, 1996; Oumeraci *et al.*, 2000);

(c) *Wave overtopping – a neglected cause of damage:* Wave overtopping is generally considered an important aspect in the functional design, but it is often overlooked that it may be an important cause of structural damage. Although the failure mechanisms and the loading associated with wave overtopping are still not fully understood, it has been shown that when excessive overtopping occurred breakwaters tilted seaward instead of shoreward (Oumeraci, 1994a);

(d) *Geotechnical and other 'hidden' progressive failures:* Sufficient attention has not yet been paid in the codes of practice to failure modes, associated with instability of the foundation and of the seabed, which very often remain hidden until collapse occurs. Even the PIANC-Committee of 1976 did 'not feel competent to examine the soil mechanics' problems involved' (PIANC, 1976). Considerable efforts have since then been devoted to this issue (De Groot *et al.*, 1996; PIANC, 2001) and it has been shown that conventional bearing capacity calculations are not sufficient. Seabed scour is another neglected design issue (see 'Other important design loads', later in this chapter). The complexity of the transient and cyclic phenomena involved in the wave–structure–foundation interaction, as well as the accumulation of irreversible soil deformations require a deeper insight into the incremental weakening of the seabed and its application in design, construction and maintenance;

(e) *Importance of three-dimensional effects:* Total wave forces, for stability analysis of vertical-face breakwaters, are generally expressed per unit length of breakwater, implicitly assuming two-dimensional conditions. However, due to effects of wave diffraction from the breakwater head, variations of wave forces occur along the breakwater. Due to the phase shift between the waves on the two sides of the breakwater, the variation in wave force per unit length is larger than the corresponding wave height variation. This may cause an increase in wave force per unit length of up to 40 per

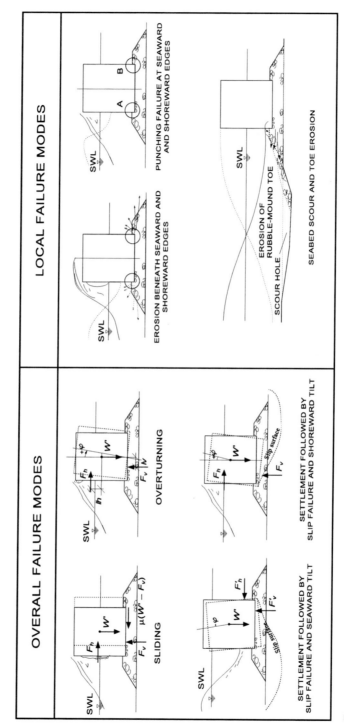

Fig. 7.29 Main modes of vertical breakwater failure

cent compared to the wave force on an infinitely long breakwater (Larras, 1942; Goda, 2000). The evidence of such effects was confirmed by prototype observations showing an undulating line of damage (displaced caissons, etc.) along the breakwater, which is often called 'meandering damage' (Ito and Tanimoto, 1972; Oumeraci, 1994a). On the other hand, oblique and short-crested waves would result in a decrease of wave force per unit length as compared to head-on and long-crested waves (see 'Wave loading of caissons with plain front walls', later in this chapter), which means that the two-dimensional assumption is rather conservative. The same applies for soil stresses in rupture surfaces, which may be 40 per cent higher than those for the actual three-dimensional surfaces (Oumeraci *et al.*, 2001); and

(f) *Need for an integrated design based on dynamic and probabilistic analysis:* Hydraulic, geotechnical, morphological and structural processes are all dynamic and stochastic in nature, and very often interact in a truly complex manner. This generally leads to incremental weakening of foundations and other components, before a collapse occurs. Therefore, the present static and deterministic design approaches are too simplistic to describe actual failure modes. Among the failure modes, which have been identified, those shown in Fig. 7.29 are the most relevant. Other failures' modes, identified by using dynamic and probabilistic analysis, are described by De Groot *et al.* (1996), PIANC (2001) and Oumeraci *et al.* (2001). In the next section, however, only traditional design approaches will be addressed, since the time does not seem ripe for practical application of the new probabilistic design tools, which have been developed (Oumeraci *et al.*, 2001).

Design of caisson breakwaters

Outline of design procedures and general considerations

The main function of a breakwater is to provide sufficient protection against waves. Wave transmission or its effect on the functionality (ship motion at berth, etc.) of the sheltered area must be reduced to an acceptable level, which strongly depends on the purpose of this area, the types of vessels, mooring conditions, etc. Besides wave transmission, wave reflection and overtopping may also be of importance (see the next section). However, before the breakwater fails completely in fulfilling its main function, a number of other failures, associated with loss of stability of the foundation and/or loss of structural integrity, have often taken place. Therefore, it is one of the main design tasks to properly analyse such failure modes. For this purpose, the design wave parameter at the structure, the design wave loads and the associated responses of the

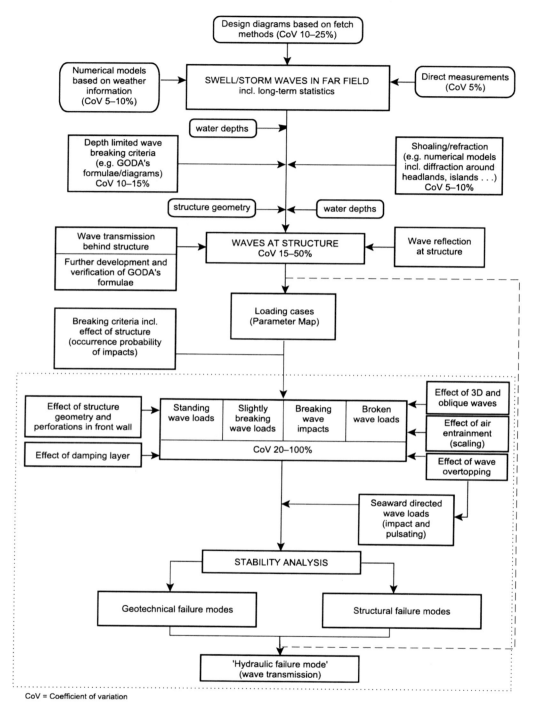

CoV = Coefficient of variation

Fig. 7.30 Flow chart of overall design procedure (adapted from Oumeraci et al., 2001)

structure and its foundation should be predicted. An overview of the overall design procedure, which is far from being exhaustive, is shown in Fig. 7.30.

The uncertainties associated with the predicted wave heights, forces, etc. are indicated by a coefficient of variation (CoV), which is defined as the standard deviation divided by the mean value of the variable considered. The following sections in this chapter focus on wave loads and stability analysis, preceded by a brief description of other aspects of hydraulic performance such as wave reflection, wave overtopping and associated wave transmission over the breakwater. A further classification of structures from a wave loading perspective is shown in 'Wave loading of caissons with plain front walls', later in this chapter.

In addition, the following items must also be addressed in the design process:

(a)   How and where to produce the caissons;
(b)   How the caissons will be transported and placed;
(c)   How to ensure scour protection during and after construction;
(d)   How to monitor, inspect, maintain and repair the breakwater components; and
(e)   How and when to perform scale-model tests as an integral part of the detailed design.

Hydraulic performance

Wave reflection

Wave reflection may cause detrimental effects to the manoeuvring of smaller vessels, to the stability of the seabed (increased scour) and neighbouring beaches (increased erosion). It is therefore one of the main design tasks to determine the reflection coefficient K_r, defined as the ratio of the reflected to the incident wave height $(K_r = H_r/H_i)$ and to keep it at an acceptable level. Since wave reflection is closely linked to wave transmission and particularly to wave energy dissipation, which of course depends on the type of structure, the determination of K_r is hardly amenable to theoretical analysis. Scale-model tests are generally required to quantify the reflection behaviour of specific structures. Alternatively, a number of empirical order-of-magnitude reflection coefficients may be used for the main caisson breakwater types shown in Fig. 7.27:

(a)   *Vertical breakwater type*: For non-breaking to slightly breaking waves and non-overtopping conditions $K_r = 0.7$ to 0.95. Lower values must be adopted for heavily breaking wave conditions such as $K_r = 0.5$ to 0.7 (see Allsop and McBride, 1994; Takahashi, 1996);
(b)   *Composite breakwater type*: For berms with moderate height $K_r = 0.5$ to 0.7 represent a reasonable range. K_r decreases with

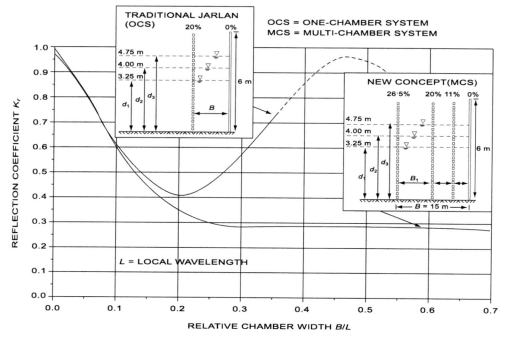

Fig. 7.31 Reflection from perforated caisson breakwater (adapted from Bergmann and Oumeraci, 1999)

increasing breaker index H_s/h_s and with decreasing relative free-board R_c/H_s (Takahashi, 1996). For a HMCB much lower values of $K_r = 0.3$ to 0.5 can be achieved (Muttray et al., 1998).

(c) Jarlan-type breakwater: For a traditional one-chamber system, the reflection coefficient strongly fluctuates between 0.3 to 0.7 with the relative width B/L ($L = $ local wavelength) of the dissipating chamber and thus with the period of the incident waves. To keep K_r constant irrespective of the wave period, a multi-chamber system as shown in Fig. 7.31 can be used (Bergmann and Oumeraci, 1999). Other parameters are the height and width of the berm (Takahashi, 1996);

(d) Armoured caisson breakwaters (horizontally composite breakwater): A range of $K_r = 0.3$ to 0.6 may be adopted. The smaller values corre-spond to larger values of the covering width of the dissipation mound. Other parameters are the breaker index H_s/h_s and the relative freeboard R_c/H_s. The minimum width at the crest generally corresponds to two or three armour units. Details are given by Tanimoto et al. (1987) and Takahashi (1996);

(e) Single perforated walls (Fig. 7.31): Reflection, transmission and dis-sipation are essentially governed by a kind of dynamic porosity

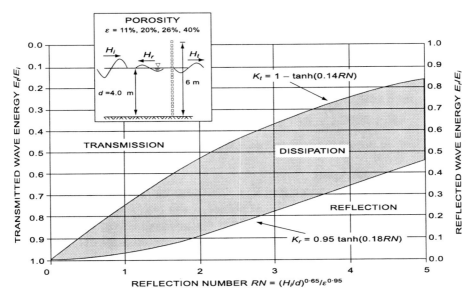

Fig. 7.32 Hydraulic performance of single perforated wall (adapted from Bergmann and Oumeraci, 1999)

parameter $RN = (H_i/d)^{0.65}/\varepsilon^{0.95}$ (Fig. 7.32) which depends on both the structure porosity ε and the hydraulic conditions (H_i/d).

Wave transmission

Waves may be transmitted through the harbour entrance (diffraction), through the rubble foundation and over the breakwater (overtopping). The first is a classical diffraction problem. The second is only relevant when both the relative height of the rubble foundation h_b/h_s and the relative freeboard R_c/H_s are large (see Box 7.1). In this case, it is rather a problem of energy dissipation and wave damping through rubble-mound structures. Apart from transmission through the harbour entrance, transmission is mainly caused by overtopping. Therefore, the governing parameter is the relative freeboard R_c/H_s. Empirical formulae and diagrams are available (Goda, 2000; Takahashi, 1996; Oumeraci et al., 2001) to predict the transmission coefficient, defined as the ratio of transmitted to incident wave height $K_t = H_t/H_i$ associated with wave overtopping. This, however, does not take into account the fact that such transmitted waves have much smaller periods, in the range of 1/3 to 1/2 of the incident wave periods and that the characteristics of these transmitted waves change considerably as they propagate away from the breakwater. For this reason, scale model tests are required for detailed design. For preliminary design, the use of simple empirical formulae for determination of transmission coefficients associated with wave overtopping, suggested by Kondo and Sato (see Goda, 2000) may be considered sufficient:

$$K_t = 0.3\left(1.5 - \frac{R_c}{H_s}\right) \quad \text{with} \quad \frac{R_c}{H_s} = 0 - 1.25$$

for a vertical caisson breakwater, and

$$K_t = 0.3(1.1 - R_c/H_s) \quad \text{with} \quad R_c/H_s = 0 - 1.75$$

for an armoured caisson breakwater.

Wave overtopping

Outline of the design problem   The crest level of the breakwater must be high enough to prevent excessive wave overtopping. Too much over-topping may cause unacceptable wave transmission, leading to damage to berthed ships, installations, vehicles and pedestrians on and behind the structure as well as to interruption of cargo-handling operations. Wave overtopping may also influence breakwater stability in the sense that the wave loading changes as compared with non-overtopping condi-tions. In order to apply the principle of overtopping-based design for determination of the crest elevation more efficiently, it is necessary to have reliable tools to predict the expected overtopping quantities and to decide on allowable limits for overtopping with regards to structural and functional integrity.

Moreover, in view of the increasing demand for sustainable develop-ment of coastal zones, the reduction of spray generation at coastal struc-tures is expected to become an important aspect in the design process. Spray may disturb traffic, cause damage to buildings and other facilities through salt corrosion and to agricultural and coastal environments through salt deposition.

Determination of wave overtopping   The governing parameter is the relative freeboard R_c/H_s. Other parameters are the breakwater type, the geometry of the superstructure, the foreshore topography, the direction of short-crested waves, wind, etc. The degree of wave overtopping may be measured in two ways:

(a)   as a *mean overtopping discharge q* per time unit and per unit length of breakwater (l/s/m), which is particularly useful for design of the drainage system; and

(b)   as an *individual overtopping volume V* per wave and per unit length of the breakwater (m³/m), which is required to assess structural damage and impacts on vehicles, pedestrians, etc.

Although the single-wave volume *V* represents damaging effects of wave overtopping much better, the available formulae and allowable limits only include the mean discharge q. Due to the complexity of the processes involved and the large number of parameters, wave overtopping conditions are best determined by scale-model tests. However, an excellent practical approach, which can be used for feasibility studies and preliminary design, is described in Chapter 5 of Goda (2000). It provides simple design diagrams for foreshore slopes of 1 on 10 and 1 on 30 to determine mean overtopping discharges for plain vertical walls with and without protection by concrete armour blocks. An alternative approach was developed by a research group (Franco and Franco, 1999). It is generally less conservative than Goda's and accounts for the reductions associated with oblique waves, wave directionality and structural measures to reduce wave overtopping (see Box 7.1). Moreover, a statistical method is proposed to calculate the individual overtopping volume V from the mean overtopping discharge q.

Users of experimental results and empirical formulae or diagrams must always be aware of the wide range of uncertainties involved, particularly when the overtopping discharge becomes small (Fig. 7.33). Depending on the type of structure, the scatter may vary from 50 per cent to 200 per cent of the actual values for $q\sqrt{gH_s^3} = O(10^{-2})$ and from 5 per cent to 1000 per cent for $q\sqrt{gH_s^3} = O(10^{-5})$. However, the range of variation of the crest elevation determined on the basis of such overtopping discharges may be less than ± 20 per cent (Goda, 2000).

Allowable wave overtopping discharges The probability that the design storm conditions will be exceeded can never be excluded. Therefore, it would be unreasonable from both economical and managerial points of view to adopt a zero-overtopping criterion for the determination of the crest elevation of the breakwater. This means that limits of overtopping quantities must be determined based on their possible impacts on the structural integrity of the breakwater itself as well as on the operations, persons and installations on and behind the breakwater. The extent of damage, which might be caused by exceeding design conditions, will be much less than if a zero-overtopping design concept would be employed because both the designer and the port operator will be aware of overtopping and its possible impacts, and they will therefore be prepared to manage it. Unfortunately, useful data on overtopping limits are scarce for both the structural integrity of the breakwater and the safe use of the areas on and behind the breakwater. Generally, the limits for the latter case are lower than those for the structural integrity of the breakwater. Therefore, only indications are given below of possible failure modes associated with the structural integrity.

Large overtopping water masses plunging into the harbour basin can generate large seaward directed forces (Oumeraci *et al.*, 2001). This

Box 7.1 Wave overtopping formulae

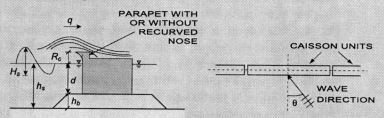

Fig. B1 Definition sketch

1 Average overtopping rate q [m³/s m]

$$\bar{Q}_* = \frac{q}{\sqrt{gH_s^3}} = 0.082 \exp\left(-3.0\frac{R_c}{H_s}\frac{1}{\gamma_\theta \cdot \gamma_S}\right)$$

γ_θ = corrective coefficient for wave obliquity and directionality

q = average overtopping rate [m³/s m]

Long-crested waves		Short-crested waves	
$\gamma_\theta = \cos\theta$	$\gamma_\theta = 0.79$	$\gamma_\theta = 0.83$	$\gamma_\theta = 0.83\cos(20° - \theta)$
For $\theta = 0$–$37°$	For $\theta \geq 37°$	For $\theta = 0$–$20°$	For $\theta \geq 20°$

γ_s = corrective coefficient for structural measures to reduce q

Structural alternatives	γ_s
• Caisson with plain impermeable front wall and parapet without nose (reference case)	1.00
• Caisson with plain impermeable front wall and parapet with re-curved nose	0.70
• Jarlan-type caisson breakwater with closed top	0.75
• Jarlan-type caisson breakwater with open top	0.60

2 Individual overtopping volume V [m³/m/wave]

$$P(V) = \exp\left[\left(-\frac{V}{A}\right)^{3/4}\right] \Rightarrow V = A\left[-\ln P(V)^{4/3}\right]$$

and

$$V_{max} = A(\ln N_{ow})^{4/3}$$

where

- $P(V)$ = exceedance probability of overtopping volume V (Weibull distribution)

- A = scale parameter [m^3] with $A = 0.84qT_m/P_{ow}$
- T_m = mean wave period [s]
- $P_{ow} = N_{ow}/N_w$ = ratio of number of overtopping waves (N_{ow}) to total number of incoming waves (N_w) (Rayleigh distribution):

$$P_{ow} = \exp\left[-\left(\frac{R_c/H_s}{0.91}\right)^2\right]$$

- V_{max} = max overtopping volume generated by one of the total number of overtopping waves N_{ow} [m^3/m]

might explain the seaward tilting of many breakwaters (Oumeraci, 1994a). The plunging water mass may also cause collapse of the deck slab and of other components of the superstructure (Oumeraci *et al.*, 2001). Some overtopping limits, which are believed to be too strict and related more to seawalls, dykes and revetments than to breakwaters, are given in CIRIA (1991). For the safe use of the areas behind the parapet, limits for vehicles and personnel are given in terms of mean overtopping discharges. Mean discharges larger than the order of 1 l/s/m become dangerous to personnel and impassable for vehicles. When using such mean values, one should always keep in mind the inherent large variation of overtopping (see Fig. 7.33). Peak values can exceed mean values by a factor of up to 100. Within port areas in Japan a mean value in the order of 10 l/s/m has been adopted for the safe use of the areas behind break- waters and jetties (Goda, 2000). Based on such limits, the required crest elevation is then determined by using the empirical formulae in Box 7.1 or the diagrams in Goda (2000) for wave overtopping. However, the final decision must be taken on the basis of scale-model tests and engineering judgement.

Measures to reduce overtopping and spray For a variety of reasons (the aesthetics of the seascape, functionality, etc.) the crest elevation deter- mined by overtopping criteria may be considered too high. In such cases measures associated with the geometry of the breakwater front and superstructure, such as permeable front walls and wave chambers, re-curved parapets, dissipating basins and other structural elements on the caisson deck, should be considered (see 'Crest structures' and 'Inno- vative caisson breakwaters' later in this chapter). Moreover innovative shapes and crowns of the caissons may also be devised to reduce wave splash and spray height. Spray can be transported by storms up to 30 km inland and a flux of salt spray up to 400 μg/m^2/s might result. Large quantities of salt water transported inland and dispersed over

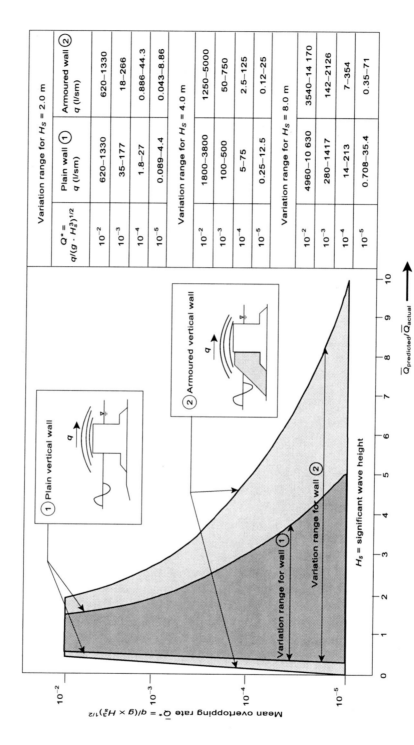

$Q^* = q/(g \cdot H_s^3)^{1/2}$	Plain wall ① q (l/sm)	Armoured wall ② q (l/sm)
Variation range for H_s = 2.0 m		
10^{-2}	620–1330	620–1330
10^{-3}	35–177	18–266
10^{-4}	1.8–27	0.886–44.3
10^{-5}	0.089–4.4	0.043–8.86
Variation range for H_s = 4.0 m		
10^{-2}	1800–3800	1250–5000
10^{-3}	100–500	50–750
10^{-4}	5–75	2.5–125
10^{-5}	0.25–12.5	0.12–25
Variation range for H_s = 8.0 m		
10^{-2}	4960–10 630	3540–14 170
10^{-3}	280–1417	142–2126
10^{-4}	14–213	7–354
10^{-5}	0.708–35.4	0.35–71

Fig. 7.33 Scatter of predicted mean overtopping discharge compared with actual discharge (based on figures from Goda, 2000)

wide coastal areas may have detrimental short-term effects (disturbance of vehicular traffic, electric power supply, port operations, etc.) as well as long-term effects (salt corrosion of buildings and other facilities, damage to agriculture and the environment, etc.).

Wave loading of caissons with plain front walls

Wave load classification

The local wave climate, the foreshore topography as well as the type and geometry of the structure including the dimensions of its rubble foundation, govern the wave loading of the structure (caisson and superstructure). The parameter map in Fig. 7.34, which was developed for the European PROVERBS project (Oumeraci *et al.*, 2001), allows the designer to distinguish between:

(a)   *breaking wave impact loads* (loading case 3) for which both magnitude and variation with time (load-time history) are required to perform the dynamic analysis of the response of the structure, and which therefore needs to be handled with special care; and

(b)   *'pulsating' wave loads* (loading cases 1, 2 and 4) which are assumed to be 'quasi-static', so that standard wave-load formulae and quasi-static stability analysis can be applied.

The principle of the parameter map in Fig. 7.34 is the use of three non-dimensional input parameters, each of which allows the designer to take a decision at the corresponding level.

At the first decision level the relative height of the rubble foundation $h_b^* = h_b/h_s$ (h_s being the water depth at the toe of the mound) governs the type of structure from the wave loading point of view:

(a)   *vertical breakwater* (Fig. 7.27) for $h_b^* < 0.3$ (bedding layer);

(b)   *composite breakwater*, further subdivided into *low mound* for $h_b^* = 0.3$ to 0.6, *moderate mound* for $h_b^* = 0.6$ to 0.9 and *high mound* for $h_b^* = 0.9$ to 1.0; and

(c)   *crown wall on a rubble-mound breakwater* (see 'Superstructure' earlier in this chapter): any monolithic concrete structure on a mound exceeding a relative height of about 1.0.

At the next two decision levels the loading case is determined by using:

(a)   the relative wave height $H_s^* = H_s/h_s$ (H_s is the significant local wave height), which represents a breaker index; and

(b)   the relative berm width $B^* = B_{eq}/L$ (L is the local wavelength), which accounts for the effect of the berm width on wave breaking conditions.

In summary, the parameter map enables the designer to identify quickly and simply the conditions which might lead to breaking-wave impacts.

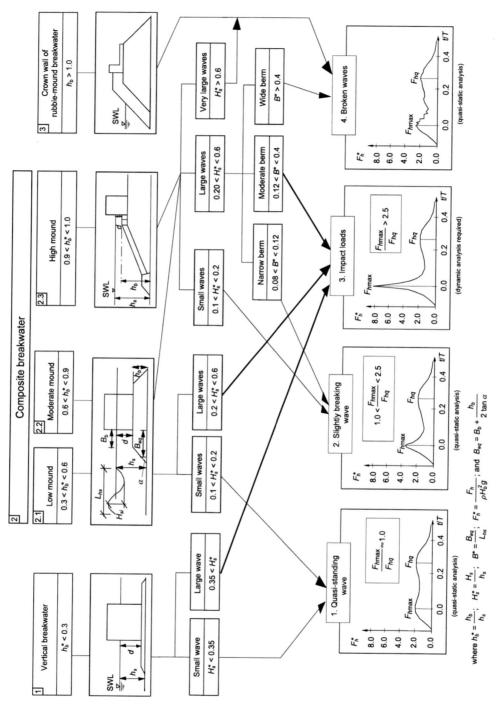

Fig. 7.34 Parameter map for classification of loading cases. Reproduced from Oumeraci et al., 2001

The latter may occur for:

(a)   vertical breakwaters with large waves ($H_s^* > 0.35$); and
(b)   low mound composite breakwaters with moderate berm widths ($0.12 < B^* < 0.4$) and large waves ($0.2 < H_s^* < 0.6$).

Further refinements may be needed for detailed design by including, for instance, the following:

(a)   improved breaking criteria accounting for the slope of the foreshore and the reflective properties of the breakwater (Oumeraci et al., 1993, Muttray and Oumeraci, 2000 and Goda, 2000); and
(b)   a procedure for estimating the likely percentage of breaking waves causing impacts. Tentative formulae proposed by Calabrese and Vicinanza (see Oumeraci et al., 2001) can be used in preliminary design for this purpose (see Box 7.2).

The loading cases 1, 2 and 4 for vertical breakwaters and low and moderate mound composite breakwaters, as shown in Fig. 7.34, are in the following called 'pulsating' wave loads in contrast to impact loads. The Goda formulae are suggested for determining the horizontal and uplift 'pulsating' wave loads directed landward (wave crest at the front wall). For the wave forces directed seaward (wave trough at the front wall) another method has to be used (see Box 7.3). For the high mound composite breakwater, a new tentative approach is suggested, because it functions differently, compared to the other structure types (see Box 7.11).

For impact loads the Goda formulae are invalid. Therefore, a new method is tentatively proposed (see Box 7.6).

Pulsating wave loads

(a)   Shoreward wave loads and associated uplift (wave crest at front wall): The horizontal pressure and force as well as the associated uplift pressure caused by non-breaking (loading case 1), slightly breaking (loading case 2), and broken waves (loading case 4) are calculated by using the Goda formulae in Box 7.3, which have been adopted in many codes of practices (e.g. BSI, 1991) and technical standards (e.g. OCADIJ, 1991). Some remarks, which might be useful, are:

(i)   The Goda formulae have been developed on the basis of extensive hydraulic model testing, theoretical considerations and analysis of failed and non-failed prototype breakwaters. Further verifications were performed by many European hydraulic laboratories (Oumeraci et al., 2001), so that uncertainties associated with these formulae may be quantified in

173

Box 7.2 Occurrence probability of impacts (supplement to parameter map in Fig 7.34)

    1 Breaker height H_b in front of reflective structure

$$H_b = L_{pi}\left[0.1025 + 0.0217\left(\frac{1-K_r}{1+K_r}\right)\right]\tanh 2\pi k_b \cdot \frac{h_s}{L_{pi}}$$

- L_{pi} = wave length in water depth h_s corresponding to peak period T_p [m]:

$$L_{pi} = L_o\left[\tanh\left(2\pi\frac{h_s}{L_o}\right)^{3/4}\right]^{2/3} \quad\text{and}\quad L_o = \frac{gT_p^2}{2\pi}$$

- K_p = reflection coefficient with $K_p = 0.50\text{--}0.70$ for composite breakwater with high mound and $K_r = 0.70\text{--}0.95$ for vertical breakwater with low mound (see section 'Wave reflections', earlier in this chapter for more details)
- k_b = empirical correction factor:

$$k_b = 0.0076\left(\frac{B_{eq}}{d}\right)^2 - 0.1402\left(\frac{B_{eq}}{d}\right) + 1$$

with the equivalent berm width

$$B_{eq} = B_b + \frac{h_b}{2\tan\alpha}$$

$\tan\alpha$ is the steepness of the seaward berm slope and B_b, h_b are the berm width and height, respectively (see Fig. 7.34), d is the water depth at the caisson wall.

2 Probability of impacts
- Percentage of breaking and broken waves

$$P_b(\%) = \exp\left[-2\left(\frac{H_b}{H_{si}}\right)^2\right]100\%$$

where H_{si} is the incident significant wave height in front of the structure [m].
- Critical wave height describing transition between broken waves and impact breaking waves:

$$H_{bs} = 0.1242 \cdot L_{pi}\tanh\left(2\pi\frac{h_s}{L_{pi}}\right)$$

- Percentage of impacts:

$$P_i(\%) = \left\{ \exp\left[-2\left(\frac{H_b}{H_{si}}\right)^2\right] - \exp\left[-2\left(\frac{H_{bs}}{H_{si}}\right)^2\right] \right\} 100\%$$

For very low value of $P_i(\leq 1\%)$ the loading can be considered as pulsating, so that a quasi-static approach can be used (see the next section).

terms of coefficients of variations $COV_x = \sigma_x/\bar{x}$ (σ_x = standard deviation of variable x, $\bar{x}$ = mean value;

(ii) The range of uncertainties associated with the calculated horizontal force F_h, the uplift force F_u and the corresponding moments M_{F_h} and M_{F_u} around the heel of the caissons and the corresponding uncertainties, which would result from scale-model tests (in parentheses), are as follows:

$$COV_{F_h} \approx 20\% \ (5\%) \quad \text{and} \quad COV_{F_u} \approx 20\% \ (10\%),$$

$$COV_{M_{F_u}} \approx 40\% \ (10\%) \quad \text{and} \quad COV_{M_{F_u}} \approx 40\% \ (15\%);$$

(iii) The design wave height $H_D = H_{\max} = 1.8H_s$ recommended for breakwaters located seaward of the surf zone corresponds to the 0.1 per cent exceedance value for a Rayleigh wave height distribution ($H_{\max} = H_{1/250}$ = arithmetic mean of the 0.4 per cent highest waves). This is, however, only a recommendation, and it is left to engineering judgement to choose a more or less strict statistical value;

(iv) The definition of the design wave height H_D, at a distance of five significant wave heights seaward of the berm toe, was recommended by Goda in order to account for the load of broken waves on the structure. Of course refinements of this simple criterion could be introduced to take into account the plunging distances and other factors; and

(v) The effect of the wave period on pressure is taken into account by a coefficient α_1, the effect of the height of the rubble foundation by α_2 and that of the wave obliquity by $\cos^2\theta$ in α_3 as well as by the reduction factor $0.5(1 + \cos\theta)$. The effect of wave overtopping on the reduction of horizontal wave loads is included in the formulae for p_4, while it is considered to have no significant influence on uplift pressure.

(b) Seaward wave loads (wave trough at the front wall): Non-breaking wave forces, directed seaward, occur when the trough of the incident wave is directly at the front wall. Thus the force on the front

Box 7.3 Goda's formulae for wave pressure with wave crest at the front wall

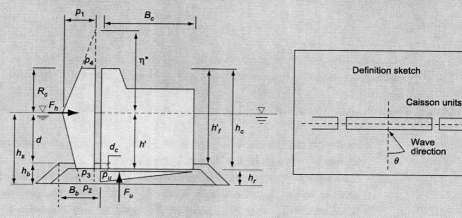

Fig. B2 Definition sketches

1 Design wave height

$$\eta^* = 0.75 \cdot (1 + \cos\theta) \cdot H_D$$

- H_D = design wave height, corresponds to the highest at a location just in front of the breakwater [m]:
  - *for location seaward of surf zone:* $H_D = H_{max} = 1.8H_s$ (Rayleigh distribution)
  - *for location in the surf zone:* $H_D = H_{max}$ taken from a random wave-breaking model at a distance of $5H_s$ from the toe of the berm
- H_s = significant wave height [m] and
- θ = angle of wave incidence relative to line normal to breakwater

2 Horizontal wave pressure and force

- Horizontal wave pressure [kPa] and force [kN/m] on vertical front wall:

$$p_1 = \tfrac{1}{2} \cdot (1 + \cos\theta) \cdot (\alpha_1 + \alpha_2 \cos^2\theta) \cdot \rho_w g H_D$$

$$p_2 = \frac{p_1}{\cosh(2\pi h_s / L)}$$

$$p_3 = \alpha_3 p_1$$

$$p_4 = \begin{cases} p_1(1 - R_c/\eta^*) & \text{for } \eta^* > R_c \\ 0 & \text{for } \eta^* \leq R_c \end{cases}$$

$$F_h = \tfrac{1}{2} \cdot (p_1 + p_3)h' + \tfrac{1}{2} \cdot (p_1 + p_4)R_c^*$$

where:

$$\alpha_1 = 0.6 + \frac{1}{2}\left(\frac{4\pi h_s/L}{\sinh(4\pi h_s/L)}\right)^2$$

$$\alpha_2 = \min\left(\frac{h_D - d}{3h_D}\left(\frac{H_D}{d}\right)^2; \frac{2d}{H_D}\right)$$

$$\alpha_3 = 1 - \frac{h'}{h_s}\left(1 - \frac{1}{\cosh(2\pi h_s/L)}\right)$$

$R_C^* = \min\{\eta^*; R_C\}$; $h_D =$ water depth at the location of design wave H_D [m]; $L =$ wave length [m] corresponding to $T_{1/3} \approx 1.2T_m$ ($T_m =$ mean wave period [s]); $\rho_w =$ density of seawater [kg/m^3] and $g =$ acceleration of gravity [m/s^2].

3 Wave uplift pressure and force
- Wave uplift pressure [kPa] and force [kN/m]:

$$p_u = \tfrac{1}{2}\cdot(1 + \cos\theta)\alpha_1\alpha_3\rho_w g H_D$$

$$F_u = \tfrac{1}{2}p_u B_c$$

4 Moment around caisson heel
- Moment around heel of caisson due to F_h and F_u [kNm/m]:

$$M_{F_h} = \tfrac{1}{6}\cdot(2p_1 + p_3)h'^2 + \tfrac{1}{2}\cdot(p_1 + p_4)h'R_C^* + \tfrac{1}{6}\cdot(p_1 + 2p_4)R_C^{*2}$$

$$M_{F_u} = \tfrac{2}{3}\cdot F_u \cdot B_C$$

wall becomes smaller than the hydrostatic force of the still water level behind the breakwater (see sketch in Box 7.4). In deep water ($h_s/L > 0.25$), the net seaward force may even become larger than the shoreward force in the previous sections. This and other circumstances under breaking and overtopping waves (see the next section), which are still not fully understood, may explain the many cases of damage with seaward tilted breakwaters (Oumeraci, 1994a), Therefore, the occurrence of such seaward forces must be considered in the design. For non-breaking waves plunging into the harbour basin, the seaward forces are considered to be 'quasi-static' and may be approximated by the Sainflou method described in Box 7.4. For deep-water waves ($H_s/h_s < 0.6$) experimental results suggest an increase of F_h, calculated from the formulae in Box 7.4, by a factor of about 1.3. This was confirmed by Goda (2000), who proposed a design diagram showing that the seaward forces become

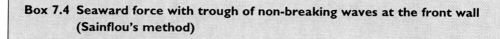

Box 7.4 Seaward force with trough of non-breaking waves at the front wall (Sainflou's method)

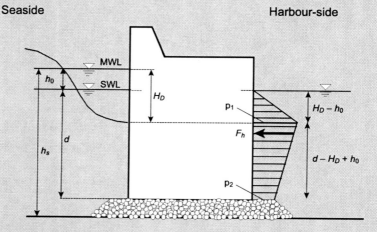

Fig. B3 Definition sketch

$$h_0 = \frac{\pi H_D^2}{L} \coth \left(\frac{2\pi h_s}{L} \right)$$

$$p_1 = \rho_w g (H_D - h_0)$$

$$p_2 = \rho_w g H_D / \cosh \left(\frac{2\pi h_s}{L} \right)$$

● Seaward force [kN/m]:

$$F_h = \tfrac{1}{2} \cdot [p_1 (H_D - h_0) + (p_1 + p_2)(d - H_D + h_0)]$$

where $H_D = H_{1/250}$ is recommended as design wave height.

larger than the shoreward forces as soon as the relative water depth $h_s/L = 0.25$ is exceeded.

Breaking wave impact loads

(a) Effects of breaking wave impacts and general considerations: From the analysis of more than 40 damaged vertical breakwaters world-wide, breaking wave impacts were found to be one of the principal reasons of the observed catastrophic failures (Oumeraci, 1994a).

In contrast to the pulsating wave loads described in the previous section, impact loads are characterised by a much shorter duration

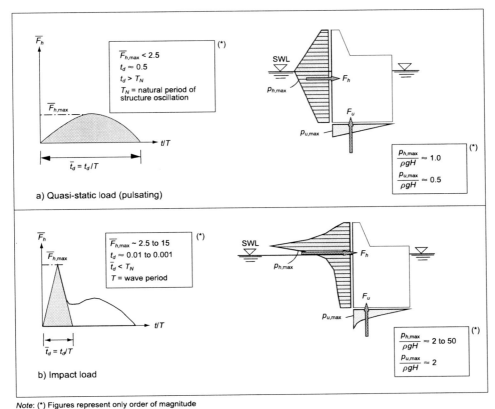

Note: (*) Figures represent only order of magnitude

Fig. 7.35 Characteristics of pulsating and impact wave loads

(less than natural oscillation period T_N of the breakwater, and a much higher magnitude of local impulsive pressures up to $50\,\rho gH$ and of total horizontal forces up to $15\,\rho gH^2$ (Fig. 7.35).

Despite their short duration, impact loads are not only critical for the structural integrity of the smaller components of the caissons (superstructure), but may also affect the overall stability of the structure and its foundation due to the induced oscillatory motion and incremental permanent displacements (Fig. 7.36). Present codes of practice do not consider impact loads to be critical for the overall stability (e.g. BSI, 1991).

In view of the effects depicted by Fig. 7.36 and the effects on the structural integrity of other components of the structure, it is generally not advisable to build a traditional caisson breakwater at a location where frequent severe breaking wave impacts are expected. Here another breakwater type, such as a rubble-mound or a high-mound-composite breakwater, would be more appropriate

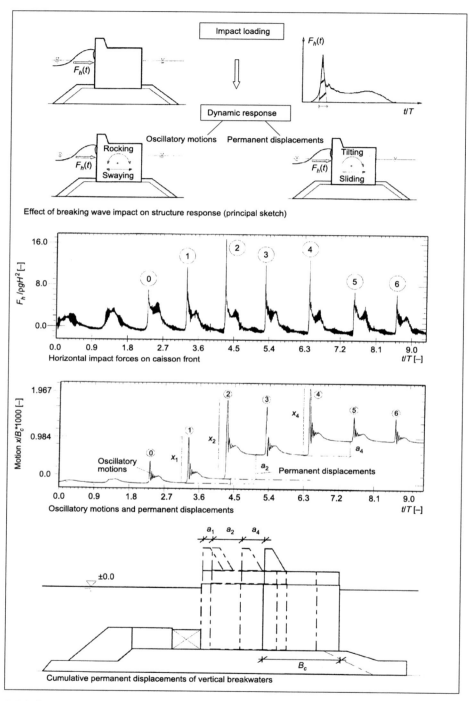

Fig. 7.36 Cumulative effect of repetitive impacts on breakwater stability

(see 'Wave loads on armoured caisson breakwaters', later in this chapter).

(b) Methods of prediction of impact loads: For frequent occurrences of breaking wave impacts (see parameter map in Fig. 7.34 and the last equation in Box 7.2) hydraulic model tests are highly recommended. For preliminary design two methods are suggested in Boxes 7.5, 7.6 and 7.7:

(i) An extension of the Goda formulae is given in Box 7.5. It is based on laboratory tests of sliding stability of caisson breakwaters (see Takahashi, 1996; Goda, 2000). This method consists of the introduction of a new coefficient into the Goda formula for pressure p_1 at still-water level (the second equation in Box 7.5), which accounts for both the occurrence and the intensity of impact pressure. It is briefly described in Box 7.5 and in more detail by Takahashi (1996). The pressures obtained by this method, which is mainly used in Japan, are effective pressures in the sense that they can be directly applied to static stability analysis, e.g. sliding stability;

(ii) A new and more generic method has been developed on the basis of large- and small-scale model tests as well as on theoretical and statistical considerations by the European Research project PROVERBS. The method is briefly described in Box 7.6 for horizontal impact loads and in Box 7.7 for uplift impact loads. For details see Oumeraci et al. (2001), where further references are provided. Unlike the loads obtained by the extended Goda formulae, the loads determined by this method are time dependent and therefore applicable to dynamic analysis of geotechnical and structural stability and the calculation of equivalent static loads (see 'Other important design loads', later in this chapter, and Box 7.8).

(c) Further important aspects related to wave impacts:

Model scale problems: Using Froude's law of similitude for laboratory results generally results in too high impacts and too short duration due to scale effects associated with less air entrainment/entrapment than under prototype conditions. Moreover, the residence time of the larger air bubbles in fresh water (used in the model) is less than the smaller bubbles in salt water (prototype), thus resulting in more air content in salt water during impacts. Tentative recommendations for alternative scaling, corrective factors and further references are given by Oumeraci et al. (2001).

Variation of impact loads along the breakwater: The present methods are based on the implicit assumption that the impact force for a unit length of the caisson [N/m] remains constant over an infinite length

Box 7.5 Extension of Goda's formulae for impact loads (Takahashi, 1996)

- Definition sketch and Goda's formulae are in Box 7.3
- Instead of α_2 in equation for pressure p_1 (see Box 7.3), a new coefficient α_* to account for impact loads is introduced:

$$p_1 = \tfrac{1}{2}(1 + \cos\theta)(\alpha_I + \alpha_* \cos^2\theta)\rho g H_D$$

$\alpha_* = \max(\alpha_2, \alpha_1)$ with α_2 according to Goda in Box 7.3 (the eighth equation); α_I = impact pressure coefficient taking into account two effects: the effect of the related wave height (α_{IH}) as well as the effect of the relative berm width and height (α_{IB}):

$$\alpha_I = \alpha_{IH} \times \alpha_{IB}$$

$$\alpha_{IH} = \begin{cases} \dfrac{H_D}{d} & \text{for } H_D \leq 2d \\ 2.0 & \text{for } H_D > 2d \end{cases}$$

$$\alpha_{IB} = \begin{cases} \dfrac{\cos\delta_2}{\cosh\delta_1} & \text{for } \delta_2 \leq 0 \\ \dfrac{1}{\cosh\delta_1 (\cos\delta_2)^{0.5}} & \text{for } \delta_2 > 0 \end{cases}$$

$$\delta_1 = \begin{cases} 20\delta_{11} & \text{for } \delta_{11} \leq 0 \\ 15\delta_{11} & \text{for } \delta_{11} > 0 \end{cases}$$

$$\delta_2 = \begin{cases} 4.9\delta_{22} & \text{for } \delta_{22} \leq 0 \\ 3\delta_{22} & \text{for } \delta_{22} > 0 \end{cases}$$

$$\delta_{11} = 0.93\left(\frac{B_B}{L} - 0.12\right) + 0.36[(h_s - d)/(h_s - 0.6)]$$

$$\delta_{22} = -0.36\left(\frac{B_B}{L} - 0.12\right) + 0.93[(h_s - d)/(h_s - 0.6)]$$

- The coefficient of impact pressure is set to $\alpha_I = 2.0$ for $B_B/L = 0.12$, $d/h_s = 0.4$ and $H_D/d \geq 2$. For $d/h_s > 0.7$, α_I is always close to zero and less than α_2, i.e. impulse pressure rarely occurs. The uncertainty for the calculation of the total horizontal impact force compared to experiments is in the range of $\sigma'_{FH} = 40\%$.

of the breakwater. As shown in Fig. 7.37, a substantial reduction of the impact load may result for longer caisson units (see also Fig. 7.48). The effects of oblique and short-crested waves are also shown.

Seaward impact forces induced by wave overtopping: Wave impact on the rear face of the breakwater, by overtopping waves plunging into the harbour basin, may represent one of the main reasons

---

Box 7.6 PROVERBS' method for determining horizontal impact loads (Oumeraci et al., 2001)

1 Calculate breaker height H_b (see Box 7.2)

The total horizontal force in a non-dimensional form is:

$$F_{h,\max}^{*} = \frac{F_{h,\max}}{\rho g H_b^2}$$

2 Determine the non-dimensional force $F_{h,\max}^{*}$ assuming a generalised extreme value distribution (GEV):

$$F_{h,\max}^{*} = \frac{\alpha}{\gamma}\left\{1 - [-\ln P(F_{h,\max}^{*})]^{\gamma}\right\} + \beta$$

where:

$P(F_{h,\max}^{*})$ = non-exceedance probability which may be taken as 90 per cent

α, β and γ = parameter of the GEV distribution of $F_{h,\max}^{*}$, which was described by Allsop and Kortenhaus (in Oumeraci et al; 2001) as a function of the bed slope.

Table B1 GEV-distribution parameter

Bed slope	α	β	γ
1 : 20	3.745	7.604	−0.295
1 : 50	1.91	3.268	−0.232

3 Determine simplified impact force history

The actual and more complicated force history is reduced to an equivalent triangular force history having the same peak value $F_{h,\max}$, the same force impulses $I_{r,Fh}$ (momentum of water mass involved in the impact) and $I_{d,Fh}$ (momentum of the total mass in the breaking wave), but different rise time (t_r) and total impact duration (t_d) shown in Fig. B4.

4 Determine equivalent rise time t_r

The actual rise time $t_{r,Fh}$ is derived from solitary wave theory and impulse theory:

$$t_{r,Fh} = 8.94 \times k \frac{\sqrt{\dfrac{d_{\text{eff}}}{g}}}{F_{h,\max}^{*}}$$

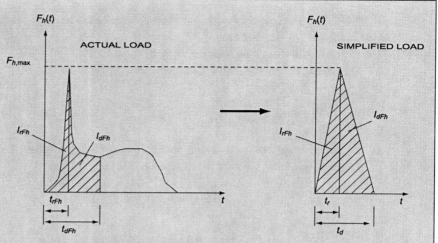

Fig. B4 Simplified force time history

where d_{eff} is the effective water depth identical to the breakwater depth just in front of the structure.

$$d_{\text{eff}} = d + B_{rel} \times m_{rel}(h_s - d)$$

where B_{rel} and m_{rel} are the parts of the relative berm width (B_b/L_{pi}) and of the berm slope $(m = \cot \alpha)$, respectively, which actually affect the breaking depth:

$$B_b = \left\{ \begin{array}{ll} 1 & \text{for } \dfrac{B_b}{L_{pi}} \leq 1 \\[2ex] 1 - 0.5\dfrac{B_b}{L_{pi}} & \text{for } \dfrac{B_b}{L_{pi}} > 1 \end{array} \right\}$$

and

$$m_{rel} = \left\{ \begin{array}{ll} 0 & \text{without any berm} \\ 1 & \text{for } 0 < m < 1 \\ \dfrac{1}{\sqrt{m}} & \text{for } m \geq 1 \end{array} \right\}$$

k is the proportion of the water mass M_{imp} of the breaking wave mass M_{wave}, which is involved in the impact and causing the force impulse $I_{r,Fh}$ in Fig. B4. It depends on the breaker type, but can be estimated to be:

$$k = \frac{M_{imp}}{M_{wave}} = \frac{I_{r,Fh}}{I_{d,Fh}} = 0.16 \text{ to } 0.25$$

The equivalent rise time t_r for a deterministic approach is

$$t_r \approx (0.5 \text{ to } 1.0) \times t_{r,Fh}$$

For a probabilistic approach, the uncertainties in k and in the relationship $t_r = f(t_{r,Fh})$ are better considered together through the following relationship to calculate directly the equivalent rise time t_r:

$$t_r = k' \cdot \frac{\sqrt{\dfrac{d_{\text{eff}}}{g}}}{F^*_{h,\text{max}}}$$

where the factor k' is described as a log-normal distribution with a mean value $\overline{k'} = 0.086$ and a standard deviation $\sigma_{k'} = 0.084$.

5 Determine equivalent total duration t_d

For a deterministic approach the upper bound of t_d/t_r is described by:

$$t_d = t_r \left[2.0 + 8.0 \exp\left(-18 \frac{t_r}{T_P} \right) \right]$$

For a probabilistic approach t_d is calculated statistically by:

$$t_d = \frac{-c}{\ln(t_r)}$$

where c is an empirical parameter $[-s\ln(s)]$ normally distributed with a mean value $\overline{c} = 2.17$ and $\sigma_c = 1.08$ (derived from large scale model tests).

6 Calculate force impulses $I_{r,Fh}$ and $I_{d,Fh}$

The force impulse over the rise time t_r and over the total duration t_d are (see definition in Fig. B4)

$$I_{r,Fh} = \tfrac{1}{2} F_{h,\text{max}} \cdot t_r$$

$$I_{dFh} = \tfrac{1}{2} F_{h,\text{max}} \cdot t_d$$

7 Determine simplified impact pressure distribution

Based on analysis of about 1000 impact pressure distributions recorded in large-scale-model tests the simplified distribution at the time of maximum impact $F_{h,\text{max}}$ has been derived (Fig. B5). Four parameters are required to describe this distribution:

(i) Possible maximum elevation of pressure distribution for an infinitely high wall:

$$\eta_* = 0.8 H_b$$

(ii) Bottom pressure p_3:

$$p_3 = 0.45 p_1$$

(iii) Impact pressure p_1 at design water level (DWL): it is based on the equivalent triangular shape of $F_h(t)$ (in Fig. B4). First, a high

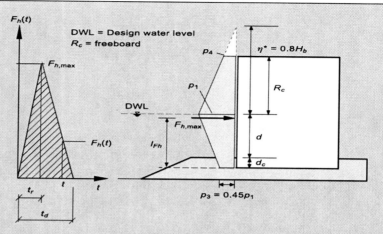

Fig. B5 Simplified impact pressure distribution

freeboard $(R_c > \eta_*)$, so that $F_{h,\max}$ or another value of $F_h(t)$ at any time of the equivalent triangular force history is described by the corresponding area of the pressure distribution:

$$F_h(t) = \tfrac{1}{2}p_1(t) \cdot \eta^* + (d + d_c)p_3 + \tfrac{1}{2}(d + d_c)(p_1(t) - p_3)$$

Substitute η^* and p_3 from the 13th and 14th equations:

$$p_1(t) = \frac{F_h(t)}{0.4H_b + 0.7(d + d_c)}$$

(iv) Pressure at the crest of the structure p_4:
- Force reduction due to overtopping is taken into account by cutting off the pressure distribution at the crest level:

$$p_4 = \left\{ \begin{array}{ll} 0 & \text{for } \eta^* < R_c \text{ (no overtopping)} \\[2mm] \dfrac{\eta^* - R_c}{\eta^*} & \text{for } \eta^* \geq R_c \end{array} \right\}$$

- In the case of overtopping, the horizontal force $F_h(t)$ in the 15th equation must therefore be reduced by:

$$\Delta F_h(t) = -\tfrac{1}{2}(\eta^* \cdot R_c) \cdot p_4$$

- Determine lever arm of the horizontal impact force l_{Fh}:

$$l_{Fh} = \frac{p_1\eta_{ov}^2 + 3p_1 d'\eta_{ov} + 3p_4 d'\eta_{ov} + 2p_1 d'^2 + p_3 d'^2}{6F_h(t)}$$

where

$$\eta_{ov} = \min\{\eta^*, R_c\}$$

and $d' = d + d_c$ (see Fig. B5).

Box 7.7 PROVERBS' method for determining uplift impact loads (Oumeraci et al., 2001)

1 Calculate breaker height H_b

Same as in Box 7.6, but for

$$F^*_{u,\max} = \frac{F_{u,\max}}{\rho g H_b^2}$$

2 Determine $F^*_{u,\max}$

Use the same formula (the second in Box 7.6) adapted for $F^*_{h,\max}$ and the following GEV distribution parameter obtained from large-scale model tests with a 1 on 50 foreshore slope:

$$\alpha = 2.17, \ \beta = 4.384 \quad \text{and} \quad \gamma = -0.1$$

3 Determine simplified uplift force history

Same procedure as in Fig. B4. For triangular uplift force $F_u(t)$ in Fig. B6:

- t_{ru} is used for the rise time
- t_{du} is used for the total duration

4 Determine equivalent rise time t_{ru} of $F_u(t)$

- *Deterministic approach*: $t_{ru} = 1.1 t_{r,Fu}(t)$ with the actual rise time of the uplift force $t_{r,Fu}$ determined by the second and eighth equation in Box 7.6, adapted to the uplift force F_u instead of horizontal force F_h.
- *Probabilistic approach*: use the ninth equation in Box 7.6, adapted to uplift, but with $\overline{k'_u} = 0.16$ and $\sigma'_{ku} = 0.17$.

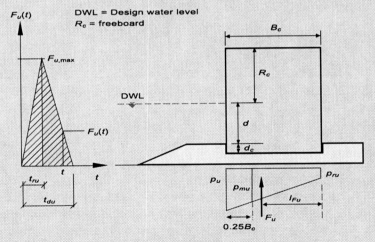

Fig. B6 Simplified uplift pressure distribution

5 Determine equivalent total duration t_{du} of $F_h(t)$
- *Deterministic approach*: use the ninth equation in Box 7.6, adapted to F_u
- *Probabilistic approach*: use the sixth equation in Box 7.6, adapted to F_u but with $\overline{c'_u} = 1.88$ and $\sigma'_{cu} = 0.99$

6 Calculate uplift force impulses over rise time and total duration

$$I_{r,Fu} = \tfrac{1}{2} F_{u,max} \cdot t_{ru}$$

$$I_{d,Fu} = \tfrac{1}{2} F_{u,max} \cdot t_{du}$$

7 Determine simplified uplift pressure distribution
Based on the analysis of more than 1000 impact pressure distributions recorded from large-scale-model tests, the simplified pressure distribution corresponding to the maximum uplift force $F_{u,max}$ has been derived (Fig. B4). Only two parameters are required to describe the distribution:

(i) Pressure at shoreward heel of caisson p_{ru}:

$$p_{ru} = \rho g H_b \left(\frac{H_b}{h_s} - 0.1 \right)$$

(ii) Pressure at seaward toe of caisson p_u:

$$p_u = \frac{2F_{u,max}}{B_c} - p_{ru}$$

8 Lever arm of uplift force l_{Fu}

$$l_{Fu}(t) = \frac{(p_{ru} + 2p_u)B_c^2}{6F_{u,max}}$$

responsible for failures associated with seaward tilting (Fig. 7.38). Experimental and theoretical evidence exists for the generation mechanisms (Oumeraci *et al.*, 2001), but no practical formulae are yet available for these impacts. The same applies for impacts due to overtopping waves or to a mass of green water impinging on the deck of the caisson (structural integrity of the deck slab).

Effect of violent impacts in confined spaces: breaking waves impinging on cavities (cracks, open joints, etc.) generate very high impact pressures and rapid flows, which may result in important degradation (Peregrine and Kalliadasis, 1996).

Box 7.8 Calculation of equivalent static force

1 Equivalent static force

$$F_{stat,equ} = \nu_L \cdot F_{dyn,max}$$

where

$F_{stat,equ}$ = static equivalent force
$F_{dyn,max}$ = maximum dynamic force (see the first equation in Box 7.6)
ν_L = dynamic load factor (estimated as given in steps 2 to 4)

2 Determination of masses and stiffness

Total mass: $m_{tot} = m_{cai} + m_{hyd} + m_{geo}$

Mass of caisson: $m_{cai} = \rho_c B_c h_c l_c$

Hydrodynamic mass: $m_{hyd} = 1.4 \rho_w d^2 l_c$

Geodynamic mass: $m_{geo} = \dfrac{0.76 \rho_s R^3}{(2 - \Gamma)}$

where

$$R = \sqrt{\dfrac{B_c l_c}{\pi}}$$

B_c, h_c and l_c are the width, height and length of the caisson, respectively, d is the water depth in front of the structure, G is the shear modulus of the soil underneath the structure, R is the equivalent radius of the caisson bottom, ρ_c, ρ_s, and ρ_w are the densities of the caisson, the soil and seawater, respectively, and Γ is the Poisson ratio.

3 Approximation of natural period

The natural period of the structure can be estimated using the natural period of the structure-foundation-system with one degree of freedom (1DoF), which is the swaying motion (horizontal motion). This is generally much larger than the rotating motion of the structure (2nd degree of freedom — 2DoF). The following calculation for the 1DoF-system may be used:

$$T_{NS,appr.} \sim 2\pi \sqrt{\dfrac{m_{tot}}{K_{11}}}$$

$$K_{11} = 3G\sqrt{B_c l_c}$$

where K_{11} is the first term of the stiffness matrix of the foundation system.

The natural period of the 2DoF-system can then be estimated using the following equation and Table B2:

$$T_{NS} = T_{NS,appr.}/k_a$$

Table B2. Approximation factor k_a.

h_c/B_c	0.5	0.75	≥ 1.0
k_a	0.95	0.90	0.80

4 Dynamic load factor

The dynamic load factor ν_L is determined from Fig. B7 as a function of the relative load duration t_d/T_N ($T_N = T_{NS}$ = natural period from the above equation).

Alternatively the dynamic load factor can be estimated by the following equation valid for k values between 2 and 5:

$$\nu_L = 1.4\left[\tanh\left(\frac{t_d}{T_N}\right)^{0.55}\right] + 0.25\sin\left[2\pi\left(\frac{t_d}{T_N}\right)C\right]$$

where

$$C = 0.55k^{-0.63}$$

$$k = t_d/t_r$$

t_d is the load duration and t_r is the rise time of the impact load.

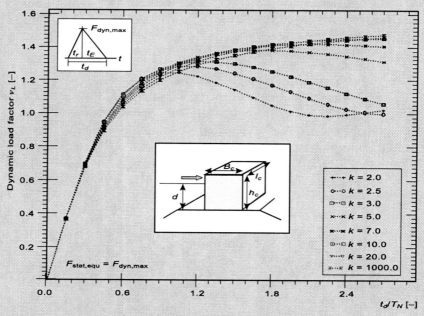

Fig. B7 Determination of dynamic load factor

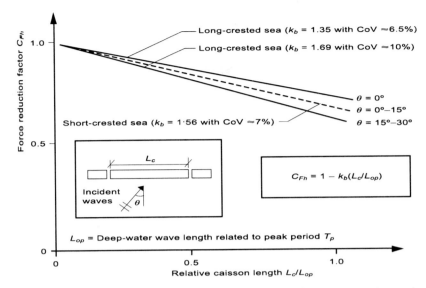

Fig. 7.37 Effects of caisson length and wave direction on wave impacts (Oumeraci et al., 2001)

(d)   Measures to reduce wave impact loads, their occurrence frequency and effects on breakwater stability.

   More than the intensity of the impact loads it is rather their frequency of occurrence, which affects the overall stability of the breakwater (see cumulative effects in Fig. 7.38). Therefore, appropriate measures must be undertaken to keep this frequency as low

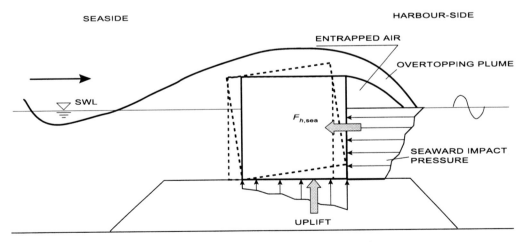

Fig. 7.38 Seaward impact loading induced by wave overtopping

as possible ($P_I < 1$ per cent suggested in Box 7.2), but also to reduce the intensity of impacts and their likely effects on the breakwater stability, if it is not technically and/or economically feasible to use another breakwater type, such as a rubble mound with a concrete crown wall (see Fig. 7.34) or a high-mound composite breakwater (see 'Wave loads on armoured caisson breakwaters', later in this chapter).

Since oblique waves ($\theta > 20°$) not only decrease substantially the frequency of occurrence of impacts, but also their intensity as compared to normal wave attack (see Fig. 7.37), it is suggested to align the breakwater whenever possible so that it cuts across the bottom contour lines (Goda, 2000). Alternatively or in addition, a more appropriate breakwater face, including wave dissipating chambers (see Fig. 7.27), wave dissipating armour blocks and other measures may be used to reduce the wave impacts and their detrimental effects (see 'Innovative caisson breakwaters' later in this chapter). This, however, should not be done at the expense of other aspects of the hydraulic performance such as wave reflection and over-topping, etc. For instance providing a sloping face on the top of the breakwater (a Hanstholm-type breakwater), see Agerschou et al. (1983), may substantially reduce the effect of impact loading on the overall stability. However, the substantial increase of wave overtopping may become a critical issue. A further possibility is to use caisson units longer than 0.5 to 2.0 times the caisson width. This is often limited by the capacity of the production yard and by structural considerations. The effect of the relative caisson length L_c on the impact force reduction is shown in Fig. 7.37.

Wave loading of perforated Jarlan-type caissons

The perforated caisson-type breakwater with a single chamber (Fig. 7.27) was first used at Baie Comeau, Canada, for a 305-m breakwater extension. Since then many have been built worldwide with various shapes (plane, curved, circular with sloped and stepped top), porosity of the front wall of 15 to 30 per cent, relative widths ($B_c/L = 1/10 - 1/4$), one to three wave chambers, berm heights and widths as well as with and without ceilings in the wave chambers. The wall porosity consists of vertical/horizontal slits or circular/rectangular holes. Its applications range from breakwaters in deep and shallow water to quay walls, jetties and seawalls with and without multi-purpose usage (promenades and other facilities for recreational activities, etc.). Its main purpose is often to reduce wave reflection, but may also be to reduce wave overtopping, splash and wave loads. The total wave forces acting on such perforated caissons are generally considered 'pulsating loads' and strongly depend on the number and dimensions of the wave chambers, the caisson geometry, the superstructure and the rubble foundation. The occurrence of impact loading, however, cannot completely be excluded. In view of the numerous possible alternatives (see 'Innovative caisson breakwaters', later in this chapter) for adapting its shape and other characteristics to any specific needs, it is not possible to provide generic formulae for the wave loads. This underlines the necessity of hydraulic-model testing.

Total wave forces

There is a pronounced relationship between hydraulic performance (e.g. wave reflection) and wave loading of perforated caisson breakwaters. Comparison of Fig. 7.31 and Fig. 7.39 shows for example that the variation of the total wave force on the perforated caisson $F_{h1/10}$ (related to $F_{h1/10}$ of a front wall with 0 per cent porosity under the same incident wave conditions) is similar to that of the reflection coefficient of the corresponding single chamber or multi-chamber caisson. This would suggest using the Goda formulae for a plain wall, as described in Box 7.3, with the design wave height H_D reduced by the reflection coefficient K_r:

$$(H_D)_{\text{perforated}} = \frac{1 + K_r}{2} H_{D(\text{Goda})}$$

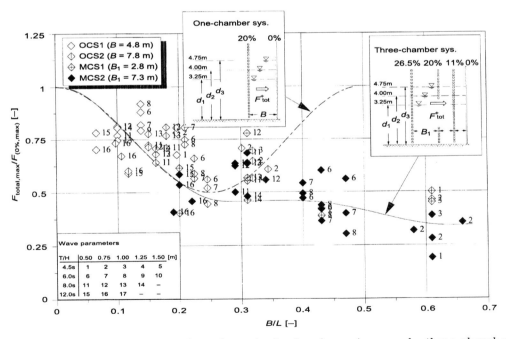

Fig. 7.39 Total horizontal wave force for a single-chamber caisson and a three-chamber caisson (Bergmann and Oumeraci, 2000)

with

$$K_r = 18.6(B/L)^2 - 7.3(B/L) + 0.98$$

where (B/L) is the relative chamber width.

This very rough approximation was first introduced by Rundgren (CERC, 1984) to reduce the Sainflou wave forces in the case of partial wave reflection. It is generally conservative. For steep waves it results in forces up to 30 per cent higher than the actual forces and should only be used for feasibility studies.

For preliminary design, the method described in Box 7.9 is preferable. It was proposed in Takahashi (1996) as an extension of the Goda formulae, including the impact pressure coefficient (see Boxes 7.3 and 7.5). This method distinguishes between three wave crest loading phases (landward forces) and three wave trough loading phases (seaward forces) as shown in Box 7.9. It also applies for typical perforated caisson breakwaters with a relatively high bottom level of the wave chamber $(d'/d < 10)$, d' being the water depth in the wave chamber.

For each of the three crest phases the Goda pressure formulae are extended by introducing modification factors λ_1 (pulsating pressure), λ_2 (impact pressure) and λ_3 (uplift pressure), which are associated with the pressure distribution on each structural member of the caisson (λ_{S1}, λ_{S2} = slit wall, λ_{R1}, λ_{R2} = rear wall; λ_{M1}, λ_{M2} = chamber bottom; λ_{U1}, λ_{U2} = uplift). The λ parameters obtained from Box 7.9 are used (see also Boxes 7.3 and 7.5) to calculate the pressure distributions on each structural member. These pressures occur simultaneously, so that the resultant horizontal and vertical wave forces can be calculated for the pulsating load case (extension of Goda formulae in Box 7.3) and the impact load case (extension of the Takahashi method in Box 7.5).

During the wave crest phases the most critical of the three phases for the overall stability depends on whether pulsating or breaking wave impact loads occur. The parameter map in Fig. 7.34 can be used as a first estimate to identify the loading type.

For impact loads the most critical case for stability is always 'Crest IIa', irrespective of the relative water depth d'/d in the wave chamber. The total force may reach up to 70 per cent of the total force calculated for a plain front wall by using the method in Box 7.5.

For pulsating loads of low-mound-composite breakwaters, both ratios d'/d and H_s/h_s are important for determination of the most critical of the loading phases. For $d'/d \approx 1.0$ and $H_D/h_s < 0.3$, 'Crest IIb' governs and the total horizontal force may reach about 80 per cent of that on a plain front wall as calculated by the Goda formulae (Box 7.3). For

Box 7.9 Extension of Goda's formulae for Jarlan-type caisson breakwaters (Takahashi, 1996)

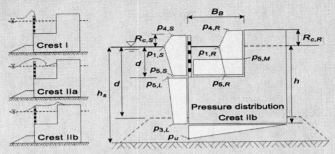

Fig. B8 Definition sketches

$$\eta_j^* = 0.75 \cdot (1 + \cos\theta)\lambda_{j,1} H_D \quad \text{(see also Box 7.3)}$$

$$p_{1,j} = 0.5 \cdot (1 + \cos\theta)(\lambda_{j,1}\alpha_1 + \lambda_{j,2}\alpha^* \cos^2\theta)\rho_W g H_D$$

$$p_{2,j} = \frac{p_{1,j}}{\cosh(2\pi h_s/L)}; \qquad p_{3,j} = \alpha_3 p_{1,j}$$

$$p_{4,j} = \left\{ \begin{array}{ll} p_{1,j}(1 - R_{c,j}/\eta_j^*) & \text{for } \eta_j^* > R_c^* \\ 0 & \text{for } \eta_j^* \le R_c^* \end{array} \right\}$$

$$p_{5,j} = \alpha_5 p_{1,j}$$

$$p_u = 0.5 \cdot (1 + \cos\theta)\lambda_u \alpha_1 \alpha_3 \rho_W g H_D$$

$$\alpha_I = 0.6 + 0.5 \cdot \left(\frac{4\pi h_s/L}{\sinh(4\pi h_s/L)} \right)^2$$

$$\alpha_2 = \min\left\{ \frac{h_D - d}{3h_D}\left(\frac{H_D}{d}\right)^2 ; \frac{2d}{H_D} \right\}$$

$$\alpha_3 = 1 - \frac{h'}{h_s}\left(1 - \frac{1}{\cosh(2\pi h_s/L)} \right)$$

$$\alpha_5 = 1 - \frac{d'}{h_s}\left(1 - \frac{1}{\cosh(2\pi h_s/L)} \right)$$

$$\alpha^* = \max\{\alpha_2, \alpha_I\}; \quad \alpha_I = \alpha_{I,0} + \alpha_{1,I} \text{ (see Box 7.5)}^{(*)}$$

$$R_{cj}^* = \min(\eta_j^*, R_{cj})$$

Structure member	λ_i	Crest I		Crest IIa		Crest IIb	
Slit front wall	λ_{Si}	0.85		0.7		0.3	
$p_{i,j} = p_{i,S}$	λ_{S2}	0.4	$(\alpha^* \leq 0.75)$	0		0	
		$0.3/\alpha^*$	$(\alpha^* > 0.75)$				
Lower plain wall	λ_{L1}	1		0.75		0.65	
part $p_{i,j} = p_{i,L}$	λ_{L2}	0.4	$(\alpha^* \leq 0.5)$	0		0	
		$0.2/\alpha^*$	$(\alpha^* > 0.5)$				
Impermeable rear	λ_{R1}	0		$20B_B/3L'$	$(B_B/L' \leq 0.15)$	1.4	$(H_D/h_s \leq 0.1)$
wall of chamber				1.0	$(B_B/L' > 0.15)$	$1.6 - 2H_D/h$	$(0.1 < H_D/h_s < 0.3)$
$p_{i,j} = p_{i,R}$						1.0	$(H_D/h_s \geq 0.3)$
	λ_{R2}	0		0.56	$(\alpha^* \leq 25/28)$	0	
				$0.5/\alpha^*$	$(\alpha^* > 25/28)$		
Chamber bottom	λ_{M1}	0		$20B_B/3L'$	$(B_B/L' \leq 0.15)$	1.4	$(H_D/h_s \leq 0.1)$
slab $p_{i,j} = p_{5,M}$				1.0	$(B_B/L' > 0.15)$	$1.6 - 2H_D/h$	$(0.1 < H_D/h_s < 0.3)$
						1.0	$(H_D/h_s \geq 0.3)$
	λ_{M2}	0		0		0	
Uplift on caisson	λ_{U1}	1		0.75		0.65	
base $p_{i,j} = p_u$	λ_{U2}	0		0		0	

Note: (*) in the calculation of the impact pressure coefficient α_I for the rear wall of the chamber α_I' should be used in the formulae given in Box 7.5 by replacing d by d', L by L', B_B by B_B' (d' = water depth in wave chamber; L' = wave length at depth d' and $B_B' = \max\{B_B - (d - d'); 0\}$).

$d'/d = 0.5$. However, both of the load phases 'Crest I' (for $H_D/h_s < 0.2$) and 'Crest IIa' (for $H_D/h_s > 0.3$) may also govern.

During the wave trough phases 'Trough II', which corresponds to the lowest water level at the front wall, is clearly governing. The seaward forces on the back of the caisson are calculated as described in 'Pulsating wave loads' earlier in this chapter.

Further aspects and remarks

The wave chamber may be topped by a ceiling slab with venting holes (Fig. 7.27). Depending on the distance between the water level and the slab bottom, the incident wave height, the size of the venting holes and the run-up velocity of the water surface in the wave chamber, severe uplift pressure on the ceiling slab may result. For a caisson a design uplift pressure of about $160\,\text{kN/m}^2$ ($\approx 2\rho g H_D$) acting uniformly on the slab was determined from scale-model tests (Takahashi, 1996). Based on the thickness of the entrapped layer of air, the run-up velocity and the venting characteristics, the pressure resulting from the compressed air can be calculated using a piston model with air leakage (Mitsuyasu, 1966).

Until better and more generic load formulae, which are under development (Bergmann, 2001), become available, hydraulic-model testing remains the sole alternative for detailed design of perforated caisson breakwaters.

Wave loads on armoured caisson breakwaters

(a)   *General considerations*: Armoured caisson breakwaters (often called horizontally composite breakwaters) as shown in Fig. 7.27 are essentially used to prevent impact loads, but pulsating wave loads and wave reflection are also reduced. They are widely used in Japan and are therefore sometimes called 'Japanese type' break-waters. Generally, the dissipating mound consists of concrete armour blocks such as Tetrapods, with a minimum width of two blocks at the crest and a slope of 1.3 to 1.5 on 1.0. The armour blocks should be designed to ensure both hydraulic stability and structural integrity. The efficiency of the dissipating mound in reducing impact loads to pulsating loads has been demonstrated by large-scale model tests. The results are shown in Fig. 7.40, which also shows that the damping ratio may reach about 80 per cent for horizontal forces and about 60 per cent for uplift forces. The damping ratio μ should be as defined in Fig. 7.40 and 7.41.

For pulsating loads the damping ratio may reach about 50 per cent for horizontal forces and about 40 per cent for uplift forces (Fig. 7.41).

(b)   *Wave load calculation*: A simple method based on the results of large-scale model tests and theoretical considerations (Kortenhaus and Oumeraci, 2000) is proposed in Box 7.10 to estimate the reduced wave forces on the armoured caisson from the corresponding wave forces on an unprotected caisson. The latter are calculated by using the method in Box 7.3 for pulsating loads and the methods presented in Boxes 7.5, 7.6 and 7.7 for impact loads. The method is applicable for the typical cross section given in Fig. B9 of Box 7.10.

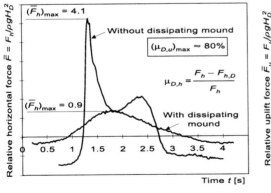
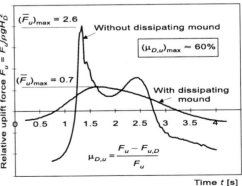

Fig. 7.40 *Damping of pulsating wave impact loads by armouring (Kortenhaus and Oumeraci, 2000)*

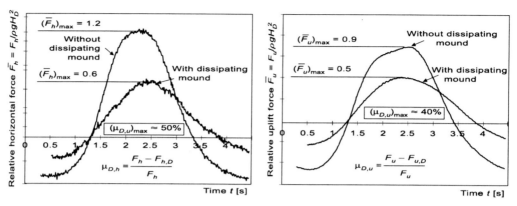

Fig. 7.41 Damping of pulsating wave loads by armouring (Kortenhaus and Oumeraci, 2000)

Box 7.10 Wave force reduction by armouring

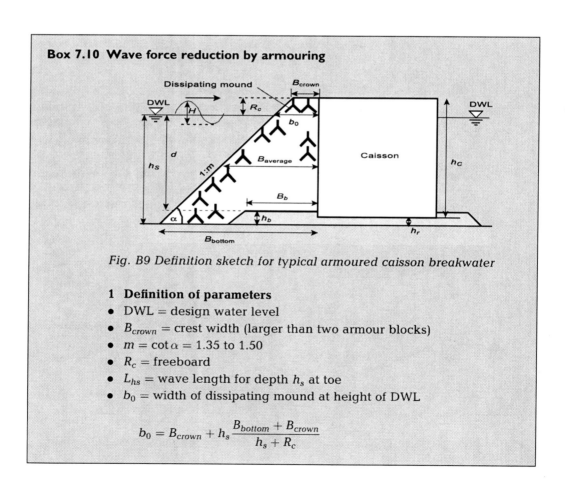

Fig. B9 Definition sketch for typical armoured caisson breakwater

1 Definition of parameters

- DWL = design water level
- B_{crown} = crest width (larger than two armour blocks)
- $m = \cot \alpha = 1.35$ to 1.50
- R_c = freeboard
- L_{hs} = wave length for depth h_s at toe
- b_0 = width of dissipating mound at height of DWL

$$b_0 = B_{crown} + h_s \frac{B_{bottom} + B_{crown}}{h_s + R_c}$$

2 Damping ratio of horizontal forces F_h

$$\mu_{d,h} = 1 - \exp\left(-2\pi \cdot C_{d,h} \cdot \frac{b_0}{L_{hs}}\right)$$

$$\mu_{D,h} = \frac{F_h - F_{h,D}}{F_h}$$

- F_h = horizontal pulsating wave force (see Box 7.3) or impact force (see Boxes 7.5 and 7.6) for non-armoured caissons.
- $F_{h,D}$ = damped horizontal wave force.
- $C_{d,h}$ = damping coefficient for horizontal forces depending on type of wave breaking and properties of mound material:
  - *pulsating load*: $C_{d,h} = 0.4$
  - *impact load*: $C_{d,h} = 1.5$

3 Damping ratio of uplift forces F_u

$$\mu_{d,u} = 1 - \exp\left(-2\pi \cdot C_{d,u} \cdot \frac{b_0}{L_{hs}}\right)$$

$$\mu_{D,u} = \frac{F_u - F_{u,D}}{F_u}$$

- F_u = pulsating uplift force (see Box 7.3) or impact uplift force (see Box 7.7) for non-armoured caissons.
- $F_{u,D}$ = damped uplift force.
- $C_{d,u}$ = damping coefficient for uplift forces depending on type of wave breaking and properties of mound material:
  - *pulsating load*: $C_{d,u} = 0.4$
  - *impact load*: $C_{d,u} = 1.0$

(c)   *Further aspects and concluding remarks*: The required size of the armour units is generally calculated by using the Hudson formula. If the mound is homogeneous and consists only of Tetrapods the Hudson-like formula by Hanzawa *et al.* (1996) has to be used, but the structural integrity should also be ensured.

Generally, the dissipating mound is too costly, if its main purpose is to eliminate too severe impact loads. However, it might be advantageous for reinforcing damaged caisson breakwaters, especially when their inside is used as berths. An alternative method for estimating wave loads, which is based on small-scale model tests and which is more conservative is given by Takahashi (1996).

Hydraulic model testing is highly recommended for detailed design and whenever the solution adopted is very different from the typical cross section shown in Box 7.10.

Wave loads on high-mound-composite breakwaters (HMCB)

(a) *The HMCB concept and general considerations*: The HMCB is not a new concept. It was used in the nineteenth century in Cherbourg, France, and Alderney, UK (Fig. 7.42). The traditional HMCB has a superstructure with an impermeable front wall, while modern HMCBs may have a permeable front wall (Fig. 7.42).

The main characteristics of the HMCB is the high mound which:

(i) is completely submerged for design water-level conditions, i.e. higher than that of a common composite caisson breakwater, but lower than that of a traditional rubble-mound breakwater;

(ii) has a relatively flat seaward slope (typically 1 on 3 or less) to cause the largest waves to break before reaching the superstructure and thus to prevent serious impact loads. The flatter slope and its complete submergence also reduce the required armour-block sizes, the wave reflection and seabed scour; and

(iii) has a superstructure, which is much smaller than the caisson of a low-mound-composite breakwater, but larger than the crown wall of a rubble-mound breakwater (see Fig. 7.34).

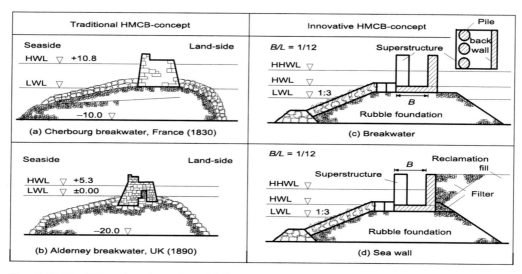

Fig. 7.42 Traditional and modern high-mound-composite breakwaters

In addition, the *modern HMCB* may be characterised by:

(i)   a permeable front wall, typically a slit wall, with a porosity of about 30 per cent, consisting of cylindrical piles. A conical parapet may be added on each pile to reduce splash and spray height (see (Fig. 7.43);

(ii) an impermeable rear wall which may also be built with a parapet at the crest to reduce wave overtopping and splash height (see Fig. 7.43); and

(iii) a wave chamber with a relative width $B/L = 1/12$ which is smaller than that of a traditional Jarlan-type caisson, which typically has $(B/L = 1/10$ to $1/4)$.

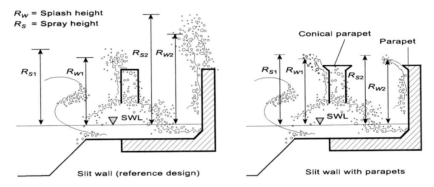

Fig. 7.43 Reduction of splash and spray height by parapets

The main purpose of this HMCB configuration is, besides the obvious reduction of wave reflection, overtopping, splash and spray height, a further reduction of the total wave load on the structure. The wave hitting the slit wall is split and substantially damped in the chamber before reaching the rear wall and a phase lag exists between the forces on the front and rear walls. This is particularly advantageous, when a larger wave breaks heavily at the wall. This may be very rare but cannot be excluded. In addition, the downward force on the bottom of the chamber has a stabilising effect.

(b)   *Calculation of wave loads*: A practical method is given in Box 7.11 for calculation of the wave pressure distribution and force on each member of the superstructure as well as the total resulting forces. This method is essentially empirical, based on extensive large-scale-model tests. A more detailed description of this method and the testing conditions is given by Muttray *et al.* (1999) and Oumeraci *et al.* (2000).

(c)   *Further aspects and concluding remarks*: The improvements over the traditional HMCB by the modern HMCB are summarised in Table 7.7.

Box 7.11 Wave loads on high-mound-composite breakwater

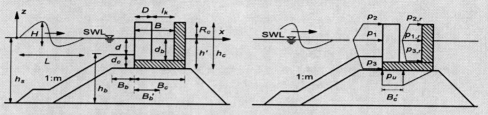

(a) Wave and structural parameters   (b) Wave pressure parameters

Fig. B10 Definition sketches

1 Critical wave heights H_{nb} and H_{bb} for the front wall load

- H_{nb} = critical wave height for transition between non-breaking and breaking waves at front wall.
- H_{bb} = critical wave height for transition between breaking and broken waves at front wall.
- $k = 2\pi/L$ = wave number associated with wave length at depth h_s and with peak period T_p.

Table B3 Wave load classification

Relative wave length			H_{nb}/d		H_{bb}/d	
Short:	$0.9 \leq kh_s < 1.2$	Non-breaking	1.2	Breaking	2.0	Broken
Intermediate:	$0.6 \leq kh_s < 0.9$	waves	1.0	waves	1.8	waves
Long:	$0.3 \leq kh_s < 0.6$		1.7		2.5	

2 Critical wave heights $H_{nb,r}$ and $H_{bb,r}$ for the rear wall load

Due to wave reflection and the increased berm width the values for the front wall are decreased to 72 per cent:

$$H_{nb,r} = 0.72 \cdot H_{nb} \qquad \text{and} \qquad H_{bb,r} = 0.72 \cdot H_{bb}$$

3 Wave pressure at SWL

The first and most important step in the calculation of pressure distributions at the front and rear walls is to determine the pressures p_1 and $p_{1,r}$ at still-water level. The calculation procedure is shown by the flow chart in Fig. B11 for the front wall, including a traditional HMCB with an impermeable front wall.

For pressure $p_{1,r}$ at SWL, on the rear wall of the slit-type structure, the same procedure as in Fig. B11 is adopted with the input critical wave heights $H_{nb,r} = 0.72 \cdot H_{nb}$ and $H_{bb,r} = 0.72 \cdot H_{bb}$ and by replacing

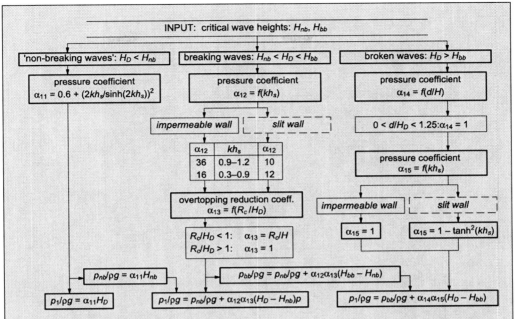

Fig. B11 Flow chart for the design of an HMCB

α-coefficients given in Fig. B11 for the front wall by the following α_r-coefficients for the rear wall:

- Non-breaking waves: $\alpha_{11,r} = \alpha_{11}$
- Breaking waves: $\alpha_{12,r} = 11$ for short waves $(0.9 \leq kh_s < 1.2)$
  $\qquad\qquad\qquad 6$ for intermediate waves $(0.6 \leq kh_s < 0.9)$
  $\qquad\qquad\qquad 2$ for long waves $(0.3 \leq kh_s < 0.6)$
- Broken waves: $\alpha_{15,r} = \min\{1, \alpha_{15}/2\}$

4 Wave pressure at top and bottom of front and rear wall

- At top of front wall (solid and slit type):

$$p_2 = (3k\alpha_2)^2 \cdot \rho g H_D \qquad \text{with } \alpha_2 = \max\left\{0; \ 1.5 - \frac{R_c}{H_D}\right\}$$

- At top of back wall (slit type):

$$p_{2r} = \rho g \alpha_{2r} H_D \qquad \text{with } \alpha_{2r} = \max\left\{0; \ 1.5 - \frac{R_c}{H_D}\right\}$$

- At bottom of front wall (for $-1 < z/h' < 0$), solid and slit type:

$$p_3 = p_1\left(1 + \alpha_3 \cdot \frac{z}{h'}\right) \qquad \text{with } \alpha_3 = 0.8\left(\text{for } \frac{d}{H_D} > 0\right)$$

- At bottom of rear wall (slit type):

$$p_{3,r} = p_{1,r} \cdot \left[1 + \frac{z}{h'}\right] - H_D \frac{z}{h'} \qquad \text{for } h' \leq z \leq 0$$

5 Vertical wave pressure on bottom slab

The uplift pressure has been assumed to act uniformly with $p_u = p_3$ over the stretch $B'_c = B_c[1 - \tanh(2kh)]$ on the seaward side and then decreasing linearly to zero at the land side (see Fig. B10(b)).

Inside the chamber $p_{3,r}$ is assumed to be constant over the entire chamber width B.

6 Resultant wave forces from pressure distribution

Since the pressures shown in Fig. B10(b) occur simultaneously, the total horizontal forces are obtained through pressure integration and super-position.

Forces on traditional HMCB (with impermeable front wall)

- Horizontal force for non-breaking and broken waves:

$$F_h = \tfrac{1}{2}h'(p_1 + p_3) + \tfrac{1}{2}R_c(p_1 + p_2)$$

- Horizontal force for breaking waves (account for time lag between p_1 and p_2):

$$F_h = \frac{1}{2}h'(p_1 + p_3) + \frac{d}{4}p_1$$

- Uplift force:

$$F_u = \tfrac{1}{2}p_u(B'_c + B_c)$$

Forces on new HMCB (with permeable slit wall)

(a) Horizontal force on permeable (ca. 30 per cent porosity) front wall consisting of piles with diameter D:

- For non-breaking and broken waves:

$$F_{h,f} = \tfrac{1}{2}[d_b(p_1 + p_3) + R_c(p_1 + p_2)]D\alpha_p n + (h' - d_b)p_3$$

with $\alpha_p = 0.3$ (reduction factor due to wall porosity and force on the back side of the front wall) $n =$ number of piles per metre length of the wall

- For breaking waves:

$$F_{h,f} = \tfrac{1}{2}d_b(\tfrac{3}{2}p_1 + p_3)D\alpha_p n + (h' - d_b)p_3$$

(b) Horizontal force on back wall:

$$F_{h,r} = \tfrac{1}{2} d_b (\tfrac{3}{2} p_{1,r} + p_{3,r})$$

(c) Total horizontal wave forces on slit type structure:

$$F_h = F_{h,r} + \tfrac{1}{2} F_{h,f}$$

(d) Total uplift forces on slit type structure:

- For non-breaking and slightly breaking waves:

$$F_v = F_u + \tfrac{1}{2} \rho g H_D l_k \qquad \text{with } F_u \text{ from the ninth equation}$$

- For breaking and broken waves:

$$F_v = F_u + \tfrac{1}{2} \rho g d_b l_k \qquad \text{with } F_u \text{ from the ninth equation}$$

Table 7.7 Improved performance over traditional HMCB by modern HMCB (without splash reduction)

Hydraulic performance	$\bar{r}_x$	COV [%]
Transmission coefficient K_t (through rubble foundation)	1.0	50
Reflection coefficient K_r	0.75	30
Mean overtopping rate $\bar{q}$	0.5	30
Maximum overtopping rate q_{max}	0.6	25
Number of overtopping waves N_{ow}	0.7	25
Splash and spray height	0.5	60
Required crest level above SWL	0.6	30
Wave forces and stability		
Horizontal forces	0.4	50
Uplift forces	0.6	40
Required weight for sliding stability	0.5	50

Notes: $\bar{r}_x$ = mean reduction factor = $\bar{x}_{\text{NewHMCB}} / \bar{x}_{\text{OldHMCB}}$; COV = coefficient of variation = $\sigma_x / \bar{x}$.

A parapet at the top of the back wall would reduce the overtopping by about 30 per cent without significant increase in total wave force on the structure. If a conical parapet is added to the top of each pile of the front wall (Fig. 7.43) a substantial reduction of the splash and spray height may also be achieved (Hayakawa et al., 2000).

The formulae proposed in Box 7.11 can be used for preliminary design of a traditional HMCB with an impermeable front wall as well as for a HMCB with a slit cylindrical pile front wall of about 30 per cent porosity. For detailed design, hydraulic model tests are recommended also for the stability of the armour units of the berm. In particular, the reduction of the required block weight as calculated from the Hudson formula should be determined.

Other important design loads

Governing for the overall stability of caisson breakwaters are generally the wave loads as described in the previous sections. Except for very soft ground, where pile foundations or other measures are necessary, the dead weight of the caissons must be sufficient to resist sliding. Against overturning, which is actually a bearing capacity failure of the foundation, the eccentricity of the dead weight should be optimised to resist both shoreward and seaward directed wave forces.

Very often the governing loads for structural design occur during towing, floating and placing of the caissons on the rubble foundation (see 'Production and transportaion of caissons', later in this chapter). The loads on the caisson walls from sand fill should be considered and calculated as earth pressure at rest, if the dynamic forces are unable to produce vibration compaction. The differential water pressure due to tides should also be considered if the internal sand fill is not drained to the harbour side.

Depending on the location and specific circumstances of the construction of caisson breakwaters, the following loads may also be relevant:

(a) *Seismic loads* The resistance against the cyclic mobility and liquefaction of foundation soil is governing. Calculations are often based on local earthquake regulations (see OCADIJ, 1991);

(b) *Ice loads* Load type and magnitude greatly depend on ice cover and ice motion characteristics at the site (see Christensen *et al.*, 1995);

(c) *Service loads* If the caisson is also used for berthing, the mooring forces, berthing impacts of ships, crane loads, etc. also be considered. (EAU, 1996); and

(d) *Earth pressure from back fill* If the caisson is used to protect a reclaimed area earth pressure should be considered (EAU, 1996).

Stability considerations

The design of a caisson breakwater requires stability analyses of the caisson and other components of the breakwater, including:

(a) the overall stability of the monolithic caisson structure against sliding and overturning;

(b) the hydraulic stability of the rubble foundation; and

(c) the stability of the foundation as a geotechnical failure may lead to the stability losses mentioned above under (a) and (b).

Generally, static stability analyses are performed using the wave loads calculated by the methods proposed in Boxes 7.3, 7.4, 7.5, 7.9, 7.10 and 7.11. For a more refined design, when severe impact loading cannot be avoided, dynamic stability analysis is recommended by using the methods in Boxes 7.6 and 7.7 to determine the impact forces, instead of the extended Goda method in Box 7.5. Equivalent static loads are calculated by multiplication of these peak forces with a dynamic amplification factor (see Box 7.8), which needs as input parameters:

(a)   the dynamic characteristics of the impact load as determined from Box 7.6 and 7.7 (peak, rise time and impact duration); and

(b)   the natural period T_N of the oscillations of the structure, which is in the range of 0.2 to 0.5 seconds for common sizes of caisson break-waters (De Groot et al., 1996).

Where impact loading is expected to occur frequently, dynamic analysis is to be preferred at the detailed design stage for the stability of the caisson breakwater and its foundation as well as for its structural integrity. Design methods, guidance and methods for this purpose are provided by De Groot et al., (1996), Oumeraci and Kortenhaus (1994) and Oumeraci et al. (2001).

Overall stability of monolithic structures

It is customary to consider overturning around the shoreward heel and seaward toe of the caisson (Fig. 7.29 and Fig. 7.45) as an overall failure mode of the caisson. However, it is considered here as a direct conse-quence of the failures associated with the hydraulic instability of the rubble foundation (see 'Hydraulic stability of rubble foundation and scour protection' next in this chapter) or with the loss of the bearing capacity of the foundation (see 'Geotechnical failure modes and founda-tion stability', later in this chapter). Therefore, in this section only sliding on the rubble foundation is addressed.

The monolithic caisson structure (caisson and superstructure) must be designed to have a safety factor of at least $\gamma_s = 1.2$ against sliding over the rubble foundation:

$$\gamma_s = \frac{\mu(W_c - F_u)}{F_h}$$

where μ is the coefficient of friction between the concrete caisson and the rubble foundation ($\mu = 0.6$); F_h and F_u are the resulting horizontal and vertical wave forces calculated by the methods described in Boxes 7.3, 7.4, 7.5, 7.9, 7.10 and 7.11 [N/m]; and W_c is the weight of the monolithic structure including the sand fill and the superstructure [N/m].

The following mass densities are suggested for various parts of the structure if no better data are available for the case under consideration:

(a)   for sand-filled caisson part below SWL $\rho_c = 1.1 \, \text{t/m}^3$;

(b)   for sand-filled caisson part above SWL $\rho_c = 2.1 \, t/m^3$; and

(c)   for reinforced concrete superstructure: $\rho_{c_g} = 2.3 \, t/m^3$.

Particularly in deep water, the safety factor against sliding toward the seaward side should be checked (see 'Seaward wave loads', earlier in this chapter). The seaward load is calculated by the method in Box 7.4. The forces used in the above equation are calculated for 1 m of the break-water length and are therefore conservative, when a caisson unit longer than the usual range of 0.5 to 2.0 times the caisson width is used.

Hydraulic stability of rubble foundation and scour protection

(a)   General considerations: The caissons are placed upon a levelled bedding layer of granular material (in the case of the vertical break-water type) or upon a well-prepared rubble foundation (in the case of a composite breakwater). In both cases wave action and longitu-dinal strong currents will cause severe turbulence, which may lead to serious erosion of the seabed and/or of the berm of the rubble foundation, thus undermining the monolithic structure (Fig. 7.44). Because of the potentially serious damage and because additional costs of reinforcing these sensitive parts of the breakwater are small compared to the costs of the whole structure, particular atten-tion should be paid to the design of the rubble foundation and its protection against scour.

(b)   Hydraulic stability of rubble foundations including toe protection blocks: Too large displacements of the armour blocks of the berm must be prevented for both normal and oblique wave attacks. The most critical zones are located at the transition between the slope and the horizontal section of the berm and around the corners of the caisson, particularly at the breakwater head. Therefore, suffi-cient dimensions and weight of the armour blocks on the berm

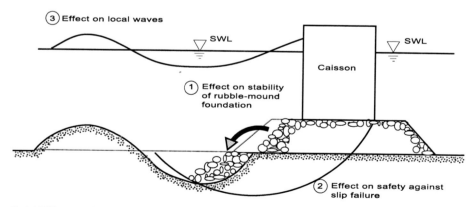

Fig. 7.44 Effects of seabed and berm erosion on caisson stability

and of the toe protection blocks are required under normal and oblique wave attack. A review and analysis of available stability design for armour blocks has been made by Oumeraci (1994c). Based on this, a state of the art method to determine the required weight of the armour blocks and the thickness of the concrete toe protection blocks are presented in Box 7.12. For more details see Tanimoto *et al.* (1982 and 1987), Kimura *et al.* (1996 and 1998), Madrigal and Valdes (1995) and Takahashi (1996). Where frequent wave breaking is expected, both the stability of the blocks and the extent of scour should be checked by suitable scale-model testing.

(c)   Seabed scour protection: Seabed scour constitutes an important source of failure and most of the damage experienced by vertical breakwaters occurred during construction (Oumeraci, 1994a). Scour protection should keep pace with the progress of work, despite the extra costs involved. Protection of incomplete works should be carefully planned, particularly at the breakwater head. Application of protection measures is described, for instance, by Lundgren (1985).

Although the possible effects of seabed scour are well known (Fig. 7.44), the specification of threshold values for the extent of scour which is about to threaten the stability of the structure is still an unsolved problem. No practical design formulae are yet available to reliably predict the depth and extent of seabed scour in front of vertical structures subject to normal and oblique storm waves. The rough guidance generally found in codes of practice and textbooks to the effect that scour protection should consist of a blanket of 0.2 to 0.5 t stones, covering the sea bottom to a distance of one quarter to three quarters of the design wave length in front of the vertical breakwater, is certainly not the most economical solution. Instead it is suggested that the designer should try to understand the underlying processes (see Oumeraci, 1994b) and based on this understanding and engineering judgement devise an adequate scour protection. Such scour protection may be (see Agerschou *et al.*, 1983) special prefabricated mattresses, geotextile sheets loaded with proper rock material and plastic bags filled with sand or concrete (Terrett *et al.*, 1979).

(d)   Further aspects and concluding remarks: The last caisson at the breakwater head should have a semi-circular end to prevent the generation of large eddies and flow acceleration at corners, which cause erosion. A round breakwater head is also favoured from a navigational viewpoint (Agerschou *et al.*, 1983).

The design of the rubble foundation and toe berm (width, height, block size and slope steepness) generally results from an optimisation procedure, which accounts for a number of conflicting factors

Box 7.12 Hydraulic stability of rubble foundation

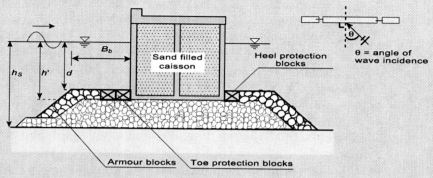

Fig. B12 Definition sketch

1 Required average weight of armour blocks W_{n50}

$$W_{n50} = \frac{\rho_s g}{N_s^3 \Delta^3} H_s^3$$

- N_s = stability number of the armour block, which is calculated by the following formula proposed by Tanimoto *et al.* (1987) and modified by Takahashi (1996):

$$N_s = \left\{ \frac{1.8}{1.3 \frac{1-\kappa}{\kappa^{1/3} H_s} + 1.8 \left[\exp\left(-1.5 \frac{(1-\kappa)^2}{\kappa^{1/3}} \frac{h'}{H_s}\right)\right]} \right\}$$

L, L' = wave length corresponding to significant wave period at depth h_s and h' respectively, H_s = significant wave height.
- Δ = relative mass density.

$$\Delta = \frac{\rho_S}{\rho_W} - 1$$

ρ_S, ρ_W = mass density of block and water, respectively.
- κ = semi-empirical parameter given below for the trunk and the breakwater head and for both normal and oblique waves:
  – *for breakwater trunk (along the breakwater):*

$$\kappa = \frac{4\pi h'/L'}{\sinh(4\pi h'/L')} \cdot \kappa_2$$

with

$$\kappa_2 = \max \left\{ \begin{array}{c} \alpha_s \sin^2 \theta \cos^2 \frac{2\pi x}{L'} \cos \theta \\ \cos^2 \theta \sin^2 \frac{2\pi x}{L'} \cos \theta \end{array} \right\}$$

$\alpha_S = 0.45$ (slope coefficient after Kimura *et al.* (1994)). $x = B_b$ for normal wave incidence ($\theta = 0°$), so that:

$$\max \kappa_2 = \sin^2 \frac{2\pi B_b}{L}$$

while for oblique waves (($\theta \neq 0°$)x varies between 0 and B_b to obtain max κ_2 for a given θ. According to Kimura *et al.* (1994) the greatest damage occurs for $\theta = 0°$ or $\theta = 45°$.

– for breakwater head:

Here the greatest damage occurs for $\theta = 60°$ rather than for $0°$ or $45°$. Overall, larger blocks are required at the head than along the trunk. Therefore, instead of the fourth equation in this box, the following formula has been proposed by Kimura *et al.* (1994):

$$\kappa = \frac{\pi h'/L'}{\sinh(4\pi h'/L')} \cdot \alpha_s \cdot \tau^2$$

with

$$\tau = \begin{cases} 1.4 & \text{for } \theta = 0° - 45° \\ 2.5 & \text{for } \theta = 60° \end{cases}$$

2 Required thickness of toe and heel protection blocks

Particularly in Japan, it is customary that two concrete blocks are placed at the seaward toe and one at the landward heel of the caisson (Fig. B12). The blocks, with a mass varying between 6 to 60 t and with thicknesses between 0.8 m and 2.2 m, are rectangular (length of 2.5 to 5.0 m and width of 1.5 to 2.5 m) and have two rectangular holes, providing 10 per cent porosity to reduce uplift pressure. Since the most relevant resistance characteristic is the weight per unit area, an empirical formula, validated by prototype observations, has been proposed by Kimura *et al.* (1994) to calculate the required thickness of the perforated rectangular blocks (10 per cent porosity):

$$t = \alpha_f \left(\frac{d}{h_s} \right)^{-0.79} H_s$$

with

$$\alpha_f = \begin{cases} 0.18 \text{ for trunk section} \\ 0.25 \text{ for breakwater head} \end{cases}$$

The ninth equation has been developed for $d/h_s \geq 0.4$ and $\theta = 0° - 60°$.

Remark: If strong currents along the breakwater and around the head occur, some guidance on the size of the blocks required may be obtained from the Isbash or Isbash-like equations for protection of channel beds, which can be found in any hydraulic engineering textbook.

and effects such as hydraulic stability of the armour units and toe protection blocks, scour protection, wave breaking and construction aspects. Therefore, the methods and guidance given in this section are primarily intended for preliminary design. For final design hydraulic-model testing with irregular waves still remains the most feasible and least costly alternative.

Geotechnical failure modes and foundation stability

(a)   General considerations: Foundation analysis is much more important for monolithic caisson breakwaters than for rubble-mound break-waters, which can better adjust to soil deformations. The bearing capacity of the foundation soil is generally decisive for the dimensions of the caisson, where the subsoil consists of weaker material (clay, loose and medium-dense sand). Soft soils might even eliminate caisson breakwaters from consideration, unless special measures (pile foundations, soil substitutions and improvement, etc.) are undertaken.

Extensive soil investigations and detailed bearing capacity analysis are essential, even more than for other breakwater types. The foundation must be designed for both single maximum wave loads, including impact loads, and cyclic wave loads. Depending on the location of the breakwater, earthquake loading may be a decisive design parameter.

Traditionally overturning around the caisson heel (Fig. 7.29) has not been considered a foundation failure, because the foundation was implicitly assumed to be rigid. Actually, the caisson is resting on a yielding foundation. Therefore overturning is a directly visible consequence of foundation failure. The bearing capacity is accounted for by comparing the stresses transmitted by the wave load with an allowable value for the foundation material of, e.g. $\sigma_{min} = 500\,kN/m^2$ (Goda, 2000).

This traditional design procedure is too simplistic and unrealistic as it completely ignores bearing capacity failure of the rubble foundation and/or the sub-soil foundation, which may lead to settlements as well as to shoreward or seaward tilting of the caisson (Fig. 7.45).

The failure modes may be caused by physical processes involved in the dynamic structure–foundation interaction, such as impact loading and load amplification, soil degradation due to repetitive wave loading, instantaneous and residual excess pore pressure effects, etc., which have also been more or less ignored in traditional design. Therefore a technical framework for caisson foundation analysis has recently been developed (De Groot et al., 1996 and Oumeraci et al., 2001). This allows practical foundation design

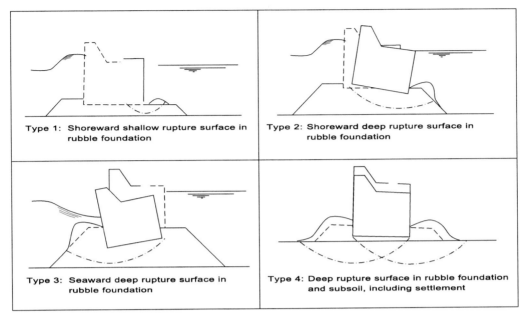

Type 1: Shoreward shallow rupture surface in rubble foundation

Type 2: Shoreward deep rupture surface in rubble foundation

Type 3: Seaward deep rupture surface in rubble foundation

Type 4: Deep rupture surface in rubble foundation and subsoil, including settlement

Fig. 7.45 Types of bearing capacity failure (adapted from Oumeraci et al., 2001)

according to the simplified flow chart and a strategy for soil investigations at different design levels presented in Fig. 7.46.

(b) Soil parameters: Extensive soil investigations in the marine environment are among the most time consuming and costly tasks of the design process. To avoid waste of time and money at an early stage, before any final decision has been made on the feasibility and location of a caisson breakwater, it is essential to identify the type and minimum amount of soils data required at each of the three design levels suggested in Fig. 7.46. Details on the soil parameters, the required investigations and practical recommendations are given by De Groot *et al.* (1996) and Oumeraci *et al.* (2001).

(c) Dynamic response of the breakwater foundation to single-wave loads: *Impact loads*: Due to the short load duration and to the comparatively larger natural oscillation period of the structure, which depends on the total mass of the caisson, including the adjacent water (hydrodynamic mass) and soil (geo-dynamic mass) as well as on the stiffness of the foundation (see Box 7.8), the peak value $F_{dyn.max}$ of the impact load cannot be used directly for static calculation of the stability of the structure and its foundation. Instead, an equivalent static load $F_{stat.eq.} = \nu_{LF_{dyn.max}}$ should be used with the dynamic amplification factor ν_L calculated according to Box 7.8. The impact load is most critical, when its duration approaches the

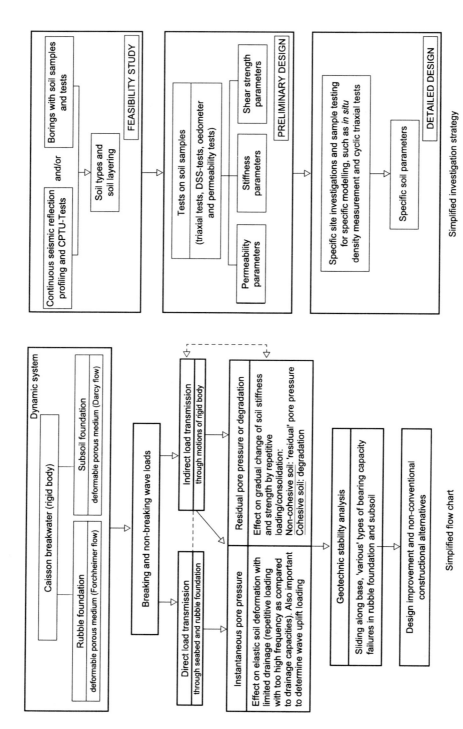

Fig. 7.46 Foundation analysis and soil investigation strategy for design of caisson breakwaters (adapted from Oumeraci et al., 2001)

natural oscillation period of the structure. Stiffer soils are more sensitive to impact loads, because they cause the natural period to approach the load duration. Also for stiffer foundations sliding over the rubble foundation may become more critical.

Instantaneous pore pressure: The pressure is generated through direct or indirect transmission of the wave loads into the rubble and subsoil foundation (Fig. 7.47). Within the rubble foundation the wave uplift force is strongly affected, and thus the sliding stability. The porous flow in the rubble foundation will cause additional seepage forces on the skeleton of the rubble foundation, which has to be considered in the calculations of the bearing capacity (see Fig. 7.46). For more details see Oumeraci *et al.* (2001).

Pore pressure in sandy subsoil may also affect the bearing capacity. For medium-dense sand and silty sand, undrained shear strength conditions may be assumed at least at the feasibility and preliminary design level. For detailed design of important projects, scale-model testing using a centrifuge is recommended.

(d) Stability of subsoil subject to cyclic wave loads: Cyclic wave loading of clay and sand may result in reduction of soil stiffness and of effective shear stresses caused by excess pore pressure. The latter will build up if the load frequency is too high compared to the drainage capacity of the soil. As a result, the effective stress is reduced, causing a reduction of the shear strength and the shear modulus. Clay is generally undrained during the entire duration of a storm and possibly even during several successive storms. In contrast, sand is subject to a certain degree of drainage, which mainly depends on the permeability of the sand (grain size distribution and degree of compaction) and on the drainage boundary conditions. Care should be taken when attempting to make use of experience from soil liquefaction induced by earthquakes to study liquefaction induced by storm waves. The fundamental difference between the two types of cyclic loading must always be kept in mind. During earthquakes undrained conditions generally prevail, even in sandy soils, whereas at worst only partially drained conditions occur for storm waves. This is essentially due to the differences in the loading frequency, which is more than an order of magnitude higher for earthquake shear stress ($f = 1\,\text{Hz}$) than for oscillatory excess pore pressure induced by storm waves ($f = 0.05$ to $0.1\,\text{Hz}$). On the other hand, the number of storm waves is much larger (in the order of 10^5 to 10^6) than the number of earthquake load cycles, which rarely exceed 20. Therefore, soil liquefaction induced by earthquake typically occurs as a result of progressive decrease of shear strength caused by build-up of excess pore pressure. For storm waves, however, the excess pore pressures have rather an oscillatory character, which results in a cyclic increase and decrease of the shear strength (Zen Yamazaki, 1991). For these reasons, earthquake induced liquefaction typically represents a sudden and dramatic event, while in the case of storm waves partial liquefaction and compaction may be transient and repeated.

Pre-loading during smaller storms prior to the design storm is beneficial for both sand and normally consolidated clay and should therefore be accounted for in laboratory testing.

For feasibility and preliminary design, diagrams for stiffness reduction, based on finite element calculations and centrifuge tests, are provided in Oumeraci *et al.* (2001) and De Groot *et al.* (1996). Recommendations for design storm characteristics to be adopted for cyclic loading of caisson foundations are given by Oumeraci in ASCE Task Committee (1995) and in De Groot *et al.* (1996). Measures for improving caisson stability and that of its foundation, together with related recommendations are given in Oumeraci *et al.* (2001).

Structural design aspects

(a)   General considerations and design sequence: Modern vertical breakwaters consist of reinforced concrete cellular caissons. The caissons are topped by a crest structure, which may have any shape, depending on the requirements for the specific use of the breakwater (wave overtopping reduction, traffic, recreational activities, etc.). The caissons can be designed as floating or non-floating caissons.

Floating caissons are generally multi-cellular and can be constructed to (almost) any size and shape compatible with ground conditions, construction requirements and intended specific use. Non-floating caissons are generally smaller because they have to be placed by cranes. They usually consist of a single cell and generally do not have a bottom. The gaps between the caissons are usually closed and sometimes have keyed joints for load transfer between caissons.

For caissons, the loading conditions expected during construction (pre-service conditions) as well as after construction (in-service conditions) must be considered. A detailed review and analysis of the structural design methods for caisson breakwaters has shown that no single code of practice currently exists, which covers the full range of structural design of the reinforced concrete components forming a caisson breakwater (Martinez *et al.*, 1999). The three main phases of structural design are briefly summarised in Fig. 7.47.

In the first phase the caisson type and general characteristics are selected according to the intended use, the available capacity and

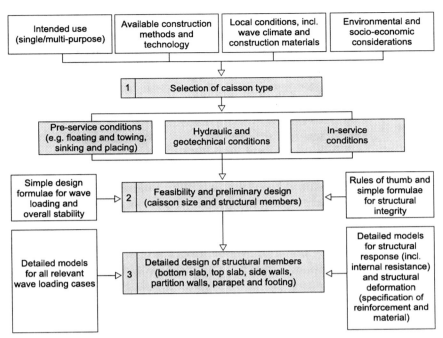

Fig. 7.47 Simplified sequence for structural design (adapted from Oumeraci et al., 2001)

technology for the casting of the caissons and the breakwater construction, the availability of construction materials, skills, etc.

In the second phase the caisson size and the approximate dimensions of the structural members are assessed by using very simple formulae and rules of thumb.

In the last phase the structural integrity of each component is assessed for the full range of possible loading cases by using more sophisticated models and formulae. As shown in Fig. 7.47, the caisson size and geometry, as well as the dimension of structural members, will essentially be determined by the hydraulic conditions and the geotechnical conditions during and after construction. Not shown in Fig. 7.47 is the consideration of the durability of concrete in the marine environment and the importance of a proper maintenance plan based on regular inspection and monitoring, which must be an integral part of detailed design ('Monitoring, inspection, maintenance and repairs', later in this chapter).

(b) Caisson shape and structural members: Floating multi-cellular caissons are generally rectangular in plan, while the face of each cell can be either straight or curved in plan. The cells, usually no more than 5 by 5 m in plan, are separated by walls up to 25 cm thick and have a bottom slab up to 70 cm thick. The outer walls are designed for internal pressure (water and fill material). In addition, the seaward walls are designed for wave loads. Curved walls are favoured because the thickness and reinforcing steel requirements are lower than for straight walls. Even the total wave impact forces are reduced, but local high-pressure concentration may occur between two adjacent caissons.

If the seaward cell of a floating caisson is provided with a perforated face and a wave chamber (Jarlan type), additional stability problems associated with floating and sinking during construction have to be accounted for. During recent decades there has been an increasing trend towards more innovative forms of the caissons and their crest structures ('Crest structures' and 'Innovative caisson breakwaters', later in this chapter).

The caisson width, generally less than 30 m, is governed by wave-load stability. The caisson length is usually governed by constructional aspects rather than by overall stability and structural integrity requirements (see also 'Caisson dimensions and structural members' later in this chapter).

Floating caissons, with a length of 100 m or more, may constitute an alternative which will substantially reduce the required caisson width (Fig. 7.48). This, however, will require much more development work to ensure that bending and torsion during towing, sinking and placing of the caissons as well as during in-service conditions (differential settlements) are within allowable limits.

Besides hydraulic performance (overtopping, etc.) the caisson height is governed by the capacity of the facilities for pre-casting, the requirements for levelling of the rubble foundation and for placing of the caissons as well as by cost considerations.

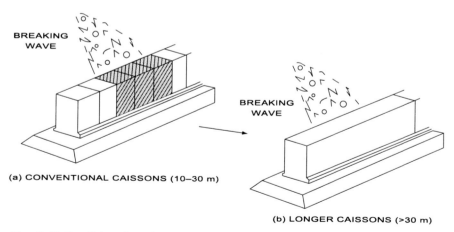

Fig. 7.48 Traditional and longer floating caissons

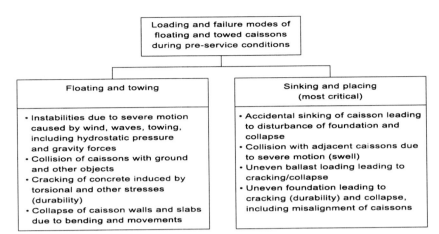

Fig. 7.49 Structural failure modes of floating caissons (Oumeraci et al., 2001)

(c)   Load conditions and structural failure modes: Structural analysis of the caisson must cover both pre-service and in-service conditions. For floating caissons, the wall and bottom loading during sinking and placing on the rubble foundation may be even more critical than the design wave loads.

Pre-service loading conditions: During casting in a dry dock or in a floating dock concentrated loads, which result from moving the caisson to a slipway or caisson platform, are the most relevant loads. The possible failure modes during floating and towing as well as during sinking and placing are briefly summarised in Fig. 7.49. However, the most critical situation may occur during sinking and placing on an uneven rubble foundation (cracking and collapse of the bottom slab). Therefore 'camel humps' and 'legs' protruding more than double the tolerance of the ground levelling are sometimes provided beneath the bottom slab to prevent too high stresses from the first contact ('Construction of caisson breakwaters', later in this chapter). Details on pre-service loads and the required structural analysis available in OCADIJ (1991) and Vrouwenvelder and Bielecki (1999).

In-service loading conditions: Permanent loads (dead weight and corresponding foundation reaction), variable loads (pulsating and impact wave loads, wave overtopping on the top slab, traffic loads, etc.) and accidental loads are usually considered. Possible structural failure modes are for instance:

(i)   bending rupture of the seaward wall, which is more critical in the splash zone due to chloride ingress, freeze–thaw cycles and wave impact;

(ii) diagonal torsion fracture of the caisson and shear dislocation between caissons due to differential settlements;

(iii) collapse of the toe and heel of the caisson due to the rocking motions caused by wave action;

(iv) shear rupture of crown wall (parapet) due to wave impact; and

(v) collapse of top slab by slamming caused by falling water masses (green water and overtopping waves).

The most critical area is the splash zone of the seaward wall, where cracks may have been caused by wave impact and chloride ingress occurs progressively. Corrosion of the reinforcing steel and spalling of the concrete cover weakens the wall, which may later collapse under wave impacts. As a result fill is washed out and the sliding safety of the caisson is decreased.

Although multi-cellular caissons are highly redundant structures and local damage will not immediately lead to a total collapse, minor structural failure such as of the top slab, may lead to a loss/ reduction of serviceability (damage to crane rail tracks, service ducts, stoppage of vehicular traffic, etc.).

Finite element modelling has become standard practice for structural analysis. However, before employing such sophisticated analysis, it is recommended to examine each structural member by using simple structural models based on slab and beam analogies. This will provide a preliminary and valuable insight into load transmission through each member down to the foundation, and thus into the possible failure modes. A detailed description of the in-service failure modes (particularly under wave impacts), of the structural simplifications to be made and of the hierarchy of models to be used at the different design stages is given by Crouch (1999a) and may also be found as an extended summary in Oumeraci et al. (2001). Further details on other in-service load cases are given in OCADIJ (1991) and BSI (1988).

(d) Durability of reinforced concrete caissons in the marine environment: High durability and long-term performance of reinforced concrete cellular caissons require not only proper structural design and construction (suitable concrete mix and arrangements to limit cracking), but also well devised maintenance plans as an integral part of the design, based on a clear strategy for instrumented monitoring, inspection and repairs. To achieve this, a deep insight is needed into the progressive degradation mechanisms and their possible consequences, which depend very much upon their precise location in the structural frame.

In addition to chloride penetration further degradation mechanisms are, for instance, sulphate attacks in the concrete matrix, salt weathering of concrete surfaces, alkali aggregate reaction, freeze–thaw action, abrasion through sediment-loaded flow, thermal cracking, plastic shrinkage and settlement as well as fatigue caused by cyclic wave loads. Details on these mechanisms, their possible locations of occurrence, preventive measures and related references are provided in FIP (1996), CEB (1992), Allen (1998), MacKechnie and Alexander (1997), PIANC (2001) and Crouch (1999b) as well as in many national codes of practice such as EAU (1996), BSI (1985,1988), ACI (1995) and OCADIJ (1991). It should also be kept in mind, that a prerequisite for a proper maintenance plan is a good understanding of the real significance of cracking and of the various stages of degradation such as the initiation thresholds, corrosion, first cracks, loss of structural integrity of members and its consequences on the structural integrity of the overall frame or on the serviceability of structural members.

Increasing the concrete cover, which is usually 50 to 80 mm thick depending on the exposure class of the reinforcing bars in a caisson member, will not necessarily improve its durability because the cracking will increase on the tensile face.

Finally, it cannot be over-emphasised that structural members of a multi-cellular reinforced concrete caisson are not necessarily designed only for extreme events (e.g. with a 100-year return period) associated with an ultimate stress state. More frequent events, e.g. with a 5- to 10-year return period, associated with durability criteria may be as and be even more relevant (Martinez et al., 1999).

Hydraulic model testing

(a) General considerations: Hydraulic-model testing represents one of the indispensable tools, which is generally used before embarking upon detailed design. The objectives of laboratory testing may be multiple:

(i) providing a general overview of the behaviour under design conditions;

(ii) determining wave forces, including impact loads on the caisson and its crest structure;

(iii) investigating the sliding stability of the caisson;

(iv) investigating the hydraulic stability of the armour of the rubble foundation, possibly including scour protection of the seabed;

(v) assessing wave overtopping conditions and associated aspects, including the effect of green water and falling water masses on and behind the structure; and

(vi) assessing the stability of the breakwater under construction.

Regarding the latter aspect, hydraulic model testing in a wave basin is an ideal tool for determining the optimum sequence for the construction phases and the optimum construction methods to be adopted in order to avoid damage.

Generally, wave-flume tests, using a Froude scale between 1 to 10 and 1 to 50, are performed at an early stage for normal wave incidence on some selected alternatives for breakwater cross sections. Depending on the wave directions, the configuration of the seabed topography and that of the breakwater head, the final design is tested and optimised exposed to both normal and oblique waves in a wave basin.

(b)   Testing requirements and procedures: The model waves should be an dequate reproduction of the prototype wave conditions. This is achieved either by generating representative wave trains or spectra recorded at the construction site or by choosing the theoretical wave spectrum (PM, JONSWAP or TMA), which best represents local conditions. However, complementary regular wave tests may also provide an improved understanding of the physical processes involved.

Tests with long-crested irregular waves are sufficient in most cases and are generally conservative compared to those with short-crested waves. The latter may be useful only for certain specific cases and for further optimisation.

Particular attention should be paid to the measurement of the waves in the model, due to the complex wave field generated by the superposition of incident and reflected waves. It is therefore suggested always to perform some preliminary wave recordings at the breakwater location. For these tests the existing seabed topography should be modelled (but without the breakwater model) for generated waves in the range of the design wave conditions. Long duration tests are necessary to obtain statistically representative values of the wave forces, overtopping rates, etc. Therefore, special care must be taken to reduce laboratory effects such as re-reflection from the wave generator and other model boundaries.

For measurement of wave loads, it is suggested to record both total forces and pressure distributions. When recording total impact forces it should always be kept in mind, that the recorded forces are not exclusively wave forces but also represent the response of the model structure including its mass, stiffness and damping properties. Pressure transducers must be used for this purpose, capable of recording high frequency impacts.

If sliding stability tests are performed the coefficient of friction between the model caisson and the rubble foundation must be determined by appropriate pulling tests. In addition, displacement meters should be used instead of accelerometers to measure the dynamic response, because the latter are much more difficult to interpret.

For measurement of wave overtopping conditions, it is suggested to record the number of overtopping waves as a proportion of the total number of incident waves as well as the individual overtopping volume of each wave. For assessment of the stability of the armour units of the rubble foundation it is important to record the number of displaced stones from a strip of a width, corresponding to one block size along the breakwater.

(c)   Data analysis and interpretation of results: Reflection analysis should be conducted with special caution because the errors on the determination of the incident wave height in front of the structure can be considerable. Another problem, requiring even more care, is interpretation of the measured impact pressure in the model, which is associated with less air entrainment/entrapment than in the prototype. Applying Froude's similitude law for converting impact pressures to prototype conditions would lead to higher values of shorter duration than the actual values. Depending on the degree of aeration and the amount of entrapped air involved in the impact process, both Froude's and Mach-Cauchy's similitude laws must be applied in combination with the different components of the impact load history (Fig. 7.50).

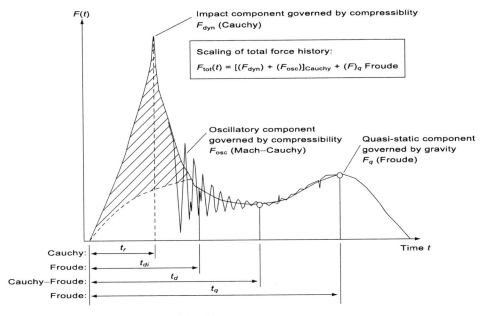

Fig. 7.50 Scaling of wave-impact loading

Details on the scaling procedure, based on separation of the different load components, are provided by Kortenhaus and Oumeraci (1999). A more pragmatic approach, based on the evaluation of a reduction factor for the impact load as well as increase factors for its rise time and duration, is suggested by Allsop and Kortenhaus in Oumeraci *et al.* (2001).

Construction of caisson breakwaters

Constructability and construction stages

To achieve a good design it is essential that the practicability of construction is known and taken into account properly. Local conditions such as the nature of the seabed, the availability of construction material, quarry output, available construction facilities for caissons, logistics support in the surrounding areas, the hostility of the sea and other environmental conditions, may adversely affect the constructability of a certain type of structure and favour another type. However, at the design stage, i.e. before tenders are invited and a contractor is selected, it is generally not certain which type of caisson and which construction method and equipment will be used. Therefore, the specifications prepared at the design stage are essentially based on experience from similar works. The specifications should allow alternatives, which are likely to provide a functional breakwater for the least total cost and within the specified time. Nevertheless, the following major stages of construction remain more or less the same for all types of caissons (Agerschou et al., 1983):

(a) production and transport of caissons to the construction site;
(b) preparation of the foundation for the caissons;
(c) sinking and placing of the caissons on the foundation;
(d) filling of caissons' cells;
(e) measures against scour and erosion during construction;
(f) securing the joints between caissons; and
(g) construction of the crest structure of the caissons (*in situ* or pre-cast concrete).

Generally, the constructability of the breakwater and the type of caisson largely depend on local conditions, such as:

(a) wind and wave conditions during construction, the tidal regime and other hydrographic and climatic conditions;
(b) the nature and complexity of the seabed (clay, sand, rock, etc.) and the availability of construction materials within a reasonable distance from the site; and
(c) the existing facilities, equipment and technology for the production, transport and placing of caissons, including logistics support in the surrounding areas.

Due to the high mobilisation costs for the facilities and equipment required to build a caisson breakwater, it is generally not advisable to choose this type of structure, if only a short breakwater is needed.

The construction time is optimum, when the average caisson production rate approaches the average placing rate. Very often the sinking and placing of caissons, including the securing of their foundations, represent the most critical phases as they require good weather conditions. It is therefore essential that rapid progress be made during calm-weather periods. Longer caissons may be favoured in this respect.

Caisson dimensions and structural members

Most of the reinforced caissons used in breakwaters consist of cells separated by walls and with a bottom slab. In addition, the outer cells have front, rear and end walls (Fig. 7.51).

The width of the caisson is essentially governed by stability considerations under wave action (see 'Overall stability of monolithic structures' earlier in this chapter). The largest caisson width used for a breakwater

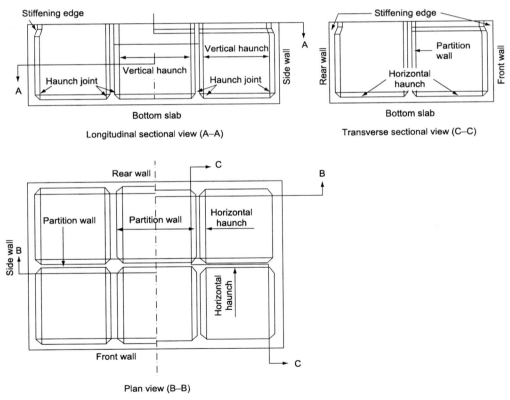

Fig. 7.51 Sketch of a multi-cellular reinforced concrete caisson (OCADIJ, 1991)

is 38 m (Hedono Port, Japan). It was designed to resist waves with $H_{max} = 17.5$ m and $T_{1/3} = 13.2$ s (Tanimoto and Takahashi, 1994).

The caisson length is very often governed by the caisson production facility as well as by considerations of the structural integrity in relation to differential settlements, etc. Generally, the caisson length is in the range of one half to two times the caisson width. The longest caisson (100 m long, 20 m high and 13.5 m wide) has been employed in Kochi Port, Japan, as a temporary breakwater to check potential problems associated with very long caissons. The caisson incorporates steel frames and has pre-stressed concrete walls. It was towed from a shipbuilding dock over a distance of 370 km (Tanimoto and Takahashi, 1994).

The optimum height of caissons, which usually exceeds 10 m, is determined by considering the capacity of the production facility, the floating stability and results of a cost–benefit trade-off analysis. The highest caissons have been used in water depths of more than 50 m (Kamashi Port, Japan, with $h_s \approx 60$ m and Santa Cruz, Tenerife, Spain, with $h_s \approx 52$ m).

The maximum size of the cells is usually about $25 \, m^2$. If larger cells are used, a detailed examination of the downward load on the top slab, caused by green water, by overtopping waves and falling objects, is required. The bottom slab is usually 50 to 70 cm thick and the partition walls are 20 to 25 cm thick. The outer walls, which are designed to sustain wave pressure and hydrostatic pressure, are usually 40 to 50 cm thick, but this can be reduced by using curved walls.

The thickness of the reinforcement cover, usually in the order of 50 to 80 mm, depends on the exposure classification of the structure. In Japan, a minimum thickness in the range of 75 mm has been used for the seaward side of the caisson walls. Details are given by PIANC (1999), JSCE (1986), Bamforth (1993), EAU (1996) and Allen (1998).

Assessment of allowable wave conditions for construction

For each construction stage, such as towing, sinking and placing of caissons, allowable wind and wave conditions have to be defined. If they are reached or exceeded the work must be interrupted. For instance, sinking and placing of caissons will require calmer conditions than floating and towing, and the wave conditions, which are going to disturb the levelled rubble foundation, may have even lower limits than those for sinking and placing. Due to the different construction methods, caisson sizes and risk levels adopted by different contractors, no generally accepted limits can be established for the respective construction stages. Once these limits have been determined, the construction can be planned based on the exceedance frequency of these limits and their duration. For the wave climate during construction, wave heights and wave periods are equally important, i.e. the wave climate should describe the exceedance probability of wave heights and periods, possibly also including wave direction and water level. Given the size of a floating structure, its draughts and natural frequency, the wave period may be even more critical than the wave height. Maximum wave periods for some operations are given by O'Hara (1994), who also describes measures to extend working conditions. A more detailed assessment of the wave conditions and their occurrence, than for rubble-mound breakwater construction, including better weather forecasts, is required, because caisson breakwater construction is more critical under exposed conditions. Moreover, the construction activities cannot so easily be adjusted to prevailing wave conditions as in the case of rubble-mound breakwaters. The risks involved in exposure of incomplete work and of equipment during construction must also be assessed and appropriate protection measures must be planned. This is best achieved through hydraulic-model testing (see 'Hydraulic model testing', earlier in this chapter). The tender documents should at least include the exceedance frequency curves for the wave climate and construction tolerance specifications, for use by contractors to assess probable weather downtime.

Caissons are brought into their final positions by appropriate means (see 'Preparation of foundation, sinking and placing of caissons', later in this chapter) to achieve the required alignment according to specified tolerances in vertical (cross section) and horizontal (plan view) directions. Tolerances defined as admissible deviations from contractors' final drawings have been suggested by Martinez et al. (1999):

> In cross section: the tolerance for every point of the top surface of the caisson is about ±0.20 m for breakwaters, but much more strict for quay walls (±0.10 m); and
> In plan view: the tolerances are in the same range as for the cross section when considering an individual caisson unit, but the accumulated deviation must remain below +5.0 m for a breakwater and −0.0 m and +1.0 m for a quay wall.

The gaps between the caissons must comply with the specified tolerances.

Production and transportation of caissons

(a) Options and methods: The production of caissons cannot be planned before the site of the facility, production equipment and the method for transporting the caissons from the production facility to the breakwater location have been chosen. Most caissons are towed to the construction site. Non-floating caissons, such as those used for the Port of Hanstholm, Denmark and for the Brighton Marina, UK, are rarely used. Their size is rather limited and their transport and placing requires a giant crane. Therefore, only floating caissons will be addressed in the following. For construction of non-floating caissons see Agerschou *et al.* (1983) and Terrett *et al.* (1979).

There are different options and facilities for producing caissons of different types and sizes:

(i) dry dock or floating dry dock;

(ii) existing shipyard with slipway for launching;

(iii) temporarily pile-supported jetties. Launching is achieved by demolition of the piling (O'Hara, 1994);

(iv) construction area with a slipway for launching;

(v) construction area with a ship lift for launching;

(vi) platform suspended in a fixed or floating installation and gradually lowered into the water as the constructed height of the caisson grows (PIANC, 1980); and

(vii) beach as construction area or excavated dock on the foreshore with slope stability maintained by well points. Flooding of the dock and float-out is achieved by dredging a channel to the open sea (O'Hara, 1994).

Each of these options has its advantages and disadvantages, and selection will depend on the specific project requirements including the size and number of caissons needed. Options (iii) and (vii) are normally most appropriate for production of one or only a small number of caissons, whereas options (i), (iv) and (v), which are usually built for the specific purpose of producing caissons, are better suited for production of a larger number of caissons.

Launching from a dry dock, a pontoon or a platform is often preferred to the slipway option. The caissons can be stored afloat in a sheltered basin, thus permitting the concrete to further cure. This would result in a considerable increase in the rate of production. Moreover, launching from a dry dock or platform would cause less cracking in caisson walls than the slipway option, provided there is a good balance between the distribution of caisson weight and that of the buoyancy. The stresses in a floating caisson may be calculated more accurately than those in a caisson on a slipway.

(b) Materials for thin-walled, reinforced concrete structures: For thin-walled structures, such as cellular caissons, exposed to an adverse marine environment, the durability requirements associated with the measures to be taken against corrosion of the reinforcement are governing for both design and construction. Deterioration mechanisms such as those caused by salt weathering, sulphate attack of the concrete matrix, freeze–thaw and abrasion should also be taken into account. Given the severity of exposure, the specifications for the cover thickness, the type and content of materials in the reinforced concrete and the curing are therefore mainly dictated by durability considerations. Since seawater contains both sulphate and chloride, the concrete has to provide both chemical resistance and protection for the reinforcing steel. The latter aspect is more important for the required concrete properties and other measures for improving durability. For a given exposure severity, the initiation and rate of corrosion mainly depend on the thickness of the reinforcement cover, the binder type and contents and the water–cement ratio.

A cover thickness of 50 to 80 mm is common. Guidance on the proper cement type, the appropriate cement content and water–cement ratio as a function of the exposure severity, is given for instance by Slater and Sharp in Allen (1998), PIANC (1990), as well as by some national and international codes of practice. However, the weakness of the existing codes is that guidance is given on concrete mixes and cover thickness, without any precise specification of cement type and exposure conditions. Therefore, it is essential that the engineer, confronted with the structural design and construction of such structures, understands the deterioration mechanisms and the significance of the associated factors such as:

(i)   the porosity structure achieved in the concrete, possibly including cracks;

(ii) the transport mechanisms for gas, water and dissolved agents, such as chloride, sulphate, etc., within the porous concrete structure, including the cracks; and

(iii) the exposure severity of the structural members.

Details on these factors are given in CEB (1992).

The quality of aggregates is particularly important with regard to resistance against abrasion, freeze–thaw and salt spalling. The shape, texture and absorbency of aggregates also affect the water content, which in turn strongly influences durability. For selection of proper aggregates refer to national standards and existing studies on local aggregates.

Guidance for reinforcing steel to be used in maritime structures (generally hot-rolled, high-yield deformed steel), including protective measures against corrosion, is provided by most national standards. Reinforcing to control early thermal tensile stress during hydration is also necessary, in addition to appropriate measures, to control the hydration heat (use of water reducing admixture, low-heat cement involving blending with pozzolan or slag, cooling, thermal curing and insulation, etc.). Methods to calculate the crack widths are given in many standard references such as BS 8007 (BSI (1987)) and CIRIA (1992). Thermal cracks recorded in caissons have shown that the reinforcement should be designed in such a way that a larger number of finer controlled cracks may occur rather than fewer and wider cracks.

(c)   Slip forming: Since most of the caissons used exceed 10 m in height, vertical slip forming is normally used for casting the walls, which takes only a few days for one unit. Of course, it is desirable to produce the full height in the initial casting area. However, for tall caissons this would require extraordinary dimensions of the production facility. Therefore, the casting in the initial area is performed to a height which ensures, that after sufficient curing, the caisson can be safely launched and towed to deeper water, where casting of the walls is continued. The concrete walls are without joints and harden free of restraints from adjacent pours. As a result, early thermal effects are minimised.

A relatively recent review of slip forming is provided by Jones and Horne (1996). Slip forming of caissons has also been described for instance by Cochrane *et al.* (1979).

The logistics' difficulties associated with curing of vertical slip-formed surfaces are described by ACI (1986).

(d)   Flotation of caissons: Flotation is achieved by providing sufficient buoyancy and hydrostatic stability. For floating stability it is suggested by OCADIJ (1991) that the metacentric height h_G, (the distance between the gravity centre G and the metacentre M) should be at least 5 per cent of the draught D of the caisson (Fig. 7.52). The floating construction stages have to consider, besides the stability aspects like trim, freeboard, wave and wind action,

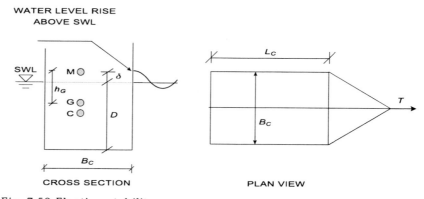

Fig. 7.52 Floating stability

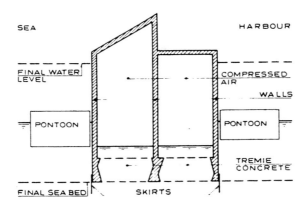

Fig. 7.53 Floating of bottomless caisson (Agerschou et al., 1983)

the effects of mooring and the ability of freshly cast concrete to sustain hydrostatic pressure as further pours cause the caisson to sink.

Even bottomless caissons, driven into the seabed, can be kept afloat by means of pontoons with the required lift capacity as shown in Fig. 7.53. Further measures and problems associated with the floating stability of bottomless caissons are discussed by Corrons *et al.* (2000).

(e)   Towing of caissons: Towing from the production site to the breakwater can be carried out only during calm weather conditions. In most cases the exact float-out dates may only be selected a few days in advance, based on the local weather forecast. Generally, the allowable conditions for float-out and towing have to be specified, so that the resulting loads are not exclusively at the contractor's risk. It is also important to keep in mind that possible damage from towing does not always become apparent during construction.

In case of very long towing distances, it may not be possible to place the caisson in the breakwater alignment immediately after its arrival. Therefore, the caisson must be properly designed in order to allow temporary grounding near the breakwater and parallel to the direction of wave propagation. Alternatively, the caisson may be anchored to ride out stormy weather. For instance, the caissons for the breakwater at Marsa el Brega, Libya (Fig. 7.54) were produced in Messina, Italy, and towed over a distance of 1,100 km across the Mediterranean Sea (Colleran and Leonard (1969)).

(f)   Loads during construction, launching and towing: The loads during construction and transport may arise from various sources such as handling with cranes during the curing period, placing on launch truck and moving to a slipway or caisson platform and launching and towing. Concentrated loads from handling, hydrostatic pressure

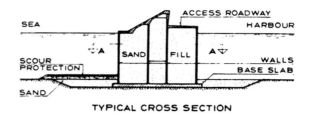

TYPICAL CROSS SECTION

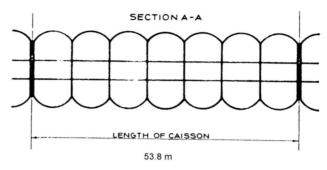

Fig. 7.54 Caisson breakwater at Marsa el Brega, Libya (Agerschou et al., 1983)

during launching, possible wave and wind action during towing, must be considered in the structural design (and floating stability analysis) as temporary design loads with an appropriate partial safety factor (e.g. $\gamma_F \approx 1.1$). If these loads cannot be anticipated at the design stage due to the lack of knowledge of details of the construction methods, the structural analysis has to be rechecked by the designer as soon as the lacking details become available. Some further indications on these loads are provided, for instance, in OCADIJ (1991) in which a safety allowance of 1 m added to the actual draft of the caisson to calculate the hydrostatic pressure on the outer walls and the buoyancy is included. The towing force T[N] is calculated as follows (Fig. 7.52):

$$T = \tfrac{1}{2}\rho_w C_D v^2 A$$

where ρ_w is the density of sea water (kg/m^3); C_D is the drag coefficient depending on the shape of the caisson; v is the towing speed (m/s); and A is the submerged area of the leading wall (m^2) with $A = B(D+\delta)$, δ is the water level rise above SWL in front of the caisson, D and B_c are the draught and width of the caisson, respectively.

In a detailed dynamic structural analysis of the towed caisson, wave pressure, heave motion due to swell, as well as roll and pitch motion due to wind and waves, and to towing and manoeuvring operations, have to be considered (Oumeraci *et al.*, 2001). A probabilistic analysis for this purpose has been suggested by Vrouwenvelder and Bielecki (1999).

Preparation of foundation, sinking and placing of caissons

Before sinking and placing of the caissons, the foundation usually has to be prepared carefully in order to avoid stress concentration in the bottom slab due to unevenness of the base as well as excessive and/or uneven settlement of the caisson. Because of the latter aspect soft soil layers must, of course, be removed.

For the support of caissons different options exist such as:

(i)   a bedding layer of gravel or shingle (Fig. 7.55);
(ii)   a rubble-mound base (Fig. 7.27);

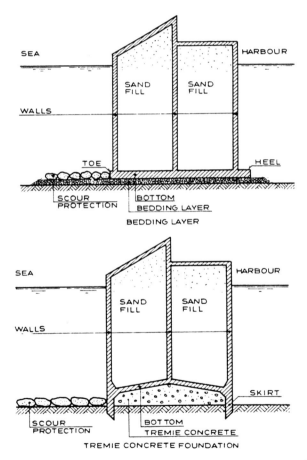

Fig. 7.55 Foundation alternatives for caissons (Agerschou et al., 1983)

(iii)   protruding outer walls driven into the seabed like skirts (Fig. 7.55);
        and

(iv)   point support method (Fig. 7.56).

The bedding layer or rubble-mound base, on which the caisson is going to rest, has to be levelled by divers and/or by special equipment, just before placing the caisson. Generally, the base has been advanced by 10 to 30 m beyond the caisson to be placed, in order to avoid erosion around the provisional end of the breakwater. Such operations are time consuming and require very calm weather conditions. For caissons driven into the seabed, the technology is similar to that used for skirt foundation of offshore gravity platforms, so that extensive use can be made of the experience available in this field of work. Generally, this method does not require diver assistance nor calm weather conditions. The caissons may be driven into the seabed by the caisson weight alone, suction generated by vacuum and seepage induced around the caisson by jets or by dredging and lowering of the water level inside the caisson (Corrons et al., 2000).

Experience using tremie concrete (Fig. 7.55) for the foundation of skirted caissons is available, for instance, from the construction of non-floating caissons for the Port of Hanstholm, Denmark (Agerschou et al., 1983). It is suggested to design the skirts for a penetration corresponding to 100 per cent water filling and 50 per cent sand filling of the caisson. Variation in soil resistance to penetration can be compensated for by adjusting the degree of filling. Additional penetration may be provided by a partial vacuum in the chamber under the caisson. Of course the bottom, the skirts and the seabed must be able to sustain the pressure difference generated by such vacuum. The skirts, made of reinforced concrete or steel, are built either under the front and rear walls only or also under the side and partition walls. As a further option, a bottomless caisson with skirts as shown in Fig. 7.53 can be used. In this case, compressed air may be used for levelling the caisson through variation of the air pressure from cell to cell, possibly including partial vacuum (Agerschou et al., 1983).

The point support, described by Lundgren in Agerschou et al. (1983) for the non-floating caissons of the Hanstholm breakwater, consisted of concrete blocks cast on the rock seabed and adjusted by divers to the correct elevations for the next caisson to be placed. As pointed out by Lundgren this method would be rather complicated for long floating caissons, while it is a natural method for smaller, non-floating caissons. To achieve better control of the first grounding of the floating caisson a number of recommendations and alternatives exist which are based on the same principle as the point support methods (PIANC, 1990). It is important to know in advance, which part of the caisson base will first touch the

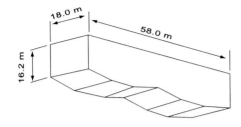

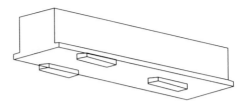

'CAMEL HUMPS' FOR CAISSONS USED IN
BROUWERHAVENSCHE GAT, NETHERLANDS

THREE SUPPORT POINTS FOR THE
FIRST GROUNDING OF CAISSON

Fig. 7.56 Grounding supports for caissons

ground, which support points will remain after placing, where the impact loads will have to be applied to the caisson base for a dynamic structural analysis and finally how to minimise the contact stresses arising from the first grounding of the caisson.

EAU (1996) in Germany suggests that a first grounding of the caisson has to be anticipated either in the middle or on two points located at two diagonally opposed edges. On the other hand, the Japanese experience suggests that an excess height of the rubble foundation should be provided in the middle. Moreover, 'camel humps', as used in the Netherlands, or other protrusions on the bottom slab (Fig. 7.56) represent further alternatives.

Experience has shown that the loading, which may occur during sinking and placing of the caisson, is more critical for structural integrity than any other loading during the life of the caisson (Vrouwenvelder and Bielecki, 1999). Such loading may include:

(i)   accidental sinking of caisson resulting in disturbance of the foundation and collapse of the bottom slab;
(ii)   collision with adjacent caisson due to waves and/or wind action;
(iii) uneven ballast loading resulting in cracking or collapse of walls; and
(iv)   uneven foundation resulting in cracking or collapse of the bottom slab as well as misalignment of the caisson.

Analysis of the accuracy of placing caissons is a very complicated problem, because the basic equations for sway, surge and yaw motion must take into account a variety of factors such as tugs, mooring lines, wave reflection and diffraction by the in-place caissons as well as the highly non-linear dynamics of the system.

Filling of caisson cells

Caissons are usually filled with sand, which may be pumped or tipped, but any other granular material, permitting calculation of the internal soil pressure with confidence, could also be used. There are generally no strict requirements for the granular material to be used. However, it should have an angle of internal friction and a saturated density, which should, at least, be equal to the value used in calculation of the internal soil pressure. Moreover, it must be suitable for underwater filling. As the density, which is achieved for sand poured into water-filled cells, is much lower than that of sand poured into open water, additional compaction by vibration (Cochrane et al., 1979, Daniels and Sharp, 1979) may be needed to achieve a secure foundation for the superstructure. Compaction will cause no significant increase of the internal soil pressure, due to the inverse relationship between fill density and internal friction angle. Japanese experience has shown that vibration caused by earthquakes does not destroy the silo effect, which is known to reduce the internal soil pressure (PIANC, 1990).

Filling should be carried out immediately after the caisson has been correctly placed. Particularly when the caisson is in an exposed location is it urgent to proceed with the filling. Therefore, any procedure requiring special caution during filling (such as restrictions on the level to which adjacent cells are filled or emptying of water-filled cells) has to be excluded. The caisson must be designed to allow any sequence of filling. Seaward cells should always be completely filled to ensure stability against wave loading. Harbourside cells can be only partially filled to reduce the bearing pressure at the caisson heel, provided the overall stability of the caisson has been considered for the seaward wave forces, which are induced, when the wave trough is at the seaward face of the caisson. If increased strength of the outer cells against impact loads is required, the cells can be filled with lean concrete. It is important to emphasise that the worst filling cases likely to occur in practice, for instance having the water level at the surface of the fill, should be considered for the structural design of both outer walls and cell partition walls.

Measures against scour and erosion during construction

Although unforeseen difficulties can never be excluded during construction, most of them can, however, be avoided through proper design and proper hydraulic model testing of the work sequences ('Stability considerations', earlier in this chapter). One of the most serious problems, which may occur during construction, is seabed scour and undermining of the prepared bed of granular material due to severe wave and/or current action at the vertical face and around the incomplete breakwater end. It is important to mention that soft rock foundations, such as chalk, can also be affected by scour. Due to the potentially serious consequences of scour and erosion in front and under the breakwater as well as around the breakwater end, the tender documents should prescribe that proper protection be placed just after the caisson has been placed in the final position. Particularly where breaking and/or oblique waves are expected to occur frequently, a detailed analysis through hydraulic model testing is a must. Due to the high costs of the structure, the protection should be designed carefully, even conservatively.

The most exposed parts are the seaward face of the breakwater and the end of the last placed caisson. The seabed and the berm in front of the seaward face are exposed to high agitation due to the strong reflection from the wall, which results in high water-particles velocities close to the bed. The situation is even more critical for oblique waves combined with currents along the breakwater. To prevent erosion of the rubble foundation by wave action only, proper design of the berm (see 'Hydraulic stability of rubble foundation and scour protection', earlier in this chapter) will be sufficient. However, if waves occur in combination with currents, it is recommended to consider the resulting combined shear stresses in calculating the required block weight. Alternatively, as suggested by Eckert (1983), the calculated block weight using the method described in Box 7.12 should be increased by about 50 per cent. In addition, it is also necessary to prevent dislocation of the berm (see Fig. 7.44). This can be achieved by various measures such as special ballasted mattresses, a quarry-run layer under and beyond the berm toe to a sufficient width, etc. Particularly in Japan, there is a revival of gravel mattresses, consisting of a thin layer of quarry run, which is allowed to be mixed by wave action with the original seabed sand, thus providing an efficient scour protection (Goda, 2000). The rough guidance found in some textbooks and codes of practice, according to which scour protection should consist of a blanket of 0.2 to 0.5 t stones covering the sea bottom up to a distance of at least 1/4 of the design wave length in front of the breakwater, is certainly not the best or cheapest solution. The apron and its extent in front of the wall should be designed carefully together with the geotechnical stability of the foundation.

The berm should also be designed and constructed to sustain the down-rush from wave splash, which can reach the bed and stir up material. In Japan considerable success has been achieved by using so-called toe protection blocks (rectangular concrete blocks with 10 per cent surface porosity for reduction of uplift) to prevent undermining of the seaward and shoreward edges of the caisson. Much more at risk during construction is the end of the last placed caisson, which is subject to larger shear stresses and vortices induced by the combined action of waves and currents. The scour and erosion potential is more pronounced for sharper corners of the caissons and for oblique waves combined with currents. Temporary protection of the incomplete breakwater end against scour is much more difficult to achieve than protection of the trunk, because it

237

has to be removed very fast and at affordable costs just before placing the next caisson. Therefore, protective elements, which are easy and fast to handle, are appropriate. Experience using geotextile containers up to water depths of 20 m is available (Pilarczyk, 1998), but other alternatives such as heavy pre-fabricated mattresses and skirted caissons with a skirt depth likely to provide sufficient protection for at least one storm can also be used. Anyhow, the contractor should always be prepared to place and remove such protection, depending on the local weather forecast. Particularly for exposed sites and important projects, hydraulic model testing is indispensable to check the efficiency and risks of various construction scenarios (see 'Stability considerations', earlier in this chapter).

The rubble foundation on the harbourside is less exposed to wave and currents than that on the seaward side. However, considerable seepage gradients resulting from waves hitting the seaward face of the breakwater may undermine the foundation or cause piping in sand. This explains why most Japanese caisson breakwaters are provided with a heel protection block on the harbourside in addition to one or two toe protection blocks on the seaside.

As mentioned by Lundgren (Agerschou et al., 1983), the underside of the caisson is rarely exposed during construction when designed properly. An exception to this rule was given by Lundgren by quoting Llewellyn and Murray (1980), who, when describing the construction of the Brighton caisson breakwater, said:

> a major incident occurred during a severe storm on September 3, 1974. The crane was standing on caissons E32 and E33 and the partially completed breakwaters were subjected to continuous south-westerly gales of force 11 for 17 h. The harbour was open on a wide front and waves were reflected from the vertical face of the north wall so that the southern arm of the east breakwater was attacked by waves from both the sea and the harbour sides. The caisson immediately behind the crane had not been filled with sand and no scour protection had been placed at that time. Seepage developed under the foundation of caissons E29 to E31 and the differential head due to waves peaking on one side with a corresponding trough on the other scoured the foundation chalk. Caissons E29 to E31 settled as a unit, sloping from zero at the joint E28/E29 to 650 mm at joint E31/32. Caissons E32 and E33 on which the crane rested were unaffected.

Securing joints between caissons

Unlike the horizontal joints, which are common in old vertical breakwaters made of block works and cyclopean blocks, vertical joints cannot be avoided. They are indispensable to ensure that differential settlements along the breakwater can occur without causing any structural damage. Despite their importance and the damage experienced by many breakwaters due to misconception of their purpose and functions, only little attention has been paid to their design and construction. It is important to stress their main functions and requirements. In addition to allowing differential settlements without structural damage, the following functions are noteworthy:

(a) making the breakwater monolithic in its longitudinal direction, so that load sharing between adjacent caissons is achieved (see Fig. 7.48). This function is much more important during the in-service life, because it is unlikely that the design wave conditions will occur during construction;

(b) preventing waves and water from rushing through the gaps with high velocities. The sealing of the joints on the seaside is therefore important during and after construction, to avoid erosion of the foundation by water jets through the gaps. Such damage was for instance experienced by the breakwater of Marsa el Brega, Libya. This experience was reported by Colleran and Leonard (1969) to illustrate that the joint must be completed or at least sealed at the seaside without delay; and

(c) allowing a flexible coupling without requiring excessive precision in caisson alignment. The usual tolerances for placing of caissons should be employed.

In addition, the joints must prevent washing out of the back fill, if any. An alternative, which offers a good compromise between load sharing and providing a flexible coupling is the use of an *in situ* gravel-filled pocket, which is plugged at both ends by concrete and sandwiched between caissons. This method, which is suggested by PIANC (1990) under the designation of 'double slot joint', is illustrated in Fig. 7.57 for rectangular caissons as used in Spain. An innovative joint for circular caissons is also shown in the figure. The benefits of shear transfer between adjacent caissons are obvious.

Another interesting alternative suggested by Lundgren and Juhl (1995) confines the shear connection to a massive bottom slab (Fig. 7.58). The shear transfer is achieved by underwater concrete, in such a manner that one caisson may rock more than another. Further examples of joints between rectangular and circular caissons are given, for instance, in EAU (1996), BSI (1988, 1991) and PIANC (1990).

Crest structures

Caissons filled with sand must be topped by a crest structure to protect the fill from being washed out by the overtopping waves. A simple top slab, as

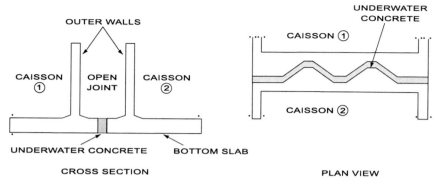

Fig. 7.57 Joints between caissons (adapted from PIANC, 1990)

widely used in Japan, (Fig. 7.59) may be sufficient. However, crest structures are very often used to achieve the required crest level of the breakwater at a minimum cost and to reduce overtopping and splash by providing a re-curved nose on the parapet (Fig. 7.59). Since such conventional parapets are often exposed to breaking waves, which may cause large overturning moments, they are sometimes substituted by so-called set-back parapets and sloping superstructures (Fig. 7.59) to reduce the wave load and its effects on the breakwater stability. The advantage of these innovative solutions is that wave forces on the crest wall occur

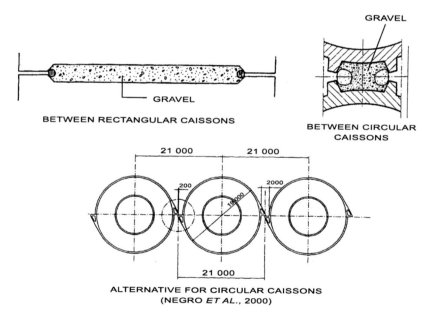

Fig. 7.58 Bottom slab joint alternatives (adapted from Lundgren and Juhl, 1995)

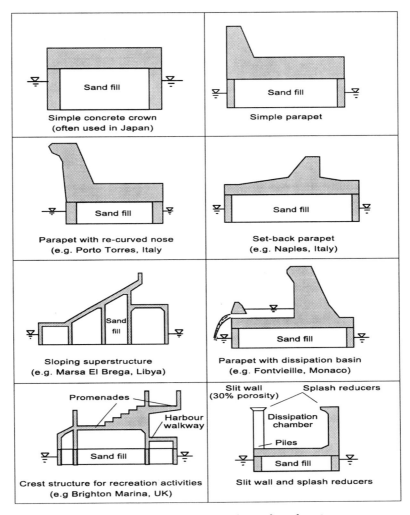

Fig. 7.59 Alternative superstructure of caisson breakwaters

later than those on the front wall. Moreover, the wave pressure on the sloping part of the crest structure has a stabilising effect in the sense that it causes a downward vertical force component. However, it tends to increase overtopping. Other innovative solutions, such as those in Fig. 7.59 can be used to substantially reduce wave loads, overtopping and splash. Crest structures of marina breakwaters and seawalls may be used for promenades and other recreation activities (Fig. 7.59). Further crest structure alternatives are described for instance by Juhl (1994) and Lamberti and Franco (1994). The crest structure can also be shaped to incorporate pipelines, facilities for port operations, and other unconventional facilities such as turbines for wave power generation (Fig. 7.60).

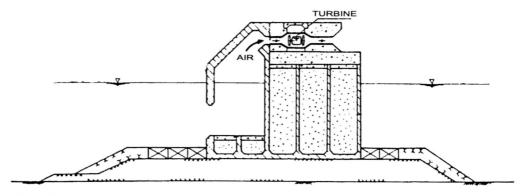

Fig. 7.60 Wave power generation

Generally, the optimal shape and height of the crest structure is obtained by means of a procedure which takes into account the required functions, the relative importance of reducing wave loads, overtopping and splash as well as constructability aspects.

For floating caissons it would of course be desirable to prefabricate the whole crest structure, because *in situ* formwork, reinforcement placing and concreting is always associated with higher costs and/or lower quality. Wave overtopping may interrupt the work and result in damage. Although most superstructures have been built *in situ* (Fig. 7.61), it would be wise to allow the contractor to propose an entirely or partially prefabricated crest structure as an alternative.

NAPLES BREAKWATER

PORTO TORRES BREAKWATER

Fig. 7.61 In situ *construction of crest structures (reproduced by permission of Società Italiana per Condotte d'acqua)*

A detailed study should be carried out of the implications of pre-fabrication of the crest structure on the floating stability of the caisson. Alternatives for floating-out the caissons, without superstructures and grounding them for completion and special provisions to increase buoy-ancy, such as pontoons attached to the caissons, should also be considered. Careful attention should also be paid to the fill beneath the top slab and whether the superstructure will be used for vehicular traffic or will it be subject to severe slamming caused by green water or over-topping water masses hitting the top slab.

Innovative caisson breakwaters

General

Development has, and is, taking place of innovative techniques to manage risks associated with coastal structures, including technology for monitoring, inspecting, maintaining and repairing. Further important technological response to the sustainability challenge in coastal and harbour engineering is the development of innovative structures (Oumeraci, 2000). The innovations should be directed towards improvements in terms of:

(a) *performance* such as reduction of wave reflection, wave over-topping, wave loads and their effects on the safety of the overall structure and its foundation. This is generally achieved by combining structural and functional design with geotechnical and constructional aspects;

(b) *total costs* over the anticipated life of the breakwater, including maintenance costs as well as possible tangible and intangible losses and/or benefits;

(c) *environmental integrity* by taking into account the effect of the structure on the water quality, the coastal ecosystem and morphology, the seascape, etc.;

(d) *construction time, quality control and standardisation* achieved through prefabrication of most of the structures;

(e) *multi-purpose use* such as recreation, wave power generation, etc. in order to achieve wider acceptance and to share the costs, which are often more than 100 million EUR per km of breakwater; and

(f) *robustness, resilience and reversibility* in regard to withstanding extreme events of low probability of occurrence without disastrous consequences. The properties associated with efficient daily performance will, however, remain equally important.

Most of the innovations of recent decades were merely directed towards improving hydraulic performance by reduction of wave reflection, over-topping, breaking wave loads and/or their effects on the overall stability of the structure and its foundation. Further innovations also emerged to cope with weak soil conditions. In recent years, however, the innovations have covered further aspects such as the environment and multi-purpose use, but also facilitating construction and reducing construction time, materials and costs.

Innovation principles and their application

(a)   Reduction of wave loads: To reduce wave loads and their effect on breakwater stability, the basic principle is based on appropriate geometry of the front of the caisson and its superstructure:

(i)   *in plan view* the front wall may consist of a series of semi-cylindrical shells (Kimura *et al.*, 2000). Depending on the depth of corners, which governs the delay of impact pressures around the shells, a reduction of the horizontal impact force up to 45 per cent may result (Fig. 7.62); and

(ii)   in *cross section* the front may have a sloping upper part to generate a stabilising, downward directed force component. This may result in a reduction of the horizontal impact force up to 60 per cent as compared to a vertical flat front (Fig. 7.62). Alternatively, a curved front can be used. A further possibility is to place the crown wall as far back as possible in order to achieve a phase difference between the breaking wave forces on the caisson front and those on the crown (Fig. 7.59). At the same time a downward stabilising force on the

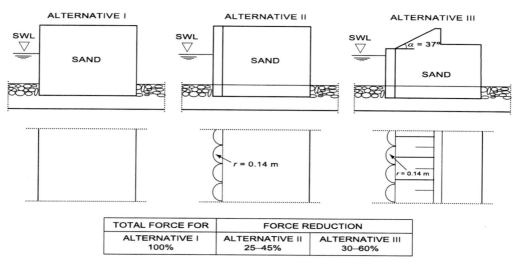

TOTAL FORCE FOR	FORCE REDUCTION	
ALTERNATIVE I 100%	ALTERNATIVE II 25–45%	ALTERNATIVE III 30–60%

Fig. 7.62 Effect of caisson front geometry on horizontal breaking wave forces (Oumeraci et al., 1993)

caisson is generated and overtopping is reduced through a dissipation basin effect.

Breaking wave impacts can also be completely avoided by providing a dissipation mound in front of the caisson (Fig. 7.27). However, this alternative is the most expensive one and can be used only under special circumstances. To increase the overall stability of the caissons against breaking, proper shear connections between the caissons should be employed (Figs. 7.57 and 7.58), longer caissons used (Figs. 7.37 and 7.48), and a curved alignment of the caisson front chosen. A more detailed description of the latter is given by Lundgren and Juhl (1995), who suggest that the head caisson should be armoured with a dissipating mound because arching is not possible at the breakwater head. This will also result in reduced wave reflection at the harbour entrance.

(b) Reduction of wave reflection and overtopping: To reduce wave reflection perforated caissons of various shapes and wave chambers have been developed as shown in Fig. 7.63. The applications described correspond to the sketches, but differ in some details.

A further reduction of wave overtopping may be achieved by providing an appropriate superstructure (see Fig. 7.59).

(c) Caissons for weak foundation soils: To cope with weak soils such as thick layers of mud, clay or silt, many caisson alternatives have been developed:

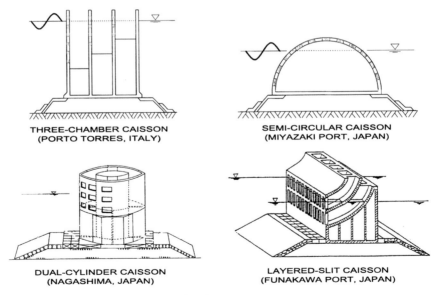

THREE-CHAMBER CAISSON
(PORTO TORRES, ITALY)

SEMI-CIRCULAR CAISSON
(MIYAZAKI PORT, JAPAN)

DUAL-CYLINDER CAISSON
(NAGASHIMA, JAPAN)

LAYERED-SLIT CAISSON
(FUNAKAWA PORT, JAPAN)

Fig. 7.63 Innovative perforated caissons

PERFORATED CAISSONS ON PILE FOUNDATION
(BOUCHET *ET AL.*, 1994)

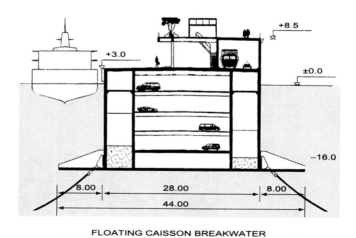

FLOATING CAISSON BREAKWATER
(PRESTRESSED CONCRETE, NEGRO *ET AL.*, 2000)

Fig. 7.64 Caisson breakwater on weak foundation soils and in deep water (alternatives for a marina in Monaco)

(i) *piled caissons* have been considered as an alternative design for a marina in Monaco (Fig. 7.64). Further piled alternatives using curtain walls are shown in Fig. 7.65;

(ii) *a floating caisson breakwater* has been adopted for the marina in Monaco. A 360 m long pre-stressed concrete caisson is going to be built in water depths up to 60 m (Fig. 7.64); and

(iii) *bottomless skirted caissons* penetrating through the weak soil layers into a bearing soil layer. This alternative has been considered, for instance, for the Malaga breakwater, Spain, on a silt layer of about 15 m thickness (Negro *et al.*, 2000).

(d) Caissons to enhance water exchange: Particularly for marinas, sufficient water exchange between the harbour basin and the open sea is essential. To achieve this goal, the following alternatives

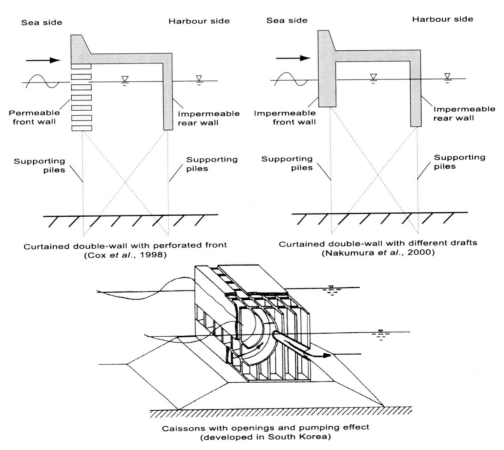

Fig. 7.65 *Alternative breakwaters enhancing the water exchange between harbour basin and open sea*

have been developed:

(i) *perforated caissons* (Fig. 7.64) or structures founded on piles (Fig. 7.65), so that the lower part of the breakwater remains open;

(ii) *openings in the lower part of a normal caisson*; which has successfully been applied, for instance, in Hokkaido, Japan; and

(iii) *special caissons* with openings in the lower part, which make use of the *pumping effect* of the waves. Such a caisson has been developed and extensively tested in South Korea (Lee and Hong, 1994). The principle is sketched in Fig. 7.65.

(e) Multi-purpose caissons: Because of the flexibility offered by prefabrication, any shape of caisson may incorporate further structural elements such as walkways and promenades for recreation, as well as other facilities. Turbines for wave power generation have already been incorporated in caisson breakwaters (Fig. 7.60) as well as promenades and walkways in marinas and reclaimed areas. Large caissons, incorporating parking garages as well as flats and office buildings on a caisson breakwater, have also been considered in Monaco (Fig. 7.64).

Perspectives on caisson-type structures

Considering the six requirements listed in 'General' earlier in this chapter and the innovation principles described in 'Innovation principles and their application' also earlier in this chapter, reinforced concrete cellular caissons and pre-stressed concrete caissons appear to be suitable for a wide scope of applications, such as breakwaters, quay walls and sea walls for protection of reclaimed areas. In addition, one should stress:

(a) their flexibility in regard to shape and to incorporating structural elements to improve the hydraulic performance and safety or to support facilities such as turbines for wave-power generation and walkways for recreation; and

(b) their suitability for prefabrication in almost any size, thus making them suitable for a wide range of water depths.

Due to the costs of the rubble foundation and the higher risk associated with sinking and placing caissons on a rubble foundation, it is likely that skirted caissons will be favoured in the future. The latter have further advantages in their built-in scour protection and adaptability to weak soils. Their use is going to require that particular attention be paid to all structural and soil mechanics' problems associated with penetration, underwater concreting, suction technology, etc. (Corrons *et al.*, 2000).

Monitoring, inspection, maintenance and repairs

Purpose and general strategy

It has been widely recognised that no design can be properly optimised without taking into account monitoring, inspection, maintenance and repairs as integral parts of the design process (Fig. 7.66). However, existing codes and guidelines for design and construction of breakwaters rarely deal adequately with these aspects. Monitoring and inspection are necessary for a variety of purposes such as:

(a) validation and early updating of design criteria and analyses for the particular breakwater as well as for general research purposes;

(b) updating of maintenance plans developed in the design phase, including equipment mobilisation and costs;

(c) identifying relevant hidden structural impairments, such as incipient reinforcement corrosion, cracks and erosion of seabed and rubble foundations, which may need urgent repairs or replacement; and

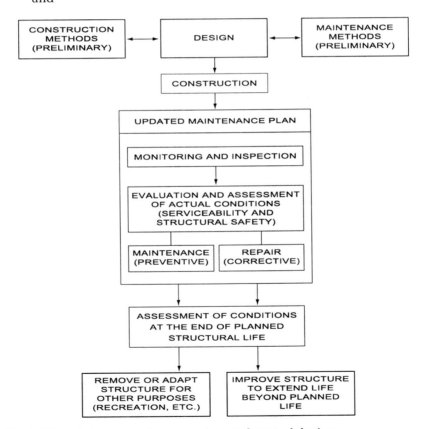

Fig. 7.66 Maintenance plan as an integral part of design

(d) providing the required basic data for analysis of the viability of short-ening or extending the life of the structure beyond the design life.

Maintenance and repair are essential to continuously ensure that the breakwater performs to an acceptable standard during its life. Although caisson breakwaters generally need much less maintenance and repair than their rubble-mound counterparts, it is advisable to establish a maintenance plan at the design stage, including a general maintenance strategy as well as practical guidance on the requirements and procedures for monitoring, inspection, maintenance and repairs. This plan must be updated during and after construction.

The maintenance plan should comprise specific practical guidance on the necessary techniques and requirements with regard to the following basic items:

(a) inspection and monitoring of environmental conditions, seabed morphology, loads and responses of the structural members and of the foundation. This should include the items to be observed/measured, the techniques and procedures to be used and the inspection frequency;

(b) definition of the type of maintenance (periodic, condition-based) and the associated time limits, before which certain action must be taken;

(c) a programme of periodic assessment of actual conditions (service-ability and structural integrity), including models for estimating the degradation until the next inspection;

(d) anticipated repairs as well as their estimated costs and time sche-dule, personnel, material and equipment needed for monitoring, inspection, maintenance and repairs, including the updating of the maintenance plan, as additional information becomes available; and

(e) the reference conditions for the breakwater recorded just after com-pletion of construction as well as during the construction guarantee period.

Monitoring and inspection

Caisson breakwaters are generally designed for a 20- to 100-year life, during which they have to perform all agreed functions satisfactorily without excessive maintenance and repairs. Each of the breakwater com-ponents will not only respond differently to the environment (waves, water levels and currents), but their malfunction may also have comple-tely different effects on the overall stability and serviceability of the struc-ture. Therefore, it is indispensable at the design stage to have an insight into all failure mechanisms, their interaction and relative contributions to overall failure in order to establish an appropriate programme for monitoring and inspection. For this purpose, probabilistic design approaches, as suggested for instance in Oumeraci *et al.* (2001), are well suited. Particularly valuable in this respect are the results of fault-free analysis with the reliability indices obtained for each failure mode.

Instrumented monitoring has basically four different purposes:

(i)   recording of environmental data such as waves and associated water levels and wind conditions, including wave loads on the structural members;

(ii)   recording of the structural response of the caisson and its super-structure, using strain gauges and of the overall behaviour of the structure, using accelerometers, displacements meters, etc.;

(iii) recording of the foundation response, including pore pressure and total stress, vibrations of the structure, total and differential settlements, etc.; and

(iv)   observation of displacement of the stones of the rubble foundation and scour of the seabed, using divers, video records, side-sonar surveys, etc.

It is essential that monitoring starts by establishing reference conditions during and just after construction, as well as during the construction guarantee period. This basic information should be documented by the contractor and delivered to the owner by the end of the guarantee period. It should include all reference measurements of alignments and levels immediately after completion, as built measurements of reinforcement concrete cover, crack and damage maps as well as video records under and above the water level. Photographs of the caissons before submergence are also an important contribution to *in situ* caisson monitoring. This baseline monitoring information should be documented in a way which allows its easy retrieval for comparison in future years of all the features mentioned in Fig. 7.67.

The monitoring and inspection frequency to be adopted will depend on a variety of local and specific conditions such as type, purpose and age of the caisson breakwater, degree of exposure and sensitivity of each structural component, including its contribution to overall failure with respect to both serviceability and structural safety, the design risk level adopted and the planned maintenance costs. However, irrespective of any specific case there are generally two basic complementary approaches to monitoring and inspection which should be used in an appropriate combination:

(a)   *scheduled routine maintenance:* Monitoring and inspection of normal conditions, which are generally associated with the serviceability limit state necessary for updating the maintenance plan established at the design stage and for routine maintenance to be

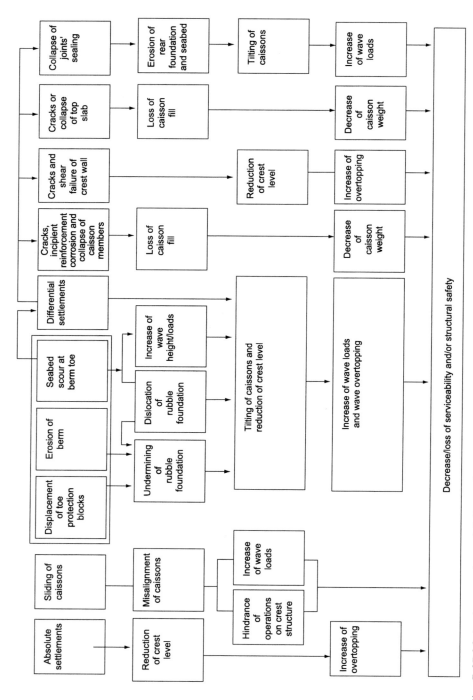

Fig. 7.67 Main damage to and failure of caisson breakwaters relevant for inspection

properly prepared. In addition to almost continuous instrumental monitoring of environmental data and structural responses, which are also used for research and for confirming/updating the initial design method, it is suggested to carry out an annual routine visual inspection. This should be documented by videos and photos of the most vulnerable items listed in Fig. 7.67, particularly those at breakwater heads and corners. This inspection may be performed above water, without divers and underwater equipment. This annual inspection is important because experience has shown that the risk of forgetting inspections is very high if they are not carried out every year. Incipient deterioration can often be detected before the occurrence of the next storm. To maintain a good overview of the general conditions of all breakwater components, and to identity significant damage at an early stage, an inspection of all items in Fig. 7.67 should be performed, say, every five years. Such an inspection may also reveal the necessity for carrying out an extended inspection by using destructive *in situ* tests, such as core drilling, or non-destructive testing, such as cover meter, rebound hammer, vibration analysis as well as underwater videos. The main goal of such extended inspections is generally to establish the real cause and extent of damage in order to evaluate appropriate measures.

A more efficient inspection strategy for preventive maintenance is suggested by Van Hijum (1998). It is based on the following three limits to be defined at the design stage, which reflect the weakening of the structure with time and approaching needed repair work:

(i)   *warning limit* at which more intensive control and higher inspection frequency are required;

(ii) *action limit* at which repairs have to be planned and carried out before failure occurs; and

(iii) *failure limit* at which the structure will fail.

Of course the time between the last two mentioned limits will strongly depend on inspection frequency and the time required for mobilisation and completion of repair works. For details on this strategy see Van Hijum (1998), De Quelerij and Van Hijum (1990) and Simm (1994).

(b)   *unscheduled inspection*: Monitoring after extreme conditions, which are generally associated with the ultimate limit state. Therefore, this task should be carried out just after each storm, which exceeds a certain level specified at the design stage. The same applies if structural damage is suspected after an accidental event such as a vessel impact. The inspection techniques to be used are generally divided into:

(i)   *overall visual inspection* along the breakwater carried out by experienced personnel to record, with the aid of photographs, the overall state of the breakwater including visible displacement, settlement, fractures, concrete spalling, etc.;

(ii)   *close visual inspection* to unveil hidden impairments in areas pre-selected, because of expected potential damage or potential structural relevance for the overall stability. This is done by divers taking high resolution still photographs, especially of cracks and incipient corrosion of reinforcement; and

(iii) *extended instrumented inspection* to determine more precisely the location, severity, dimensions, depth and causes of damage. This may include side-scan sonar recording of seabed scour and berm erosion, non-destructive testing (cover meter, rebound hammer, etc.), as well as destructive testing, e.g. core drilling of concrete. Measurements of the natural period of caisson oscillations may provide valuable information about possible changes of the stiffness of the foundation.

Each inspection report should include the environmental conditions and accidents recorded since the last inspection.

No information on the likely costs of regular monitoring and inspection is available for caisson breakwaters. It is difficult to identify clearly such costs because vessels, equipment and personnel engaged in the survey work are also used for other purposes and the cost sharing between activities is often not accounted for. Experience from many concrete off-shore-platforms has shown that annual costs for regular monitoring and inspection are less than 0.2 per cent of the total construction and installation costs (FIP, 1996). If, for some reasons, the resources for preventive inspection are very scarce, they should be allocated to the most critical areas, which have to be identified at the design stage.

(a)   *Purpose and main types*: Properly designed caissons are generally very durable. They may sustain large local structural damage without jeopardising the overall safety. However, if durability is defined to include serviceability during the design life, the structure and its members should be kept at their original safety levels. Thus maintenance and repairs become indispensable. Otherwise degradation may lead to very expensive repairs or will necessitate complete replacement. Generally one distinguishes between preventive maintenance and corrective repairs. The type, frequency and extent of maintenance work will of course depend on the results of the monitoring and inspection activities as well as on the resulting appraisal of the actual performance of the breakwater in terms of both serviceability and structural integrity.

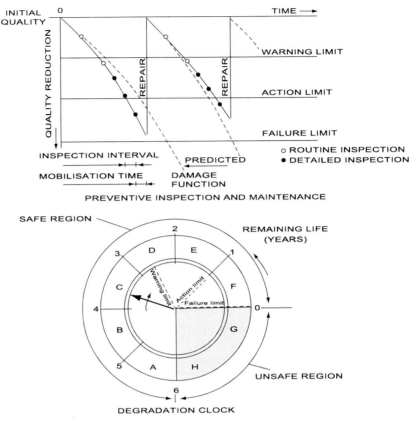

Fig. 7.68 *Maintenance strategy (adapted from De Quelerij and Van Hijum, 1990)*

(b)   *Appraisal of breakwater performance*: Each principal and extended inspection results in determination of the actual degradation level as compared to the degradation limits specified at the design stage (warning limit, action limit and failure limit as previously defined. Ideally, this would result in the assessment of updated safety and serviceability levels as well as an assessment of the remaining life of the breakwater. Given the degradation limits and the observed actual degradation levels and damage, a degradation clock, similar to that in Fig. 7.68 can be devised for each failure mode. For details on this methodology see De Quelerij and Van Hijum (1990), Simm (1994) and Van Hijum (1998). A fault-tree analysis and probability analysis (Oumeraci *et al.*, 2001) may enable assessment of the relative contribution of every failure mode to the overall safety and serviceability and thus make possible a degradation control clock panel for the whole breakwater.

255

(c) *Maintenance strategy*: Based on the performance assessment, a maintenance strategy must be developed which clearly describes the different options and their implications in terms of required equipment, personnel, costs, mobilisation time, etc., as well as the possible consequences of postponing any option. Possible options might be, for example, to await the next inspection before making any decision, to carry out additional, extended monitoring and inspection before undertaking any action, to carry out temporary or emergency repairs and postpone major repairs (in this case possible adverse effects of postponement must be assessed) and to carry out permanent and major repairs to restore the required safety and serviceability levels. Since both safety and serviceability standards change with time and since inspection techniques (by which the actual performance is determined) and prediction models (by which degradation limits are specified at the design stage) improve, updating, which keeps pace with the new developments, is necessary.

(d) *Repair techniques and procedures*: Detailed descriptions of possible repair techniques, procedures and associated equipment must be given in the maintenance plan for each type of repair activity such as:
(i)   filling scour holes with sand-filled geo-textile containers;
(ii) reinforcing rubble foundations, possibly including grouting and placing toe protection blocks;
(iii) sealing of joints between caissons;
(iv) increasing the stability of caissons against sliding, e.g. by adding wave dissipation armour;
(v)   sealing of cracks in the caisson and its superstructure, restoring concrete cover of reinforcement by epoxy, acrylic resin or cement injection; and
(vi) replacing deteriorated concrete by new concrete.

Details on repair procedures and techniques are given for instance in FIP (1996), based on experience from offshore concrete platforms and Mallick *et al.* (1999), who describe the repair works for three commercial harbours in Libya, including caisson breakwaters and quay walls.

References

ACI (1986) *Recommended Practice for Curing Concrete*, (ACI 308-1981), Detroit: American Concrete Institute.

ACI (1995) *Building Codes Requirements for Structural Concrete and Cement*, (ACI 318R-95), Detroit: American Concrete Institute.

Agerschou, H. *et al.* (1983) *Planning and Design of Ports and Marine Terminals*, Chichester: John Wiley.

Ahrens, P. and McCartney, B. L. (1975) 'Wave period effect on the stability of riprap', *Civil Engineering in the Oceans*.

Allen, R. T. L. (ed.) (1998) *Concrete in Coastal Structures*, London: Thomas Telford.

Allsop, N. W. H. and McBride, N. W. (1994) 'Reflections from vertical walls – the potential for improvement in vessel safety and wave disturbance', *Proceedings of the International Workshop on Wave Barriers in Deep Waters*, Yokusuka: PHRI.

ASCE Task Committee (1995) *Wave Forces on Inclined and Vertical Wall Structures*, New York: ASCE Monograph.

Bamforth, P. B. (1993) 'Concrete classification for RC structures exposed to marine and other salt laden environments', *Proceedings of the Conference on Structural Faults and Repair*, Edinburgh, June–July.

Bergmann, H. (2001) 'Hydraulische wirksamkeit und seegangsbelastung senkrechter strukturen mit durchlässiger front', Ph.D. thesis, Mitteilungen Leichtweiss-Institut für wasserbau der Téchnischen Universität Braunschweig, Germany (in German).

Bergmann, H. and Oumeraci, H. (1999) 'Hydraulic Performance of Perforated Structures', *Proceedings of the 5th International Conference*, COPEDEC, Cape Town.

Bergmann, H. and Oumerachi, H. (2000) 'Digue innovante en caissons multichambres', *Revue du Génie Civil*.

Bouchet, R. *et al.* (1994) 'New types of breakwaters – two projects in Monaco', *Proceedings of the International Conference for Hydro-technical Engineering for Port and Harbour Construction*, HYDROPORT '94, Yokusuka.

BSI (1985) *Structural use of Concrete BS 8110, Part 1: Code of Practice*, London: British Standard Institute.

BSI (1987) *Code of Practice for Design of Concrete Structures for Retaining Aqueous Liquids, BS 8007*, London: British Standard Institute.

BSI (1988) *Code of Practice for Maritime Structures, BS 634, Part 2: Design of Quay Walls, Jetties and Dolphins*, London: British Standard Institute.

BSI (1991) *British Standard Code of Practice for Maritime Structures, Part 7: Guide to the Design and Construction of Breakwaters, BS 6349*, (also *Part 1: General Criteria*, 1984), London: British Standard Institute.

Burchart, H. F. (1994) 'The design of breakwaters', in W. A. Abbott and W. A. Price (eds) *Coastal, Estuarial and Harbour Engineer's Reference Book*, London: Chapman & Hall.

CEB (1992) *Durable Concrete and Structures, Design Guides*, London: Thomas Telford.

CERC (1984) *Shore Protection Manual*, 4th edn, Vicksburg, Mississippi, USA.

CIRIA (1991) *Manual on the use of Rock in Coastal and Shoreline Engineering*, CIRIA Special Publication '83, CUR Report 154, London.

CIRIA (1992) *Early-age Thermal Crack Control in Concrete*, Report 91, London: Construction Industry Research and Association.

Christensen, F. T. *et al.* (1995) *Ice Loading and Pile up Against Vertical and Inclined Seawalls*, ASCE Task Committee.

Cochrane, G. H. *et al.* (1979) 'Dubai dry dock – design and construction', *Proceedings of the Institution of Civil Engineers, Part 1*, London, February.

Colleran, R. J. and Leonard, T. (1969) 'Harbour extension in Libya', *The Dock & Harbour Authority*, Vol. 50, London.

Coppons, A. (2000) 'Experimental modelling study of caisson driving into the seabed', *ASCE Proceedings of the International Conference on Coastal Structures '99*, Santander.

Cox, R. J. *et al.* (1998) 'Double walled, low reflection wave barrier', *ASCE Proceedings of the 25th International Conference on Coastal Engineering*, Copenhagen.

Crouch, R. S. (1999a) 'In-service behaviour of cellular reinforced concrete caissons under severe wave impact', *Final Report PROVERBS MAST3 – CT95 0041*, edited by Leichtweiss Institute, Braunschweig.

Crouch, R. S. (1999b) 'Some observations on the durability and repair of concrete structures in a marine environment', *Final Report PROVERBS MAST3 – CT95 0041*, edited by Leichtweiss Institute, Braunschweig.

Daniels, R. J. and Sharp, B. N. (1979) 'Dubai dry dock: Planning, direction and design considerations', *Proceedings of the Institution of Civil Engineers, Part 1*, London, February.

De Gerloni *et al.* (1991) 'The safety of breakwaters against wave overtopping', *Conference on Coastal Structures and Breakwaters '91*, London, November.

De Groot, M.B. *et al.* (1996) *Foundation Design of Caisson Breakwaters*, Vols 1 and 2, Publication No. 198, Oslo: Norwegian Geotechnical Institute.

De Querlerij, L, and Van Hijum, E. (1990) 'Maintenance and monitoring of water retaining structures', in K. W. Pilarczyk (ed.) *Coastal Protection*, Rotterdam: Balkema.

EAU (1996) 'Empfehlungen des Arbeitsausschusses, Ufereinfassungen', *Häfen und Wasserstraßen*, 9th edn (in German), Berlin: Ernst & Sohn.

Eckert, J. W. (1983) 'Design of toe protection for coastal structures', *Proceedings of Coastal Structures '83*, New York: American Society of Civil Engineers.

FIP (1996) *Durability of Concrete Structures in the North Sea. State of the Art Report*, London: Federation Internationale de la precontrainte, Seto.

Franco, C. and Franco, L. (1999) 'Overtopping formulas for caisson breakwaters with non-breaking 3D waves', *Journal of Waterway, Port, Coastal and Ocean Engineering*, ASCE, 125:2.

Fukuda, N. *et al.* (1974) 'Field observations of wave overtopping of wave absorbing revetment', *Coastal Engineering in Japan*.

Goda, Y. (2000) *Random Seas and Design of Maritime Structures*, 2nd edn, World Scientific Co.

Gravesen, H. and Sørensen, T. (1977) 'Stability of rubble mound breakwaters', *Permanent International Association of Navigation Congresses*.

Gravesen, H. *et al.* (1979) 'Stability of rubble mound breakwaters', *Conference on Coastal Structures*, VA, USA.

Hanzawa, T. *et al.* (1996) 'New stability formula for wave dissipating concrete blocks covering horizontally composite breakwaters', *Proceedings of the 25th International Conference on Coastal Engineering*, ASCE.

Hayakawa, T. *et al.* (2000) 'Wave splash height on high-mound composite seawall', *Proceedings of the International Conference on Hydrodynamics, ICHD 2000*, Balkema.

Hudson, R. Y. (1959) 'Laboratory investigation of rubble-mound breakwaters', *Journal of the Waterways and Harbours Division*, New York: ASCE.

Iribarren, R. and Nogales, C. (1954) 'Other verification of the formula for the calculation of breakwater embankment', *Permanent International Association of Navigation Congresses, Bulletin 39*.

Ito, Y. and Tanimoto, K. (1972) 'Meandering damages of composite type breakwaters', PHRI *Technical Note No. 112*, Yokosuka, Japan.

Jensen, O. J. and Klinting, P. (1983) 'Evaluation of scale effects in hydraulic models by analysis of laminar and turbulent flows', *Coastal Engineering 7*, Amsterdam: Elsevier Science.

Jensen, O. J. (1983) 'Breakwater superstructures', *Conference on Coastal Structures*, VA, USA.

Jensen, O. J. (1984) *Monograph on Rubble Mound Breakwaters'*, Danish Hydraulics Institute, November.

Jensen, O. J. (1987) 'Wave overtopping on breakwaters and sea dikes', *2nd International Conference on Coastal and Port Engineering in Developing Countries*, Beijing, September.

Jones, M. N. and Horne, R. D. (1996) 'Slip formed structures – the second century – a review and preview in concrete for infrastructure and utilities', in R. Dhir and N. A. Genderson (eds) *Concrete in the Service of Mankind*, London.

JSCE (1986) *Standard Specifications for Design and Construction of Concrete Structures, Parts 1 (Design) and 2 (Construction)*, Tokyo: Japan Society of Civil Engineers.

Juhl, J. (1994) 'A comparative study on wave forces and overtopping of caisson breakwaters', *Proceedings of the International Conference on Hydro-technical Engineering for Port and Harbour Construction*, Yokusuka: Port and Harbour Research Institute.

Kimura, K. *et al.* (1996) 'New stability formula for wave dissipating concrete blocks covering horizontally composite breakwaters', *ASCE Proceedings of the 25th International Conference on Coastal Engineering*.

Kimura, K. *et al.* (1998) 'Wave force and stability of armour units for composite breakwaters', *ASCE Proceedings of the 26th International Conference on Coastal Engineering*.

Kimura, K. *et al.* (2000) 'A slit caisson with half-circle wave chambers and its mechanism of shock pressure reduction', *ASCE Proceedings of the International Conference on Coastal Structures '99*, Santander.

Kortenhaus, A. and Oumeraci, H. (1999) 'Scale effects in modelling wave impact loading of coastal structures', *Proceedings of the Hydralab Workshop*, Hannover, edited by K. U. Evers, J. Grüne and A. Van Os.

Kortenhaus, A. and Oumeraci, H. (2000) 'Wave loads of horizontal composite breakwaters', *ASCE Proceedings of the 27th International Conference on Coastal Engineering*, Sydney.

Lamberti, A. and Franco, L. (1994) 'Italian experience on upright breakwaters', *Proceedings of the International Workshop on Wave Barriers in Deep Waters*, Yokusuka: Port and Harbour Research Institute.

Larras, J. (1942) 'La déformation ondulatoire des digues verticales', *Travaux no. 108*, June.

Lee, D. S. and Hong, G. P. (1994) 'Korean experience on composite breakwaters', *Proceedings of the International Workshop on Wave Barriers in Deep Waters*, Yokusuka: PHRI.

Llewellyn, T. J. and Murray, W. T. (1980) 'Harbour works at Brighton Marine: Construction', *Proceedings of the Institution of Civil Engineers*, London, Part 1, Vol. 66, 1979 and Part 2, Vol. 68, 1980.

Lundgren, H. (1969) 'Wave shock forces: an analysis of deformations and forces in the wave and in the foundation', *Proceedings Symposium on Research on Wave Action, II*, Paper 4, Delft.

Lundgren, H. (1985) 'Development of caisson breakwater technology', *PIANC Bulletin No. 50*.

Lundgren, H. and Juhl, J. (1995) *Optimization of Caisson Breakwater Design*, New York: ASCE Task Committee.

MacKechnie, J. R. and Alexander, M. G. (1997) 'Exposure of concrete in different marine environments', *ASCE, Journal of Materials Civil Engineer*, 9:1.

Madrigal, B. G. and Valdes, J. M. 'Results on stability tests for the rubble foundation of a composite vertical breakwater', *Final Proceedings of MAST2, MSC-Project on Monolithic Vertical Structures*, edited by Leichtweiss-Institute, Braunschweig.

Mallick, D. V. *et al.* (1999) 'Maintenance strategies of harbour structures', *Proceedings 5th International Conference COPEDEC*, Cape Town.

Martinez, A. *et al.* (1999) 'Structural design of vertical breakwaters. Limitations of Current Practice and Existing Design Codes', *Final Report PROVERBS MAST 3-CT95-0041*, edited by Leichtweiss Institute, Braunschweig.

Mitsuyasu, H. (1966) 'Shock pressure of breaking waves', *ASCE Proceedings of the 10th International Conference on Coastal Engineering*.

Muttray, M. and Oumeraci, H. (2000) 'Wave transformation on the foreshore of coastal structures', *Proceedings of the 27th International Conference on Coastal Engineering*, ASCE, Sydney.

Muttray, M. *et al.* (1998) 'Hydraulic performance of a high mound composite breakwater', *ASCE Proceedings of the 26th International Conference on Coastal Engineering*, Copenhagen.

Muttray, M. *et al.* (1999) 'Wave loads on an innovative high mound composite breakwater', *ASCE Proceedings of the International Conference on Coastal Structures '99*, Balkema.

Nakumura, T. *et al.* (2000) 'Enhancement of wave energy dissipation by a double-curtain-walled breakwater with different drafts', *ASCE Proceedings of the International Conference on Coastal Structures '99*, Santander.

Negro, V. *et al.* (2000) 'Technological innovations in the conceptual design of vertical breakwaters', *ASCE Proceedings of the International Conference on Coastal Structures '99*, Santander.

OCADIJ (1991) *Technical Standards for Port and Harbour Facilities in Japan*, Tokyo: Overseas Coastal Area Development Institute of Japan.

O'Hara, J. F. (1994) 'Construction of maritime works – contractors' methods and equipments', in M. B. Abbott and W. A. Price (eds) *Coastal, Estuarial and Harbour Engineer's Reference Book*, London.

Oumeraci, H. (1994a) 'Review and analysis of vertical breakwater failures – lessons learned', *Coastal Engineering*, Vol. 22:1/2.

Oumeraci, H. (1994b) 'Scour in front of vertical breakwaters – review of problems', *Proceedings of the International Workshop on Wave Barriers in Deep Water*, Yokusuka: PHRI.

Oumeraci, H. (1994c) 'Berm stability and toe protection of caisson breakwaters', *Paper for MAST Advanced Study Course on Probabilistic Approach to the Design of Reliable Coastal Structures*, Bologna.

Oumeraci, H. (2000) 'The sustainability challenge in coastal engineering', *Keynote Address in IAHR Proceedings of the 4th International Conference on Hydrodynamics, Yokohama*.

Oumeraci, H. and Kortenhaus, A. (1994) 'Analysis of dynamic response of caisson breakwaters', Elsevier Science Publishers B.V., *Coastal Engineering, Special Issue on 'Vertical Breakwaters'*, Vol. 22, no's 1/2.

Oumeraci, H. *et al.* (1993) 'Classification of breaking wave loads on vertical structures', *Journal of Waterways, Ports, Coastal and Ocean Engineering*, ASCE, Vol. 119:WW4.

Oumeraci, H. *et al.* (2000) 'Wave loading of a high mound composite breakwater (HMCB) with splash reducers', *IAHR Proceedings of the International Conference of Hydrodynamics*, Balkema.

Oumeraci, H. *et al.* (2001) *Probabilistic Design Tools for Vertical Breakwaters*, Balkema.

Peregrine, D. H. and Kalliadasis, S. (1996) 'Filling flows, cliff erosion and cleaning flow', *Journal of Fluid Mechanics*, Vol. 310.

PIANC (1976) *Final Report of the International Commission for the Study of Waves*, Supp. to PIANC Bulletin 25:III.

PIANC (1980) *Final Report of the 3rd International Commission for the Study of Waves*, Supp. to PIANC Bulletin 36:II

PIANC (1990) *Inspection, Maintenance and Repair of Maritime Structures Exposed to Material Degradation Caused by a Salt-water Environment*, PIANC.

PIANC (2001) *Breakwaters with Vertical and Inclined Concrete Walls*, PTC II WG 28, Final Report.

Pilarczyk, K. W. (1998) 'Composite Breakwaters Utilizing Geosystems', *ICE Proceedings of the International Conference on Coastlines, Structures and Breakwaters*, London: Thomas Telford.

Simm, J. D. (1994) 'Maintenance of coastal structures', in M. B. Abbott and W. A. Price (eds) *Coastal Estuarial and Harbour Engineer's Reference Book*, London.

Stephenson, D. (1978) 'Rockfill in hydraulic engineering', *Developments in Geotechnical Engineering 27*, Amsterdam: Elsevier.

Svee, R. (1962) 'Formulas for design of rubble-mound breakwaters', *Journal of the Waterways and Harbours Division*, ASCE, May.

Takahashi, S. (1996) *Design of Vertical Breakwaters*, Reference Doc. No. 34, Yokusuka: Port and Harbour Research Institute.

Tanimoto, K. and Takahashi, S. (1994) 'Design and construction of caisson breakwaters – the Japanese experience', *Coastal Engineering*, 22.

Tanimoto, K. *et al.* (1987) 'Structures and hydraulic characteristics of breakwaters – the state of the art of breakwater design in Japan', *Report of Port and Harbour Research Institute*, Vol. 26:5.

Tanimoto, K. *et al.* (1982) 'Irregular wave tests for composite breakwater foundation', *ASCE Proceedings of the 18th International Conference on Coastal Engineering*, Cape Town.

Terrett, F. L. *et al.* (1979) 'Harbour works at Brighton Marina: Investigations and design', *Proceedings of the Institute of Civil Engineers*, Vol. 66:1 (incl. discussions in Vol. 68:1, 1980).

Terzaghi, K. and Peck, R. B. (1967) *Soil Mechanics in Engineering Practice*, New York and London: John Wiley.

Thompson, D. M. and Shuttler, R. M. (1976) *Design of riprap slope protection against wind waves*, CIRIA Report No. 61.

US Army Coastal Engineering Research Center (1973) *Shore Protection Manual*, Fort Belvoir, VA, USA.

US Army Coastal Engineering Research Center (1977) *Shore Protection Manual*, Fort Belvoir, VA, USA.

Van der Meer, J. W. (1993) *Conceptual Design of Rubble Mound Breakwaters*, Delft Hydraulics, Publication No. 483:December.

Van Hijum, E. (1998) 'Aspects of execution and management', in K. W. Pilarczyk (ed.) *Dikes and Revetments*, Balkema.

Vrouwenvelder, A. and Bielecki, M. (1999) 'Caisson reliability during transport and placing', *Final Report PROVERBS MAST3-CT95-0041*, edited by Leichtweiss Institute, Braunschweig.

Zen, K. and Yamazaki, H, (1991) 'Field observations and analysis of wave-induced liquefaction in seabed', *Soils and Foundations*, Vol. 31:4.

Zwamborn, J. A. (1979) 'Analysis of causes of damage to Sines breakwater', *Conference on Coastal Structures*, VA, USA.

CHAPTER 8

Berth and terminal design in general. Storage facilities and cargo-handling systems

Harry Ghoos, Jens Korsgaard, Leif Runge-Schmidt and Hans Agerschou

Introduction

This chapter deals with the functional requirements of ships, cargo handling and storage facilities.

The choice of cargo-handling equipment will influence the layout and design of the terminal. Weight, load distribution, size and manoeuvrability of cargo-handling equipment may influence areas and layouts as well as design of structures, foundations and pavements.

Similarly, existing fixed facilities will influence the choice of cargo-handling equipment, while cargo units influence the choice of both cargo-handling equipment and of storage facilities.

Unless the terminal is designed for only one specific usage, the multi-purpose character of the terminal should prevail, not only in layout and design of fixed facilities but also in the choice of equipment. Storage facilities and cargo-handling systems should be as adaptable as possible. Purchases of specialised equipment should only be made when forecast future throughputs clearly justify the acquisition.

With a definite trend toward larger homogeneous lots of general cargo (break bulk) and more importantly with the growth of container through-put, the requirements have become wider quay aprons, larger transit sheds and open storage areas as well as replacement of transit sheds by open storage.

Terminals at the Churchill dock (7th harbour dock) in Antwerp, built in the late 60s, had a total land area of 6 ha. Antwerp's Delwaide dock terminals (9th harbour dock), built 10 years later, were allocated an area of 30 ha. Both of these terminals have a quay length of 600 m. This means that the depth of the terminals was increased from 100 to 500 m.

While earlier the basic layout of a terminal was rectangular and often quite narrow along the quay, with time the ideal shape was considered to be close to square. Newer terminals are again becoming rectangular, but with the quay as the short side. With forklifts of different sizes and capacities, and with better horizontal transport equipment, the distance between the quay crane and the storage site is not as important anymore.

The quality of the ships' gear has improved quite a lot. The old-fashioned derricks and winches have been replaced by rotating derricks, or more often by electric or hydraulic cranes. Cargo is picked up and landed by ships' gear and by quay cranes at a number of varying locations along ships of different length, to and from which the cargo has to be moved horizontally. A general cargo berth is, therefore, normally required to be one continuous land-connected structure for berthing, mooring and cargo handling.

At the other end of the spectrum are tanker terminals. Cargo handling usually takes place only through one mid-ship manifold. Thus, only one loading platform (this terminology is common, even if the terminal is used exclusively for unloading) of modest size is required for loading arms and hose-handling equipment. On each side of the loading platform are separate berthing and mooring platforms connected only by catwalks or other light structures. Storage tanks need not be located close to the berth as the cargo is transported inexpensively by pipeline. Thus, tanker terminals need not be close to land areas and need not be connected to land areas other than by one or more pipelines, which may be resting on the sea bottom or buried in it.

For dry-bulk terminals the requirements are that loading/unloading has to take place to and from a number of hatches along the ship. Thus more extensive loading platforms than for liquid bulk are required for the cargo-handling installations, which have to cover a good part of the length of the ship. Dry bulk is often handled by permanent conveyer systems. Storage facilities should be relatively close to the ship although the cost penalty for having the storage relatively far from the ship, say up to 1 km, is modest.

Container terminals require large areas for stacking of containers. The sheds for stuffing and un-stuffing of containers (container freight stations) should be situated behind the stacking yards. Roll-on/roll-off (ro/ro) requirements are again different. Most ro/ro vessels are fitted with quarter ramps that do not require any special berthing facility, except clear landing space on the apron for the ramp. Ships with straight stern or bow ramps need purpose-built berthing facilities. Floating platforms offer flexibility, both in regard to berth allocation and tidal variations.

Storage and equipment requirements are difficult to determine for general cargo facilities, the functioning of which cannot normally be computer simulated due to the lack of availability of the multitude of required input data.

Surface elevations for berth and storage areas are often difficult to establish. In principle, surface elevations should be determined, based on known or forecast high-water probabilities, by a trade-off analysis between, on one hand, the economic costs of increasing the surface elevation and, on the other hand, the economic value of reduced interruption of operations, and of reduced damage to cargo and equipment.

Flooding of general cargo and container berths interrupts operations, damages cargo, except perhaps containers stored on semi-trailers (which is not common anymore), and damages cargo-handling equipment such as quay cranes, which cannot be moved to higher ground before extreme high water occurs.

For bulk terminals the situation would normally be somewhat different. Bulk cargo may not be damaged by flooding and/or storage facilities may not be exposed to flooding. Loading, berthing and mooring platforms and their equipment may not be damaged by flooding, and operations may thus continue if started before flooding occurs. However, berthing of bulk carriers and start of cargo handling may not be possible after flooding has started. A trade-off analysis would thus only involve delay and possibly interruption of cargo handling.

General cargo berths

Berth surface elevations

The minimum surface elevation should correspond to a combination of high water and wave action, which is forecast to occur very infrequently, certainly not even once a year. For operational reasons a higher surface elevation, around 2 m above MSL is often chosen. Where the tidal range does not exceed approximately 2 m it is possible to use ship's gear during the full tidal cycle, except for very small ships. Where quay cranes are used a somewhat larger tidal range may be tolerated.

Apron widths

The apron width is the distance from the quay face to the transit shed or to the open storage area. In the past, cargo was commonly handled directly between ships and multi-level storage buildings, without being put down on or picked up from the apron. Therefore, it was reasonable to regard the apron as a combination of road and railway, with space for quay cranes on rails, for direct access alongside ships. Thus an apron width of about 13 m was considered desirable for one rail track on the apron (5.5 m for a roadway, 4.5 m for rail track, twice 0.75 m for crane legs, and 1.5 m from quay face to crane leg) and corresponding additional width for more than one rail track (Fugl-Meyer, 1957). This is a clear-cut rational approach. However, it is no longer valid.

The quay apron is no longer an alongside road–railway combination. It has become more of a short and wide road between ship and transit storage, serving criss-cross traffic of forklifts, tractor–trailer units, etc., as well as a buffer storage area for cargo. Railway traffic on the apron is no longer compatible with the across-the-apron cargo handling between ship and shed or open storage. Usually direct deliveries to and from trucks on the apron, although saving handling costs, negatively influence overall handling productivity and thus increase the ships' turnaround time. The latter is the highest value item involved in the loading/unloading process. With increasing use of pallets for general cargo the transfer from transit shed and open storage to onward vehicle has become less expensive.

How is the apron width determined in a rational manner for this case? There is no established approach, however, the governing considerations are that the distance between the hook of the ship's gear or the quay crane in its extreme shoreward position and the transit shed should be as short as possible, subject to the following constraints:

(a) there should be space for buffer storage of cargo;
(b) there should be space for lateral transport of cargo to and from open storage areas located behind and between transit sheds; and
(c) there should be space for access of trucks for direct unloading of inbound cargo.

In current practice, 20 m is considered the minimum acceptable apron width, and 40 m is normally considered the maximum desirable width. Apron widths within this range usually give the required flexibility for cargo-handling operations. The apron width may also be influenced by the optimum location of transit sheds and of stored cargo, in regard to their effects on the stability of the quay structure.

Total widths of land area behind quay face

With an apron width of 20 to 40 m, a transit shed or open storage area width of 60 m, and a 10- to 20 m-wide road and/or rail access area behind the storage facilities, the minimum total width of the land area behind the quay face is going to be 90 to 120 m.

When there is a need for warehouses, open storage or marshalling yards behind the transit storage facilities, the total width requirement increases considerably. Land area widths may be subject to physical and cost constraints such as basin width requirements, water depths, availability of fill, and topography (cliffs and mountain slopes).

Transit sheds

Transit storage area requirements in general are discussed in Chapter 2. The average weight of cargo stored per m^2 of the total storage area in question depends upon:

(a)   the stowage factor (m^3/tonne) for different commodities and cargo units;

(b)   the average stacking height, which may be limited by foundation conditions and by the lifting height of cargo-handling equipment; and

(c)   floor space required for cargo handling by appropriate equipment as well as for cargo access.

Stowage factors for a number of commodities and cargo units are available in Thomas (1983).

Clearly, large area requirement differences exist between, for example, a large consignment of exclusively inbound or outbound, bagged cargo and many small consignments of both inbound and outbound mixed general-cargo units. The latter may only be stacked 1 to 3 m high and require relatively large areas for equipment operation and cargo access, whereas the former may be stacked much higher and requires only small areas for equipment operation and access. Nevertheless, the literature contains suggested standard area correction factors, e.g. 1.4, to account for floor space required for equipment operation and cargo access. Similar rule of thumb correction factors from 1.2 to 1.5 are quoted for above-average cargo throughput combined with below-average stacking height. This interval is too narrow to be realistic and, if expanded by raising its upper limit considerably, only demonstrates the need for a case by case approach.

The length and width of sheds corresponding to the required area depend upon berth lengths, space for open storage, operation of cargo-handling equipment and trucks, and occasionally on the need for rail tracks between sheds.

General structural design of sheds is not within the scope of this book. We do think, however, that the following particular aspects need to be dealt with. It is not an indispensable rule that a single-span structure is required. One or more rows of interior columns may not cause serious problems. However, an area around each column cannot be used for storage. It is necessary to prevent structural damage to all columns. This may be done by raising and enlarging the concrete footings for the columns or by placing other permanent obstructions which will prevent cargo-handling equipment from colliding with columns. The choice between single-span or interior-column structures is made based on a cost comparison which will take into account the somewhat larger total area required for the latter.

It is rarely, if ever, economical to construct transit shed walls, such that cargo may be stowed against the walls. Further, access problems would

267

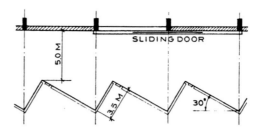

Fig. 8.1 Truck platform layout

result. Nevertheless, it is not uncommon to observe that cargo is stowed against shed walls which are not designed for the resulting lateral pressure. This cannot be prevented by physical obstructions, which would impede cargo handling along the walls of the shed, but has to be eliminated by supervision and strictly enforced disciplinary measures.

The number of lift or roll gates on each side of the shed depends upon the number of ships' holds and the distance between the centre lines of holds, typically 20 to 30 m. The gate width would usually be 5 to 6 m, to allow simultaneous passage of two forklift trucks. The height of the gates should not be less than 5 m, to allow forklift truck access.

Shed lighting should be both natural (transparent roof panels) and artificial. The recommended light intensity is normally 100 lux ±25 per cent.

If forklifts are used, which is usually the case, a platform along the back of the shed is not needed. If forklifts are not used, which is hardly ever the case anymore, there should be a platform which corresponds in height to the bed of locally used railway wagons and/or trucks. A straight platform running along the shed is needed for railway wagons. For truck platforms a saw-tooth arrangement as shown in Fig. 8.1 is recommended. This makes possible loading/unloading from the back of the truck as well as from its left side. The reason for the left-side access is that there is much better visibility for the driver while he backs his truck into the loading bay. In countries where the driver's seat is on the right side, the platform should be modified accordingly. While railway wagon beds usually are of uniform height this is not the case for trucks. The truck-bed height variation should be carefully investigated and provisions made for simple height adjustment if necessary (see Fig. 8.2).

Open transit storage

Open storage should be provided for cargo that will not be damaged by rain, sun, snow, ice and open-air temperature variations as applicable, and which is not subject to pilferage. Open storage is commonly used for vehicles, logs, structural and reinforcing steel, drums and low-value metal

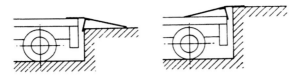

Fig. 8.2 Platform height adjustment

ingots. Area requirements are determined as for transit sheds. More flexibility exists in regard to location and shape for open storage areas than for transit sheds. Open transit storage is often placed between and behind transit sheds. Lighting which makes night work possible is indispensable. The lighting fixtures should be fastened to the transit sheds and/or to relatively few tall masts. The lighting fixtures on the transit sheds should be protected well against damage by cargo-handling equipment and the lower parts of masts should also be protected against structural damage. The light intensity requirements are the same as for transit sheds.

Pavements
Pavements for quay aprons, transit sheds and open storage areas do not present unique problems in regard to their structural design, which is not within the scope of this book.

Very even pavements are required where forklifts and other small-wheel equipment are used. Both the high front-wheel loads for forklifts, as well as the grinding effect of their turning rear wheels when the forklift is empty and the rear wheels carry the full load of the counterweight, are hard on pavements. Pavements in ports are exposed to petroleum products and other chemicals leaking from cargo-handling equipment and from damaged drums and bags. The chemicals may soften bituminous pavements and sometimes lead to their complete disintegration. Concrete pavements are more resistant to most chemicals and may be relatively easily cleaned. However, they have other disadvantages, the most important of which is that they are prone to crack when differential settlements occur, making it difficult to operate small-wheel equipment. Bituminous pavements are easier and less costly to repair.

Nowadays most terminals use relatively small concrete blocks as pavements. Depending on local conditions, the costs of the stone, gravel and sand layers required under the concrete blocks, may vary considerably. Labour-intensive block laying may be used in developing countries, whereas machines, that will place one m^2 at a time, are often used in developed countries to reduce costs. Maintenance of concrete-block surfaces, although continuous, does not require any special machinery nor outside contractors, as is the case for bituminous and concrete surfaces. Not only general cargo terminals but also most container terminals are adopting concrete-block pavements.

Cargo-handling systems

(a)   *General considerations*: Cargo-handling systems requirements in general are discussed in Chapter 2. The required number of equipment units of each category, their lift capacity, reach, and carrying capacity as applicable, depend in addition to realistic cargo-handling rate forecasts upon:

   (i)   the ships' forecast of turnaround time;

   (ii)   the statistical distribution of the total number of ship's hatches to be worked simultaneously;

   (iii)   the distribution of cargo handling between direct loading/ unloading, to and from transit shed and to and from open transit storage;

   (iv)   distributions of cargo units and of their weights;

   (v)   location of transit storage facilities;

   (vi)   storage stacking heights;

   (vii)   other equipment categories in the port, which may be used instead of the category in question, the number of units and expected availability;

   (viii) the expected availability for the equipment category in question;

   (ix)   the number of working hours per 24-hour day; and

   (x)   the labour situation in regard to unemployment, productivity, discipline and quality of supervision.

Theoretically, the handling rate of a crane or other vertical and/or horizontal cargo-handling device, which functions in cycles, may be determined based on:

   (i)   its lift capacity or carrying capacity as applicable;

   (ii)   lifting height, speed and acceleration or average speed over lifting height;

   (iii)   same as (ii) for lowering;

   (iv)   horizontal travelling (translation and/or rotation) distance, speed and acceleration; and

   (v)   (ii), (iii) and (iv) without load.

However, this is only theory. It does not account for:

   (i)   the average load, which depends upon the cargo units;

   (ii)   time required for loading/unloading of hook, forklift, trailer, etc.;

   (iii)   interruption of cargo handling because of weather, breakdown, etc.;

   (iv)   competence and condition of operator(s) (whether well rested or tired, etc.);

(v)   average availability of device (time not under repair or waiting for repair work); and

(vi) average lift, lowering and travel accelerations and speeds, which can actually be sustained over the duration of the shift, rather than those, claimed by manufacturers.

It is hardly possible to estimate very accurately the effects of all the above six factors, in particular when planning a new port. In an existing port proper collection and processing of relevant data enable determination of average cargo-handling rates for each category of equipment and each operation. Also, in an existing port, data should be available concerning most of the above six factors. This experience is clearly useful for estimating handling rates for new types of equipment and new operations.

There is no doubt that precisely the same equipment, equally well installed, will achieve widely varying handling rates which do not depend upon the characteristics of the equipment, but depend mainly upon the mix of cargo units as well as equipment operation and maintenance.

The optimum combination of categories, numbers and characteristics of equipment (cranes, forklifts, tractors and trailers, etc.) may, in principle, be determined based on time-and-motion studies followed by simulation of the entire port operation. It would be necessary to include labour performance, realistic estimates of feasible future productivity improvements and for each equipment item, its lift capacity, reach, carrying capacity, etc. To our knowledge, such has never been done successfully.

Systems should not be planned in such a manner that one component has to wait for another, e.g. ship's gear should not load cargo directly from forklifts. The cargo should be put down on the apron and picked up from there. Such temporary buffer storage saves equipment time and more important ships' alongside time. Equipment versatility is another important consideration. A larger number of multipurpose equipment units is often a better solution than much smaller numbers of several different specialised types of equipment, when capacity, required availability, capital costs, operation and maintenance are considered. The quality and price of cargo-handling equipment vary among manufacturers and from time to time. An equally important consideration is local availability of factory-organised service and supply of spare parts, which varies from country to country.

(b) *Handling between ship and apron*: Ships may be loaded/unloaded by means of portal cranes on rails, ships' gear or mobile cranes on rubber wheels. For decades port operators have debated the relative advantages and disadvantages associated with the use of portal cranes and of ships' gear. Considering recent and continuing radical improvement of ships' gear, which on relatively new ships consists of electric or hydraulic cranes, investments in quay cranes, particularly in countries where electric power supply is not always reliable let alone the quality of crane maintenance, are not advisable. Cargo-handling productivity rates vary widely from less than 10 tonnes per crane hour for mixed general cargo to more than 30 tonnes per crane hour for large consignments of pre-slung bagged cargo or other similar unitised cargo. Ships' gear supplemented by mobile cranes on rubber wheels probably constitutes the optimum solution except for extreme low-water conditions, which would cause the deck level to be considerably below the quay surface. While cranes on rails may only be used for loading/unloading of ships, mobile cranes on rubber wheels may be used to lift cargo wherever needed, inside or outside the port limits.

The operator of an ordinary mobile crane cannot see the inside of the ship's holds and has to rely upon a signalman on deck. Several manufacturers have developed mobile tower cranes with the operator's cab positioned sufficiently high up to give him a good view of the hold. However, it remains to be seen, whether real operational advantages resulting from the use of complex mobile tower cranes compensate for their high capital and maintenance costs.

(c) *Handling between apron and transit storage*: The choice of equipment for this purpose depends largely upon cargo units, distance, stacking height and consideration of equipment-intensive versus labour-intensive handling. If the distance does not exceed 100 m, forklifts are normally preferred, unless more labour-intensive cargo handling is demonstrated to be more economical. For bagged cargo manually loaded/unloaded tractor-drawn trailers are often economical, where labour costs are low, perhaps combined with the use of bag stackers (mobile conveyors) and gravity rollers.

Where forklifts are used primarily, the number required per gang is typically three or four assuming an average availability of 75 per cent.

Tractor-drawn trailers are to be preferred over longer distances to and from transit storage yards. At least three trailers are needed for each tractor to make waiting time for the latter insignificant while trailers are being loaded. One tractor and accordingly three or four trailers are required per gang, with reserve tractor units corresponding to an average availability of about 75 per cent. Tractor-drawn trailers are also used for transport of cargo from transit storage to longer-term storage within port limits. Lately the use of so-called fifth-wheel tractors has gradually taken the place of ordinary or so-called agricultural tractors. Although more expensive they are more versatile, especially for longer distances within the port area. Also the coupling and uncoupling of the plate into the kingpin of the trailer is faster and safer. When productivity and labour costs are important issues, it is worth considering the additional investment in fifth-wheel tractors.

(d) *Handling inside transit storage areas*: Forklifts, bag stackers, gravity rollers and manual labour are predominantly used in transit sheds. In transit storage yards mobile cranes and forklifts are primarily used. Yard cranes on rails cannot be recommended except for well-defined purposes, because of their high costs and limited yard coverage.

(e) *Equipment characteristics*: Table 8.1 shows characteristics, advantages and disadvantages of the common equipment categories suitable for handling and transport of general cargo. The forklift trucks are off-the-shelf items for which attachments are available for

Table 8.1 General-cargo handling equipment

Type	General characteristics	Advantages	Disadvantages
Forklift trucks	Capacity: 2–45 tonnes Lift height: 2.5–5 m	Suitable for pickup, transport over short distances, loading and stacking	High loads on front wheels
	Power: petrol diesel electric	Acceleration Long engine life No air pollution Preferred for work in ship's hold	Carbon monoxide exhaust fumes Charging of batteries, after 5 to 6 hours use, takes 12 hours
Mobile cranes on rubber wheels	Capacity: 2–40 tonnes	Versatile: may be used where needed and for all kinds of general cargo	Only for stationary lifts, not for transport of cargo. Stabilising jacks normally used during lifts
			Operators cab too low for him to look down into ship's hold. He must rely upon signalman on deck of ship
Tractors and trailers	Horse power: 50–100 Capacity: 10–20 tonnes	Inexpensive and relatively easy to maintain	Only for horizontal transport. Have to be supplemented by lift equipment

handling of bales, cartons, drums, wire and cable coils, paper in rolls, etc. Mobile cranes on rubber wheels are also mass-produced items, except those with a very high-lift capacity and/or long reach. Common or modified agricultural tractors or fifth-wheel tractors should be used (see above). Trailers should have a low bed and wide wheels.

Container terminals

General considerations

Containerisation was introduced in the US during the 1950s. Originally only for military use, its advantages were seen by the Sealand organisation, which first put it to use in domestic transport.

Apart from its vast road tractor and trailer fleet, Sealand not only owned containers but also owned and operated container vessels, as well as container terminals. Due to the rapid development in container transport, the principle of door-to-door service gained momentum and is now widely accepted. In the initial stages, the shipping lines also invested in the onward land transport facilities. Later they realised that such investments were better left to specialised operators as sub-contractors for the shipping lines. The shipping line, in most cases, remained the one organisation, responsible for the door-to-door transport. The cargo was transported under one through bill of lading, issued by the shipping line.

Container terminals are, in principle, rapid transit facilities at the interface between land and sea transport. Soon after arriving in the terminal, inbound containers ought to continue their inland journey to consignees. Nearly all of the inbound containers, at least 90 per cent, should move on within one to three days. Customs processing of the sealed containers should take place on the premises of consignees or at inland freight stations, situated relatively close to the consignees.

Outbound containers should arrive at the container terminals as fully loaded containers (FLC), not more than 5 to 7 days before the expected arrival of the vessel. Most container terminals will not accept containers more than one week in advance of the expected arrival of the vessel, at which time the stack(s) allocated for that particular vessel are declared open.

Unfortunately, procedures in many countries still lack behind the excellent intentions in regard to reduced transit storage time and improved safety of cargo and operations. Customs processing of containers, like any other cargo, is still often required to take place within the terminal, where the container has been landed. Inland freight stations for stuffing and un-stuffing containers often do not exist and customs processing at the business or factory premises, at consignees or consignors, was not foreseen in customs regulations. Customs processing will therefore have to take place in the port area if not at the container terminal itself. Such antiquated arrangements prolong the time containers spend in the terminal, while stuffing and un-stuffing inside the port goes against the door-to-door principle. As a result, the container throughput capacity of the terminal, which is inversely proportional to the average storage time for containers (dwell time), is greatly reduced.

The uniform and modular cargo consist predominantly of 20 and 40 ft containers, but also some 45 ft containers, all of which are 8 ft wide. This together with the almost identical cranes used for handling between ship and land side, the relatively few equipment items used for transport between apron and storage yard, for stacking, delivery and receiving of containers in the yards, does make container terminals suitable for systems analysis and computer simulation.

Container throughput is commonly measured in 20-ft equivalent units (TEUs). One 40-ft container is equivalent to two TEUs, etc. The standard heights are 8.0 and 8.5 ft. Today, so-called 'high-cube' containers, especially of 40- and 45-ft lengths with heights of 9 or 9.5 ft, are very common in the Far East, US and European trades. Table 8.2 shows container characteristics. The current maximum total weight of a 20 ft by 8.5 ft container is normally 20.3 tonnes, but for an extra strengthened container it is 30.5 t. For a 40-ft container it is 32.5 t. ISO International container standards are available in Jane's Freight Containers (Yearbook) and other publications.

When the cargo-stowage factor measured in m^3 per metric tonne exceeds 2.3 to 2.8 for 40 ft containers and 1.7 to 1.8 for 20 ft containers (with one exception of 1.2 for the extra strengthened container), the cargo volume capacity becomes governing. The cargo weight is thus reduced below its maximum value. Many commodities have stowage factors which considerably exceed the above break-even values, and a large percentage of the container throughput for a terminal may have weights which do not exceed half of their maximum total weight. In general container-handling rates are on a unit basis. Thus weight does not matter to the terminal operator, as long as it does not exceed the capacity of his equipment.

Apron elevation and width

The apron elevation should, in principle, be the same as for general cargo berths (see 'Berth surface elevations', earlier in this chapter). Container terminals, like general cargo berths, normally have continuous aprons alongside the ships, although this is not strictly necessary in principle. If the apron is immediately adjacent to the ship, the apron width is governed by the distance between the legs of the gantry crane, which is typically 20 to 30 m for newer cranes and about 10 to 15 m for older cranes, plus the

Table 8.2 Container characteristics

Length (ft) and (m)	Width (ft) and (m)		Height (ft) and (m)		Material	Weight of container (t)	Maximum cargo weight (t)	Maximum total weight (t)	Inside volume (m³)
20	6.10	8	2.44	8.5	2.59	Aluminium	1.9	18.4	20.3	33.0
20	6.10	8	2.44	8	2.44	Steel	2.0	18.3	20.3	31.0
20	6.10	8	2.44	8.5	2.59	Steel	2.2	18.1	20.3	33.0
20	6.10	8	2.44	8.5	2.59	Steel	2.3	28.2	30.5	33.0
40	12.19	8	2.44	8	2.44	Aluminium	2.8	27.7	30.5	63.3
40	12.19	8	2.44	8.5	2.59	Aluminium	3.4	27.1	30.5	67.0
40	12.19	8	2.44	9.5	2.90	Aluminium	3.9	26.6	30.5	75.0
40	12.19	8	2.44	8	2.44	Steel	3.4	27.1	30.5	63.0
40	12.19	8	2.44	8.5	2.59	Steel	3.6	26.9	30.5	67.0
40	12.19	8	2.44	8.5	2.59		3.8	28.7	32.5	67.0
40	12.19	8	2.44	9.5	2.90		3.9	28.6	32.5	75.0

reach of the crane behind its inside rail, which is normally 10 to 20 m.

In order to limit their draft, container vessels tend to have a relatively wide beam. The consequence for gantry cranes is the need for a longer reach. This in turn results in a wider span and a longer back reach. Horizontal transport should preferably go under the crane's span in order to reduce the distance of travel of the trolley. The area under the back reach is commonly used for temporary storage of containers that need to be 'shifted' (to give access to other containers) and will be reloaded onboard the ship. Terminal operators should avoid transporting such containers to the stacks as it may lead to errors when re-loading. When under the back reach of the crane they remain there to remind the terminal operator to re-load them.

Container-yard equipment
One of the most important questions in regard to container terminals is how large a yard area is required to accommodate a certain known or forecast throughput of inbound and outbound TEUs. The answer is intimately related to the equipment used for transport to and from the yard, for stacking and for delivery and receiving on the landside. The latter is, in the following, called yard equipment.

The following yard-equipment systems predominate at present in large container terminals:

(a) tractor–trailers: containers are placed on or picked up from trailers by quay cranes. The trailers are pulled to and from the yard by tractors. Only very specialised operators, mainly in the short sea and ro/ro trade, will keep the containers on road trailers;

(b) straddle carriers, which stack containers two or three high and also remove them from the stacks. They may also be used for transport between the apron and the yard. Figure 8.3 shows a typical straddle carrier;

(c) yard gantry cranes on rubber wheels (RTGs), normally combined with tractor–trailer transport between apron and yard. Containers are stacked three to five high. Figure 8.4 shows a typical RTG;

(d) yard gantry cranes on rails, also combined with tractor–trailer transport. Containers are stacked four to five high;

(e) large-span gantry cranes on rails for direct handling between ship and yard; and

(f) mixed yard-gantry crane and straddle carrier systems.

Table 8.3 summarises characteristics, relative advantages and disadvantages of the systems.

With front-end loaders (forklifts equipped with spreaders) or side loaders, both of which are normally only used in smaller terminals,

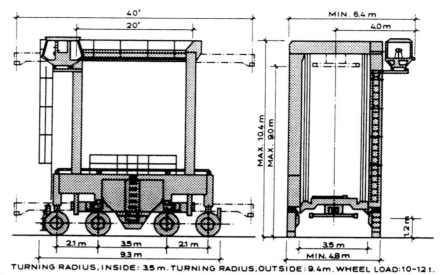

TURNING RADIUS, INSIDE: 3.5 m. TURNING RADIUS, OUTSIDE: 9.4m. WHEEL LOAD: 10-12 t.

Fig. 8.3 Typical straddle carrier

containers may be stacked two or three high and two wide, as there is only access from the sides of the containers. The space required between the rows for operation of equipment is considerably wider for front-end loaders than for side loaders because the former carry the containers perpendicular to their direction of travel and have to make 90° turns for loading or unloading.

In recent years several manufacturers have introduced the so-called reach stacker. This machine combines the fast operation of the forklift

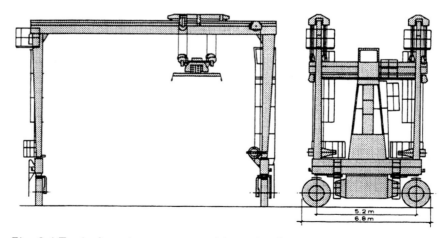

Fig. 8.4 Typical gantry crane on rubber wheels

Table 8.3 Container-yard equipment characteristics

Type	Stacking height and width measured in containers	Operational considerations	Costs and useful life	Flexibility in use	Availability
Gantry cranes on rubber tyre wheels	Three to five high. Six wide	Average number of movements per container, for access to delivery, increases with stacking height. Good safety record	Relatively low maintenance costs. Relatively long life	Run on fixed concrete beams. Not (rapidly) moved to other beams	Relatively good, with conscientious maintenance
Gantry cranes on rails	Four to five high. Six wide	See above. Highest possible accuracy for horizontal positioning	See above	Run only on their two rails. Cannot be used elsewhere	See above
Large-span gantry cranes, with overhang on rails for automatic terminal	Five to six high. Twenty or more wide	Quay-side overhang used for placing on conveyor for loading by quay crane or picking up from conveyor after unloading by quay crane. Landside overhang used for receiving from and delivering to accurately positioned trucks	See above	See above	See above
Straddle carriers	Two to three high. One wide	Poor visibility for operator. Accident prone. Poor yard area utilisation	Relatively high maintenance costs. Relatively short life	May be used anywhere, also for transport between yard and apron	Relatively poor
Forklifts	Three to seven high. Two wide*	Large space required between narrow stacks for 90° turn. Very poor yard area utilisation. Very high wheel loads	Relatively low maintenance costs. Long life	See above	Good, with conscientious maintenance
Reach stackers	Four to seven high. Four wide*	Limited visibility for operator. Accident prone		May be used anywhere, but not suitable for transport between yard and apron. Able to rotate 90 degrees	

*For direct access.

279

with the greater reach of the mobile crane; it has become the preferred type of equipment for railway operations. Because of its reach it can be used for combined vertical and horizontal transport, in most cases it will be more economical to use the tractor–trailer combination for horizontal transport. When straddle carriers are used, it is possible to stack three high in single rows, spaced 1.5 to 2.0 m. When RTGs are used, the stacks may be five high and are normally not wider than six containers. The RTG span in addition to six containers should also allow for a truck and tractor–trailer lane for receiving and delivery. Gantry cranes on rails may allow five-high stacks fifteen or more containers wide.

The choice of equipment is related to whether the terminal, at one end of the spectrum, is part of an integrated single-user, sea and land transport service, or at the other end of the spectrum, is a multi-user terminal (both on the sea-transport side and on the land-transport side). Further, it is important whether the terminal serves short-sea traffic (normally frequent service, relatively few containers per ship and relatively low, average storage time), deep-sea long-distance service (less frequent service, relatively many containers per ship and relatively high-average storage time), intermediate distance traffic, or a combination of the above.

The following are the important parameters governing the relationship between yard area and container throughput:

(a)   the average stacking height and its statistical distribution. The maximum stacking height ranges from one to five, depending upon the yard equipment. The average stacking height as a fraction of the maximum height is governed by the traffic mixture and by the organisation of the yard operations;

(b)   the open areas required for operation of equipment and for access to containers;

(c)   the number of calendar days of period considered;

(d)   average TEU storage time in calendar days and its statistical distribution.

The relationship is often expressed in the simple equation:

$$C = (L \times H \times W \times K)/(D \times F)$$

where C is the TEU throughput during the period of time considered; L is the number of TEU ground slots; H is the average stacking height in number of containers; W is the average utilisation factor for ground slots; K is the number of days of the period of time considered; D is the average TEU storage time in days; and F is a peaking factor for a combination of higher than average TEU throughput, less than average stacking height and more than average storage time.

The problems associated with this equation are caused by the use of average values for C, H and D. The peaking factor is a rather crude attempt at compensating for the variation of these parameters. The optimum yard area should be determined by trade-off between, on the one hand, the cost of additional yard area and/or stacking height, and on the other hand, the benefits from reduction of waiting time for ships and/or yard transport, which is caused by storage congestion. Queuing analysis has been suggested for solving this problem. However, it has to be assumed that the statistical distributions of equipment cycle times are exponential or that the links are processes, which may be represented by Erlang distributions. It is claimed that it is possible to get around the need for such assumptions and that a four-berth terminal model with up to ten different types of equipment systems or operations has been developed and calibrated (Frankel, 1980). Computer simulation is probably a better approach.

Yard layout and stacking patterns

Yard layout and stacking patterns are intimately related to the choice of equipment, which together with types of containers, shippers and consignees, modes of onward transport and even contents of the containers determine the nature, size and stacking patterns.

Contrary to what one may think, area is not necessarily the main determining factor in choosing a proper layout or adopting a stacking pattern, both of which should first of all be productivity driven. This means that, despite area availability, one should concentrate on providing the shortest distances between the vessels and the stacks.

Another factor that should be taken into account is convenient access to the stacks for land transport. For road transport one should try not to deviate from public road traffic rules inside the terminal, e.g. left- or right-side-of-the-road traffic.

Forklifts and reach stackers only

For traffic flow reasons, stacks perpendicular to the quay are preferred. The width between stacks depends on the equipment turning radius or length. Transfer between stacks and the apron is best done with tractor–trailers, but transfer by forklift or reach stacker may be considered.

When using tractor–trailers as transfer equipment, the forklift or reach stacker only moves backward and forward in a straight line when carrying a container and turns only without a load. Damage to pavements is therefore minimal compared with turning with a full container load in the spreader.

Reach stackers have a better weight distribution than forklifts, resulting in lower maximum wheel loads. They also allow for higher stacks and some direct access to second-row positions, which is not possible with forklifts.

When using forklifts only, it is recommended, if space is available, to make the stacks two containers wide. Thus every container row is directly accessible.

For reach-stackers four row stacks is not a problem and some terminals even use eight rows. This may, however, lead to some shifting for outbound stacks and most certainly for inbound stacks.

Straddle carriers

Straddle-carrier stacks are one-row stacks, either two or three high, depending on the straddle carrier.

Most terminals have their rows perpendicular to the quay, but 45° and parallel stacks do exist. Some layouts have been dictated by prevailing wind directions.

A straddle-carrier operation is usually not supported by transfer equipment, so that the space between rows is the minimum required for two straddle carriers to pass each other.

To improve safety a transfer point for delivery or pick-up of containers is usually located well away from the stacks.

For straddle carriers, outbound stacks should also be closest to the vessels. For unbalanced trades, mainly large numbers of empty containers being discharged, container terminals have lately introduced different types of stacks for empty and full import containers. Empty containers are being transferred by special tractor–trailers able to carry up to eight 20-ft containers, which are block stacked.

Rubber tyre and rail-mounted gantry cranes

The stack patterns of these depend on the equipment. Early RTGs spanned over only four containers plus one truck or tractor–trailer lane and could lift one container over three-high stacks. Nowadays they accommodate six container rows plus one truck or tractor–trailer lane and lift one container over four to five high. The result is often two to three different stack lengths for the same terminal. One should remember that RTGs move on specially strengthened straight beams, that cannot be altered to a different layout for new, high-performing equipment. RTG stacks are typically parallel to the quay, although perpendicular stacks do exist in some terminals. Again inbound and outbound stacks should be well separated, with the latter closer to the vessels.

Tractor–trailer transfer between yard and apron is required when yard gantry cranes are used.

The length of the stacks is debatable. It should allow smooth traffic flows for the tractor–trailer transfer, when several vessels are working simultaneously. One should not have stacks parallel to the quay, which are longer than the average length of the calling vessels. This would cause too much interference for the transfer traffic for vessels working at the same time, in front of these stacks.

Stacks should also be related to the total number of TEUs that one vessel usually loads or discharges. It may be spread over a number of stacks so that several gangs can work simultaneously on the same vessel without getting in the way of each other.

Stacks for empty containers

Stacks for empties should be separate from the stacks for containers with cargo and are designed in accordance with the equipment used.

Delivery of empty containers, once separated according to type and ownership, is mostly on a first-available basis. Containers of the same category are therefore block stacked using the cheapest type of equipment for stacking and delivery: either forklifts, reach stackers or mobile cranes of different types. Stack patterns are dependent on the type of equipment.

Stacking according to container contents

Stacking of containers with dangerous cargo, chemicals, etc. is governed by special rules. A general rule, often disregarded, is to avoid double-handling as much as possible. Additional transfer of such containers is therefore not recommended. They should be easily accessible to fire-brigade intervention in case of accident. Containers with cargo with low flash points should not be exposed to direct sunlight.

Stacking according to consignee or shipper

In order to avoid shifting, which is common in stacks for inbound containers with cargo, it is advisable to block-stack containers for the same consignee as much as possible.

Organising stacks is often dictated by the particular software program, that terminal operators use for planning of the vessels. Some of the above may therefore need to be adapted to the specific program, that is used.

Handling between ship and apron

Specialised gantry cranes equipped with spreaders for container handling, were developed in the 1960s. Their lifting capacity commonly ranges from 30 to 45 tonnes, at reaches from the quay face of 30 to 37 m. Their back reach, from the inside rail, is normally 10 to 20 m and the rail gauge is 20 to 30 m for newer cranes and 10 to 15 m for older cranes. The crane boom, which extends outside the quay face, is raised to a vertical position to allow passage of the superstructure of ships. Only first generation main-line ships and some short-sea and feeder service ships are equipped with their own container handling gear.

The time required for one operational cycle of a container crane is about two mins. Thus about 30 containers may be handled per hour. Newer cranes however have been equipped with better drives and higher rates are reached for empties and light containers. It is therefore not unusual, for a mixture of containers with cargo and empties, to achieve hourly averages of 40 containers and more. Attachments have been developed to lift two or four empties at a time, while cranes in hub terminals are fitted with two trolleys, with intermediate transfer from one to the other, thus reducing the horizontal travelling distance.

Pavements

Container-terminal pavements are normally exposed to much heavier loads than general cargo-terminal pavements, but less exposed to oil and other chemicals, except for oil leakage from equipment. The potential damaging effects on pavements of heavy traffic of straddle carriers, fork-lifts and tractor–trailer units have been assessed (Barber and Knapton, 1979 and 1980). Forklifts, which have very high front-wheel loads, were found to be much worse in this respect than the other yard systems, which carry about equal loads on all wheels. Concrete block pavements are normally preferred (see 'Pavements', earlier in this chapter). All ground slots should be clearly marked by lines and numbered. Continuous maintenance of marking lines and numbers should be carried out.

Buildings

Sheds for stuffing and un-stuffing of containers as well as office buildings should be located behind the transit storage yards. The only obstructions allowed in yards should be lighting masts and their protection against structural damage from moving equipment.

Partly and fully automatic terminals

Yard gantry cranes on rails, as well as recent RTGs, provide for accurate horizontal positioning of containers and recording of their positions. Such cranes together with a computer program, which records all movements of containers from their arrival in the terminal to their departure, as well as their storage positions, both in ground slots and vertically in the stacks, has made automation of terminals possible. Automatic terminals vary in different parts of the world, but perhaps are most advanced in Singapore, where yard cranes on concrete superstructures span twenty rows of containers stacked five to six high, with overhangs of about 20 m at both ends. The quay cranes are automatic as well. The yard crane is directed by computer to position or pick-up each container to and from the yard, to and from trucks accurately positioned under the landside overhang, as well as to and from a container conveyor under the quayside overhang. The horizontal tolerance is ± 5 cm. The yard crane gantry travels and its trolley lifts or lowers containers simultaneously at high speeds and accelerations. The container conveyor, which automatically follows the quay crane, serves as a two-way buffer storage between the yard crane and the quay crane. A secondary yard, located behind the yard described above, is equipped with a similar gantry crane as well as with RTGs, which span six rows of containers, stacked four high, and a truck lane. The movements of the RTGs are not automatic, but they do have automatic position controls.

Mixed general cargo-container terminals and general cargo terminals suitable for conversion to container traffic

Where a limited volume of container traffic exists, and is not expected to increase rapidly, a separate container berth will not be economically and/ or financially feasible and containers will have to be handled at one or more general cargo berths. The containers are likely to be handled between ship and apron by ships' gear, existing quay cranes or by mobile cranes on rubber wheels. Straddle carriers or gantry cranes will usually be too costly for yard handling of the relatively small number of containers. Heavy-duty forklifts, or reach-stackers, equipped with spreaders and tractor–trailers are likely to constitute the least cost solution. They may also be used for other purposes, when not needed for container handling. The high front-wheel loads of forklifts are likely to require pavement reinforcing, which may not be required for reach stackers.

Where there are good reasons to believe that considerable containerisation of general cargo will take place, one or more needed berths may be built for general cargo, but with provisions made for conversion to accommodate container traffic. The major provisions are foundations for container quay-crane rails, at least for the front rail, and transit-shed structures, which may relatively easily be dismantled and re-erected elsewhere as required.

Roll-on/roll-off terminals

General considerations

Conversion of an existing berth for use also by ro/ro ships, if conversion is required at all, is normally a fairly simple and inexpensive matter. New facilities exclusively for ro/ro ships, equipped with stern or bow ramps, are also inexpensive for locations with moderate tidal variations, provided that conditions are such that the ships may be moored to dolphins or buoys

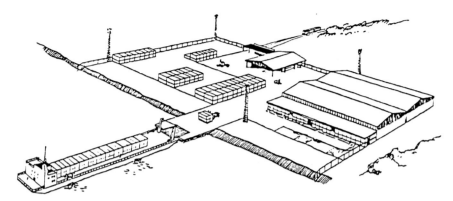

Fig. 8.5 Inexpensive ro/ro berth (reproduced by permission of MariTerm AB)

in a direction perpendicular to the terminal land area. This is illustrated in Fig. 8.5. Many ro/ro ships are equipped with their own ramps designed to reach a straight stretch of quay along which the vessel is moored. This is not only the case for ships with side ports, but also for ships with stern ports from which the ramps are lowered at an angle to the quay. The requirements for berths used by such ro/ro ships are limited to paved areas for ramp traffic, marshalling yards for trailers and vehicles as well as storage facilities for trailers and other cargo units.

Facilities for ro/ro ships without ramps or with longitudinal ramps
Ships which are not equipped with their own ramps require special port facilities. For ships with side ports it is a simple matter to provide shore ramps. While some ships are equipped with stern ramps at an angle to the quay, as described above, it is normally not possible to install similar shore ramps for ships without their own ramps, because of the lack of standardisation of ships' ports, tidal variations, etc. Thus it becomes necessary to provide adjustable shore ramps in extension of the longitudinal direction of the ships. This in turn requires fixed facilities at the stern end of the berth. The required facilities are either a platform structure extending from the end of a berth, equipped with a fixed ramp or an adjustable ramp, or a traditional ferry terminal. However, the latter may only be used for ro/ro ships, whereas, the former is a supplement to a lift-on/lift-off (lo/lo) berth, be it a general cargo, container or multipurpose berth. This latter arrangement is shown in Fig. 8.6.

The ro/ro platform has to be suificiently large to accommodate the traffic entering and leaving the ramp,

An ISO international standard (ISO 6812) concerning ro/ro ship-to-shore connections was adopted and issued as international standard during

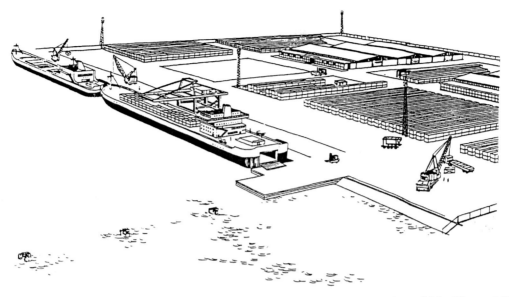

Fig. 8.6 Combination ro/ro and lo/lo berth (reproduced by permission of MariTerm AB)

1983. It may be acquired from ISO Technical Committee 8/Subcommittee 19, or from national standardisation organisations. The following summarises briefly the shore ramp standards. For both fixed and adjustable ramps:

(a)   the roadway shall have a minimum clear width of 9 m for two-lane traffic and of 5 m for one-lane traffic;

(b)   the minimum clear height above the ramp surface shall be 7 m;

(c)   design loads shall comply with road regulations in countries to and from which traffic is expected and shall allow for loads of cargo and cargo-handling equipment expected to be moved over the ramp in question; and

(d)   the changes of gradients upwards and downwards, between surfaces, shall be limited to allow the use of small-wheel cargo-handling equipment. Angles between surfaces, which are normally not to be exceeded, are recommended.

For normal water-level variations of less than 1.5 m fixed ramps are considered feasible. Two classes of fixed ramps are considereded. They are shown in Figs. 8.7 and 8.8. Class A ramps are for ships equipped with ramps which can reach levels from 0.25 to 1.75 m above the water level, in all loaded conditions of the ships. Similarly, class B ramps are for ships with ramps, which can reach levels from 1.5 to 3.0 m above the water level.

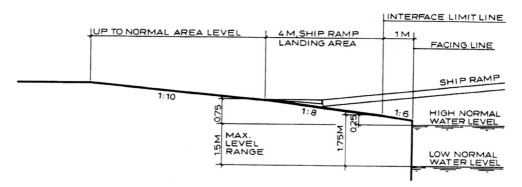

Fig. 8.7 Fixed shore ramp. Class A. ISO Standards (reproduced by permission of the ISO Secretariat)

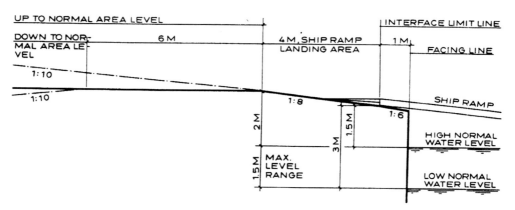

Fig. 8.8 Fixed shore ramp. Class B. ISO Standards (reproduced by permission of the ISO Secretariat)

For fixed shore ramps:

(a)   the width shall be 32 m or equal to the beam of the largest ship expected to call; and

(b)   the gradient shorewards of the normal landing area for the ship ramp shall normally be limited to 1 on 10.

The adjustable ramp characteristics are shown on Fig. 8.9. The outer end shall be wide enough to accommodate the ramp of the largest ship expected to call.

Advanced ro/ro systems

Systems for handling units larger than single 40-ft containers would for lo/lo handling require very expensive cranes because of the heavy lifts involved. However, for ro/ro handling such systems might be feasible.

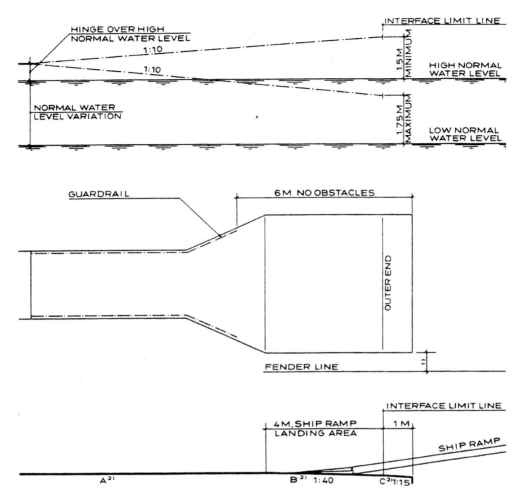

1) THE DISTANCE BETWEEN THE FENDER LINE AND THE ADJUSTABLE SHORE RAMP HAS TO BE CHOSEN WITH REGARD TO THE SHIPS WHICH ARE EXPECTED TO CALL AT THE TERMINAL
2) FOR SLOPE A = 1:10 THEN SLOPE B = 1:10 + 1:40 = 1:8 AND C = 1:10 + 1:15 = 1:6.

Fig. 8.9 Adjustable shore ramp. ISO Standards (reproduced by permission of the ISO Secretariat)

Such a system was developed in Sweden and named LUF (Lift Unit Frame) after one of its basic components. The frame is built for up to six standard 20- or 40-ft containers. Equipped with a platform, it becomes a large pallet which may carry a load of 100 tonnes of normal palletised cargo, cartons, newsprint or sawn timber. Equipped with a large container it may be used for bulk. Special frames accommodate cars, pipes, logs or other types of cargo with special lashing requirements. The frames would be loaded in the port of origin in advance of the arrival of the ship and unloaded in the port of destination. The frames are about 55 cm high and thus do not add much to the height of their cargo. If they have to be returned empty, they can easily be stacked and do not take up much space. A very low trailer with a large number of small wheels made to fit the frame is guided into the latter as it stands on the ground or on deck. The trailer, and with it the frame, is raised hydraulically to provide ground clearance for the frame. The hydraulic power has to be supplied by the towing unit. The trailer has coupling devices designed for standard terminal tractors and for special tractors built for integration with the trailers.

Several prototype units were built and tested, e.g. in the ports of Singapore and Gothenburg, but did not apparently catch on.

Ferry terminals

General considerations

While container and ro/ro terminals can be considered rapid transit facilities exclusively for a specialised type of cargo, but serving many different ships in more or less regular services, ferry terminals are reserved for a smaller number of ships in regular service. On the other hand, the cargo is much less uniform and usually includes:

(a)   vehicles (cars, buses, trucks, trailers);
(b)   trains; and
(c)   walk on/walk off (wo/wo) passengers.

Due to the large difference in loading/unloading time and other requirements, trains and cars are not normally transported simultaneously on the same ferries. Another significant aspect is that loading/unloading of the vehicles themselves is not carried out within the terminals. Neither is cargo storage provided for, except perhaps parking space for a relatively low number of trailers. Pure passenger ferry services may be carried out from any suitable ordinary quay and are not treated in the following.

Ferry terminals typically link fairly rapid road or railway traffic by a 'floating bridge' system with carrier speeds of around 15 to 20 knots.

However, in recent years a growing interest has been experienced in high-speed ferries for transport of passengers, cars and trucks, but not of rail cars. Such ferries are often catamarans and usually made from aluminium in order to reduce the weight. Their cruising speed is in the range from 32 to 40 knots. Such high speed generates special wake waves, which may cause physical damage to ships and structures in the vicinity, if proper remedial measures are not taken (see 'High-speed ferry terminals', later in this chapter).

The introduction of high-speed ferries may change the transport concept for a certain crossing completely. For a traditional ferry service, it may be optimal to reduce the crossing time by locating the ferry terminals where the crossing distance is at a minimum. For a high-speed ferry service it may turn out to be preferable to operate the service directly between the origin and destination of the bulk of the traffic, i.e. the nearest cities.

Great efforts should be made to speed up ferry operations by facilitating approach, berthing, and loading/unloading as well as by introducing new types of high-speed ferries, such as catamarans, hovercraft and hydrofoils. As high-speed ferry terminals are dealt with later in this chapter, this section will be devoted to the functional elements of traditional ferry terminals

The design of the following functional elements in a ferry terminal is of decisive importance:

(a)   marshalling yards;
(b)   passenger facilities; and
(c)   ferry berth(s) with three main elements:
      (i)   fendering system;
      (ii)   ramp(s); and
      (iii) scour protection.

The basis for the planning of these elements is clearly the traffic which is expected. The total volume of traffic, its distribution over time and its distribution over the different modes of land transport are important parameters. It should be pointed out that the forecasting of these parameters is very difficult and its accuracy quite limited. Also, once a new terminal starts to operate, it will influence the traffic pattern in ways that are difficult to predict. Fortunately, ferry routes are usually flexible systems which, within a year or two, may be modified to cope with changes in the traffic volume and pattern. New ferries may be chartered or built within 6 to 18 months, and the terminal structures may be modified over the same period of time.

The traffic

Traffic forecasts are prepared for:

(a)   vehicles arriving at the terminal for crossing, divided into:
      (i)   passenger cars;
      (ii)   trucks, including trailers; and
      (iii) buses.
(b)   vehicles delivering and picking up wo/wo passengers divided into:
      (i)   private cars;
      (ii)   taxis; and
      (iii) buses.

The forecasts are results of a multidisciplinary study involving statistics, economic growth analysis, origin–destination estimates, customs regulations where applicable, interviews with potential users, etc.

Forecasting of the passenger traffic involves evaluation of traffic generation and diversion, and determination of annual growth factors. The traffic generation and diversion may be assessed on the basis of a passenger-traffic generation model, if sufficient data are available for its calibration, using the existing road network and the present ferry service system. The travel habits will probably change in accordance with the improved connection roads and ferry services in ways which may only be judged with limited accuracy. The basis for the forecasts of the future cargo traffic is the expected cargo movements between the regions connected by the ferry route in question. In this context, it is essential to assess the competition with other modes of transport, such as conventional shipping, air freight and road/bridge transport.

Having forecast the annual volumes of different transport categories, the next step is to assess the distribution over time of the various categories. In particular, the identification of peak traffic is required for design of the functional elements of the ferry terminal. The required maximum daily transport capacity varies from one ferry service to another and may exceed the daily average by a factor of two or more.

It is often practical to express the forecast traffic categories as equivalent passenger car units (pcus). One pcu may be considered to occupy $10 \, m^2$ of deck, including space around each car. Trucks are equivalent to 4 to 7 pcus depending on their size and buses to about 4 pcus. This is used for an estimate of the transfer capacity in terms of car deck area, as well as for marshalling-yard requirements.

The turnaround time depends mainly upon the distance between the terminals, the cruising speed and the loading/unloading time. Financial analysis will establish whether a few large ferries or more smaller ferries should be preferred, the desirable cruising speed and the number of ferry berths. These options, however, will have different impacts on the traffic, because more frequent departures are likely to stimulate traffic.

Marshalling yards

In order to speed up operations, reduce bottlenecks and avoid accidents, it is essential to prepare the layout of the land areas in such a way that different traffic categories are kept separate. The departing traffic should never interfere with the arriving traffic. The latter should, after arrival at the terminal, be separated in:

(a)   passenger cars;
(b)   trucks and trailers;
(c)   buses; and
(d)   vehicles which discharge or pick-up wo/wo passengers.

Also lanes for waiting vehicles with and without reservation should be separated. A reservations system will increase the required number of lanes for waiting vehicles, but will reduce the required total length of the lanes, because vehicles with reservations spend less time waiting.

The number and lengths of the lanes for the different traffic categories are determined from the traffic forecasts and arrival distributions, combined with the capacity of the ferries. A rough idea of the size of the marshalling area may be obtained based on the assumption that waiting time during peak traffic of more than two to three hours would not be acceptable. With a departure frequency of about once per hour, this would mean a marshalling area capacity corresponding to two or three ferry loads.

Figure 8.10 shows the layout of a terminal, which has recently been designed according to the above-sketched principles. The ferry is of the catamaran type. However, the planning elements for a traditional ferry are the same.

Passenger facilities

The required passenger facilities are mainly shelter for use while waiting for departure. In addition, facilities for sale of refreshments and convenience shopping as well as toilets should be provided. The requirements vary from one ferry service to another. The total area requirement should not be less than about $1\,m^2$ per passenger. This figure must be used with caution, as it is heavily influenced by climatic conditions and local standards. For safety reasons, special gangways for wo/wo passengers should be provided.

Berth facilities

(a)   *Fendering system*: in order to allow rapid berthing, a fendering system with smooth front panels is required. The fendering system shall be designed to absorb the impact energy from the berthing ferries with reasonably low impact forces on both ship and quay. During the last 10 to 12 years significant developments in the design of fendering systems have taken place. Formerly a continuous fendering system with smooth front panels was normally used. The berths ranged from the closed type, with approximately equal lengths of fendering on each side of the ferry, to the open type with fendering along one side of the ferry only. Such arrangements were necessary, because the manoeuvrability of the ferries was limited. Recent developments in ferry design have included introduction of the shuttle propeller, able to rotate 360° around a vertical shaft. With one shuttle propeller at either end of the ferry, a traditional rudder becomes superfluous. The ferry is extremely

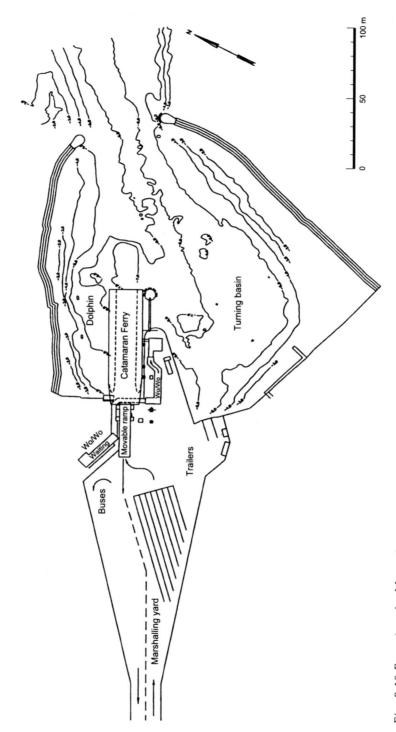

Fig. 8.10 Ferry terminal layout

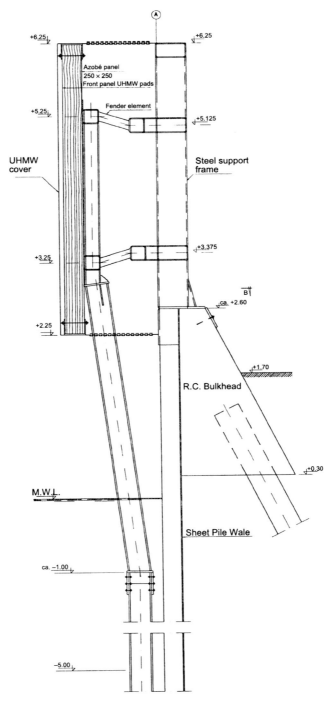

Fig. 8.11 Fender panel. Example from catamaran ferry berth

manoeuvrable, and a continuous fendering system can be reduced to a system of separate fendering elements. This concept has also been enhanced by the development of very smooth front covers for the fender panels, which reduce the friction coefficient significantly. The material commonly used is of the UHMW-PE (Ultra-High-Molecular Weight PolyEthylene) type. An example of such a system for a traditional ferry is shown in Fig. 8.11.

Normally, the mass of a catamaran ferry is less than the mass of a traditional ferry with similar dimensions. Hence, the impact energy to be absorbed by the fenders would be less. However, it should be borne in mind, that the car deck of a catamaran is situated at a higher level than a traditional car ferry deck. Further, the added hydrodynamic mass for a catamaran, which moves sideways, is higher than that for a traditional single-hull ferry.

(b) *Ferry ramps*: movable ramps are installed ashore at the inner end of the ferry berth. As a ferry is usually not equipped with ramp(s), the outer end of the shore ramp is connected directly to the bow or stern of the ferry. Ferries with two or more decks require the same number of shore ramps to provide access to the different decks. The approach road from the marshalling yard to the ramps must be designed in such a way that it is possible to divert cars to any deck. Ramps are designed to sustain the vehicle load and be able to transfer the vehicles safely also when extreme water levels occur. The length of the ramp may be determined based on high- and low-water statistics, the maximum allowable steepness of the ramp and the specifications of the vehicles to be transported. The maximum allowable steepness is normally 1 on 10. The ramps must be able to allow small roll movements of the ferry. For reasonably well protected ferry berths roll movements of more than ±3 degrees will rarely occur. An ISO standard for fixed and adjustable shore ramps in general is summarised in 'Roll-on/roll-off terminals' earlier in this chapter.

In some instances, small ferries with short crossing times of some 10 to 20 mins, are equipped with movable ramps, where fixed landing slopes are available ashore. Due to the considerable ramp length compared to the ferry dimensions, the ramps remain in a horizontal position during the crossing and are thus vulnerable to damage. Further the rationale of reducing the payload by transporting two ramps (one at either end) on a ferry during each crossing is debatable.

The shore ramp may be constructed with longitudinal I-beams and a transverse system of secondary beams, supporting the deck. The ramp is supported on land and on the ferry in a recess. The design of ramps does not normally present particular problems, except where extremely long ramps are required or other special circumstances exist. The weight of the ramp may change the trim of small ferries significantly. This may be reduced by suspending the ramp partly by means of counterweights or by submerged uplift tanks.

Figure 8.12 shows an example of a movable shore ramp for a catamaran ferry. Due to the high levels of car decks on catamaran ferries, the ramp will normally be longer than shore ramps for traditional ferries. In order to minimise its length the ramp is designed with two horizontal hinges. The ramp is supported by and fixed to the ground at its landward end. Two supplementary sets of supports are provided by pairs of hydraulic cylinders, enabling the ramp to be adjusted continuously during loading/unloading.

Scour protection

Proper scour protection against propeller erosion from traditional ferries and from the water jets of catamaran ferries is an important component of ferry terminals. Published theoretical research concerning this matter is rare, so the design is based almost exclusively on experience and/or hydraulic model tests. Scour protection may consist of:

(a)   layers of armour stones on the seabed;
(b)   a continuous layer of smaller stones encased in steel wire cages (gabions); or
(c)   a continuous concrete pavement.

For (a) it is essential to observe the filter criteria to avoid that seabed material is extracted through the protective layers. When (c) is constructed by placing concrete flagstones it is necessary to place filter layers under the flagstones. The gabions must be inspected regularly, because the cages deteriorate through corrosion. The gabion and concrete pavement solutions require the least thickness to achieve the desired protection. Fibre-reinforced concrete may be constructed directly on the sea bottom, either to seal erosion holes or as new erosion protection.

The free edge of the scour protection is very exposed to erosion. This in particular is the case for concrete pavements, which cannot by themselves close minor erosion holes, while gabions and layers of armour stones are to some extent self-repairing.

High-speed ferry terminals

General considerations

The main feature of high-speed ferries are the adverse effects caused by their speed of 32 to 40 knots or even higher. The ferries are normally of

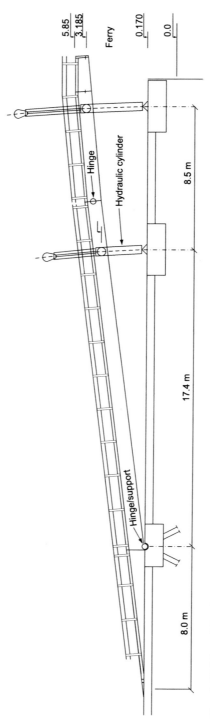

Fig. 8.12 Movable shore ramp (catamaran ferry)

the catamaran type and the descriptions in this section is confined to this type, since other ferry types, such as hovercraft, only require a simple landing ramp on the beach. This section should be read in conjunction with the previous section 'Ferry terminals', in which some highlights of the characteristics of high-speed ferry terminals have already been incorporated.

The traffic

In the earlier section also called 'The traffic', the forecasting principles have already been dealt with. These principles are, of course, also valid for high-speed ferries. It should, however, be borne in mind that transport of heavy vehicles such as trucks and buses is extremely costly on high-speed ferries. Normally the time reduction benefit for trucks on high-speed ferries does not justify the higher costs and, consequently, traditional ferries are generally competitive. The time and comfort benefits associated with transport of buses on high-speed ferries may in some few cases keep long range bus lines alive.

Marshalling yards

Since the cost and, consequently, the price of transporting a car or a passenger by a high-speed ferry is higher than transport by a traditional ferry it is essential that the customers experience all elements in the transport process as being efficient and smooth. This means that waiting time has to be reduced as much as possible, so that the customers may arrive late at the terminal, typically some 15 mins before departure. In order to assure that this is possible, an ample number of ticket boxes must be provided with appropriate associated marshalling lanes. One consequence of such a late-arrival pattern for the customers may, on the other hand, reduce the requirements to the total marshalling area, since the loading operation will be going on, while customers are still arriving at the terminal.

Similarly the unloading of vehicles and passengers must keep the same high pace. Hence, the unloading operations may cause regular high influxes to the general traffic in the destination zone near the terminal. In order to assure smooth and rapid traffic, special measures are required. Such measures may include traffic-light systems with priority to the ferry traffic through city centres. It will probably also be necessary to provide accommodation lanes for leaving cars in the immediate vicinity of the ferry terminal in order to empty the ferry as quickly as possible and, at the same time, ease the integration into the general city traffic.

A terminal inaugurated in the spring of year 2002 is shown in Fig. 8.13. The largest ferry using this terminal accommodates some 200 passenger

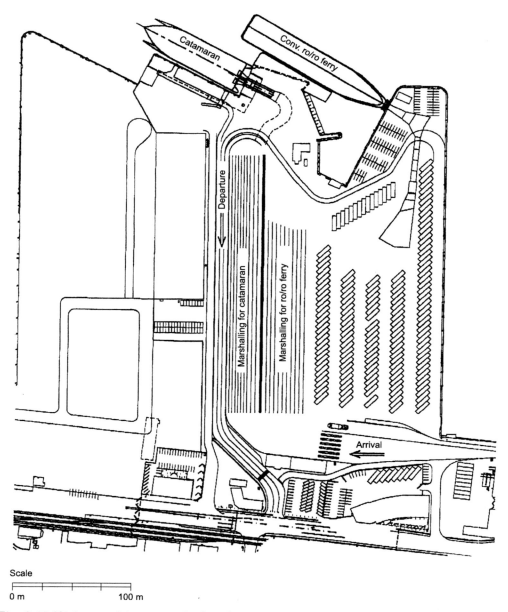

Scale

0 m   100 m

Fig. 8.13 High-speed ferry terminal and conventional ro/ro ferry terminal (reproduced with permission from Mols–Linien A/S)

cars and four buses. A total number of eight ticket boxes provide quick access to nine lanes, in which some 400 pcus are accommodated. Cars leaving the ferry may line up in three lanes corresponding to some 180 pcus before entering the city.

Passenger facilities

The passenger facilities requirements are very limited since the waiting time is very short. However provisions for toilets and limited shelter should be made. Normally, these facilities will not be used very much, since passengers arrive at the 'kiss-and-wave' area near the ferry in the last minutes and are picked up on arrival immediately.

It should, however, be emphasised that vehicles and passengers for safety reasons should be separated in the terminal area to the greatest extent possible. This may, among other measures, require installations of gangway overpasses, elevators, etc.

Berth facilities

The main aspects are presented in 'Ferry terminals', earlier in this chapter.

In order to assure quick operations it is essential that the fendering system provide a smooth and distinct berth line with sufficient fendering capacity. For quick operations, fender units with smooth UHMW-PE front panels are considered indispensable. Another measure, saving both time and operational costs, is installation of auto moorings operated by the ship's crew using a small radio transmitter (remote control unit).

The ramp is operated in the same way. When designing the ramp, careful consideration of the ship's setting by the load of the ramps should be made. Since catamarans are sensitive, especially to bow loads, it may be necessary to implement counterweight systems at the ramp, which may be accomplished hydraulically.

The catamaran ferry berths, shown in Fig. 8.13 and Fig. 8.12 allows accommodation of two different catamaran ferry types, with different widths and different fender line and car-deck levels. Therefore, a movable and adjustable fendering system and a long movable land ramp have been provided. The time required for wireless adjustment from one ferry type to another is less than five minutes.

Scour protection

The water jets of the catamarans will create extensive erosion of the seabed almost regardless of the seabed material (except rock). Therefore proper scour protection is mandatory.

Due to the heavy propensity to erosion at the free edge of the scour protection, it must be extended to cover the erosion area adequately. If the scour protection is constructed of concrete, gabions should be placed at the edges in order to provide an adjustable transition to the original seabed.

Liquid-bulk terminals

General considerations

The major cargo category in world trade is bulk crude oil and petroleum products. Other liquid-bulk cargoes, such as molten sulphur, chlorine, other chemicals, and foodstuffs constitute a small percentage of the total liquid-bulk shipments. Hence, this chapter is primarily concerned with crude oil and petroleum products terminals.

Crude oil and petroleum products are transferred between land-storage facilities and vessels by means of pipelines. Since pipelines can easily and inexpensively be placed over or under other traffic lanes, the location of liquid-bulk facilities relative to land transportation is not critical. However, the distance between storage and vessel as well as the relative elevation is of importance. A distinction must be made between loading and discharging terminals. Cargo is normally discharged by ships' pumps. This is a factor over which the port designer has no control. Typical values of pump capacities for various sizes of crude-oil tankers are shown in Table 8.4. As a general rule, additional pumping capacity is required if the storage facility is more than 5 km from the berth and/or elevated more than 20 m above sea level. The terminal provides the additional capacity in the form of so-called booster pumps. This is undesirable from an economic point of view as the booster pumps are only used a small percentage of the time. In certain cases, booster pumps are unavoidable; in others they may not be needed if a reduction in discharge rate can be accepted. The choice is made by trade-off between the value of the ship's time and the cost of boosting.

The terminal provides the energy to pump the liquid aboard the vessel. Elevated storage facilities may permit the liquid to flow by gravity alone to the vessel. Special problems arise from very low-lying storage facilities for the reason that a positive head is required at the pump suction manifold. This has, in some cases, caused pumps to be located below ground-water level.

Most liquid cargoes are hazardous in one or more ways. Most are flammable and many are toxic. This leads to requirements that the facilities be

Table 8.4 Typical pump characteristics for crude oil tankers

Size of tanker (DWT)	Pump capacity (m^3/hour)	Maximum pressure in head of liquid (m)
200 000	12 000	130
100 000	7000	110
50 000	4500	110
25 000	3500	90

located away from population centres, that special fire and clean-up installations be provided and that specially trained personnel be on standby duty. For each cargo there are separate sets of requirements, which vary from country to country. The requirements are also updated from time to time. It is beyond the scope of this book to provide details.

Conventional berths for crude oil and petroleum products
The berths often consist of a central loading platform (this term is used even if only unloading takes place) and a number of berthing and mooring dolphins. A typical layout is shown in Fig. 8.14.

The loading platform is usually just large enough to accommodate the piping and other mechanical equipment, safety equipment, and personnel access. Special consideration must be given to ships' supplies. Most tanker terminals, especially export facilities, are not equipped to supply vessels. If there is such a requirement it could be that the least cost solution is to provide the service by barge(s) tied up on the outside of the vessel. Otherwise the loading platform and possibly the trestle must be enlarged to accommodate this operation. It should be noted that there is a trend to use standard 20-ft containers for the supplies.

The cargo is transferred through flexible devices which can accommodate the motion of the vessel relative to the loading platform. Horizontal sway and surge motion is caused by wind, wave and current forces, as well as by passing ships. Vertical heave and roll motion is caused by changes in loading conditions, by tide and by wave action. Pitch motion

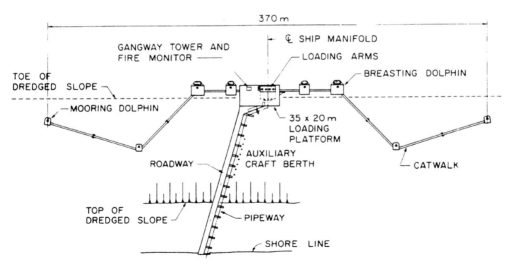

Fig. 8.14 Typical oil-tanker berth layout (reproduced by permission of Dravo Van Houten Inc.)

303

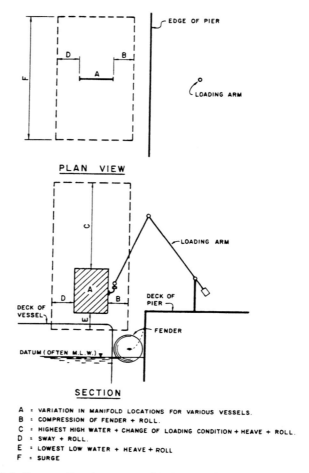

PLAN VIEW

SECTION

A : VARIATION IN MANIFOLD LOCATIONS FOR VARIOUS VESSELS.
B : COMPRESSION OF FENDER + ROLL.
C : HIGHEST HIGH WATER + CHANGE OF LOADING CONDITION + HEAVE + ROLL.
D : SWAY + ROLL.
E : LOWEST LOW WATER + HEAVE + ROLL
F : SURGE

Fig. 8.15 Relative motion between ship's manifold and loading platform

is not a problem when the manifold is located amidships. For ship motions, see Chapter 5, 'Motion and moorings'. Possible relative motion between the manifold and the loading platform is shown on Fig. 8.15. The influence of the range of vessel sizes should be noted. A voluntary standard exists for manifold locations. Virtually all tankers comply with this standard.

The flexible devices that convey the liquid between vessels and loading platforms can be divided into two basic types, namely loading arms and hoses. A loading arm is a series of rigid metal pipes connected by swivel joints in such a manner that the outboard ends can move freely within a prescribed envelope. A typical loading arm is shown on Fig. 8.16. The balancing of the loading arm is a special problem, which various manufacturers solve in different ways. The balancing may be achieved by counter weights, hydraulic cylinders, springs or a combination of the above. The

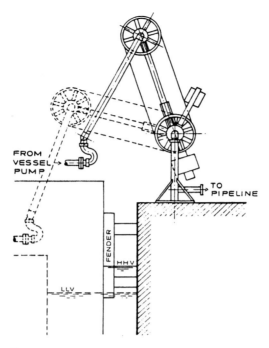

Fig. 8.16 Loading arm

arm may be manually operated, i.e. it is brought into position by pull on wires, or it may be power operated, usually hydraulically. In general, small arms (6 in diameter or smaller), at berths which are used infrequently, are manually operated, whereas arms larger than 8 in diameter are power operated.

The alternative is usually a composite rubber, steel wire and fabric hose. A number of different designs exist and manufacturers should be consulted for specific applications. Factors to be considered include the following: chemical resistance of hoses, minimum and maximum pressures, and allowable bending radius and stiffness. The hose is usually suspended from the vessel in such a manner that it does not chafe. However, if care is not taken the hose may be squeezed between the platform and the hull. If this happens the hose may be damaged and an oil spill may result.

On the loading platform provision must be made for draining hoses or loading arms, which are usually empty when handled. The drain system, which is called the slop system, also collects any other spillage which may occur during operations. It includes a slop tank, which serves as temporary storage. The slop is either removed by tank truck or re-injected into the pipelines. The practice in any particular case is dictated by the allowable contamination of the liquid in the pipeline.

305

Storage facilities for liquids

The storage facility at a typical liquid–bulk terminal consists of a series of cylindrical steel tanks, either with roofs that float on the liquid in the tank, or with cone roofs. In both cases, the purpose of the roof is to prevent contamination of the liquid by rain. The floating roof also serves to prevent evaporation of the liquid to the atmosphere. This function is provided for in most cone-roof tanks by an interior, floating barrier, which may be of simpler design as it is not required to drain rainwater. In some modern tank farms a vapour-collection system is installed. Its dual purpose is to recover valuable volatile substances lost through evaporation and/or to prevent air pollution by these same volatile substances. Aside from the air pollution aspects, it frequently makes good economic sense to equip tank farms with vapour-collection systems. The tanks in the tank farm may be switch tanks, i.e. able to store a number of different liquids (one at a time) or be dedicated to one particular liquid. The choice between switch tanks or dedicated tanks is a complicated economic problem involving the cost of cleaning and product degradation at each switch of liquid and, more important, it has a profound impact on the economic size of the tank farm. If a tank farm is made up mainly of switch tanks, the total required capacity of the tank farm is, as a rule of thumb, three to four times the largest shipload received or shipped. For exclusively dedicated tanks, the required capacity is three to four times the largest shipload received or shipped for each type of cargo. More rational determination of the required storage capacity is outlined in Chapter 2, 'Computer simulation'.

Conventional terminals for refrigerated and/or compressed liquid gases

Hydrocarbons, which are gases at normal temperature and pressure, present special transportation problems. They have to be transported in liquid state in specially built vessels. The liquid state is achieved either by compressing the gas at natural temperature or by cooling the gas to below its boiling temperature at atmospheric pressure. In some cases, a combination of the two methods is used and the gas is transported at moderate pressure and below the corresponding boiling temperature. Table 8.5 shows the boiling temperature and the critical temperature (temperature above which the gas cannot be liquefied, irrespective of pressure) for various gases. It may be noted from the table that it is not possible to liquefy methane (principal component in natural gas), ethylene and ethane at natural temperature. Hence, these cargoes can only be transported as liquids by means of cooling. They are usually transported at atmospheric pressure. Other gases are liquefied by all of the methods mentioned. However, for propane and butane the usual practice is to

Table 8.5 Boiling point and critical temperature in centigrades for gases

Type of gas	Boiling point	Critical temperature
Methane	−161	−83
Ethylene	−104	+9
Ethane	−89	+32
Propylene	−48	+92
Propane	−42	+97
Ammonia	−33	+132
Butadiene	−4	+152
N-Butane	−1	+152

employ the cooled state for long voyages in large vessels and to use the pressurised state at natural temperature for short voyages in small vessels.

The cargo is transferred in much the same way as described for liquid bulk in 'Conventional berths for crude oil and petroleum products', earlier in this chapter. The lower the transfer temperature the higher is the required degree of insulation for pipelines. Due to the difficulty of insulating the end of the loading arm it is common that large lumps of ice form here during transfer operations. Hoses made from rubber are not suitable for low temperatures; hence articulated loading arms are normally used.

There is, however, one important difference from ordinary liquid bulk. When filling the vessel excess vapour is formed due to boiling of the liquid. This vapour has to be transferred back to shore by means of a separate piping system called the vapour-return system. It is burned at the flare of the facility, or re-liquefied and re-injected into the pipelines. In either case significant expenses are incurred. Another solution is to deliver the liquid to the berth slightly below its boiling temperature and apply an ejector to suck the vapour back into the pipeline and utilise the available energy of the sub-cooled liquid for liquefying the vapour. Some modern ships have sufficient liquefaction capacity installed to be able to cope with the excess vapour. However, a vapour-return system with a flare will always be required for emergency conditions.

The berths are usually constructed in the same manner as for petroleum bulk. The requirements to distance from population centres and in regard to operations, to safety devices and access limitations during operations, are more stringent for liquid-gas berths.

Certain ships for the transport of liquefied natural gas (LNG) are not designed for sloshing of the LNG in partly full tanks. Sloshing may result in damage to the insulation with potential catastrophic consequences.

For this reason, these vessels require very calm conditions at berth during loading and discharge. Further, they cannot leave the port under any circumstances before the tanks are either full or empty. This may add considerably to breakwater expenditures, and combined with the requirement of distance to other facilities may cause very difficult terminal siting problems.

Storage facilities for liquid gases

Liquefied gases require special low-temperature and/or high-pressure tanks. Both capital and operating costs for such tanks are much higher than for ordinary oil storage tanks. The tanks, which maintain gases in liquid form at low temperature, have to be insulated and must have an associated refrigeration plant for re-liquefying vapours. Large tanks, which maintain gases in liquid form through high pressure, should be spherical. The rule of thumb for required storage capacity is two to three times the maximum shipload. For more rational determination of the required storage capacity see Chapter 2, 'Computer simulation'.

Liquefied gases in storage tanks with atmospheric pressure may develop stratification. The stratification may result in warmer, heavier mixtures in the bottom of the tank overlain by lighter and colder layers. This is because the liquefied gases typically contain a number of different gases, the separation of which, by composition and temperature, makes stratification possible. The warmer mixture at the bottom of the tank may not boil because the hydrostatic pressure at the bottom of the tank increases the boiling temperature. However, if the stratification is disturbed then explosive boiling may result. Consequently the tanks must be designed to either prevent the stratification through, for example, stirring or to contain safely the sudden pressure increase resulting from explosive boiling.

Offshore terminals

Due to the facility with which liquids are transported in pipelines special types of offshore facilities have been developed for these cargoes.

(a)   *Conventional buoy moorings*: this is the oldest type of offshore mooring. The vessel ties up at a fixed position by means of multiple chain-anchor moorings. These moorings are so arranged that the vessel maintains a fixed position and orientation. In many, but not all, cases one or two of the moorings are made up by the bow anchors with which all vessels are fitted. A typical arrangement is shown in Fig. 8.17. A major drawback of conventional buoy moorings is the difficulty of navigating into the mooring. Figure 8.17 represents the best layout in this regard. Most existing moorings are laid out in

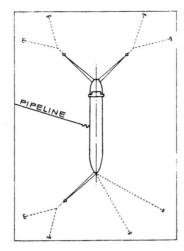

Fig. 8.17 Conventional multi-buoy terminal

a manner that makes occasional collision between the ships and one or more mooring buoys highly probable.

The cargo is transferred through a flexible rubber hose connecting the amidships' manifold on the vessel with a sub-sea manifold, which in turn is connected to land facilities by submarine pipelines. When no vessel is present the hoses are left on the seabed but attached to a marker buoy on the surface.

In most cases this is the lowest capital-cost, liquid–bulk terminal. However, even moderate wave conditions prevent the mooring of tankers and operation of the terminal, which is also prone to high maintenance costs. In principle, there is no limitation to the size of vessel which may be accommodated at a properly designed conventional buoy mooring. However, it is not common to consider this type for vessels larger than 100 000 DWT. This type of berth makes it possible to pump the cargo ashore through floating hoses and thus avoid a submarine pipeline. It may be the preferred solution in cases where very few ships are handled per year.

(b) *Single-point moorings*: at this type of berth the vessel ties up with a bowline only. Consequently, the vessel is free to weathervane around the point to which the bowline is attached, and will tend to align itself in the direction of least resistance. The vessel may stay moored under even very severe conditions. The cargo is transferred by means of rubber hoses floating on the surface connecting the vessel's amidships manifold to a swivel on the single-point mooring. Most liquid–bulk vessels only have amidships manifolds. Technically it would be advantageous to have the manifold located at the

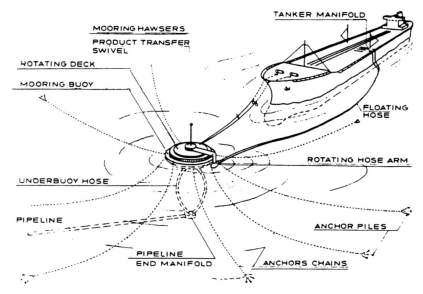

Fig. 8.18 Single-point CALM

bow of ships, which tie up to single-point moorings. This is done at some terminals, which are used only by a small number of vessels modified especially for bow transfer.

Three different types of single-point moorings are in common use:

(i)   catenary anchor leg mooring (CALM);
(ii) single-anchor leg mooring (SALM); and
(iii) tower moorings.

CALM consists of a buoy floating on the surface moored by a number of (usually five to eight) conventional chain-anchor legs. The cargo is conveyed between the submarine pipeline and the buoy by means of submarine hoses. The mooring line is connected to the cargo swivel by a lever arm. This device is used to rotate the swivel with the mooring line. Floating hoses connect the cargo swivel on the buoy at the surface to the amidships manifold. Figure 8.18 shows a CALM.

CALM is the most common single-point mooring. Its maintenance costs are fairly high. However, it is competitive with other types in relatively shallow water (30 m or less).

SALM is a buoy moored by means of one vertical chain or pipe fastened to a single anchor as shown in Fig. 8.19. The buoy must be so proportioned that even under extreme conditions the chain always remains taut, otherwise failure results due to impact loads. The cargo is conveyed from a submerged cargo swivel to the amidships manifold by means of hoses

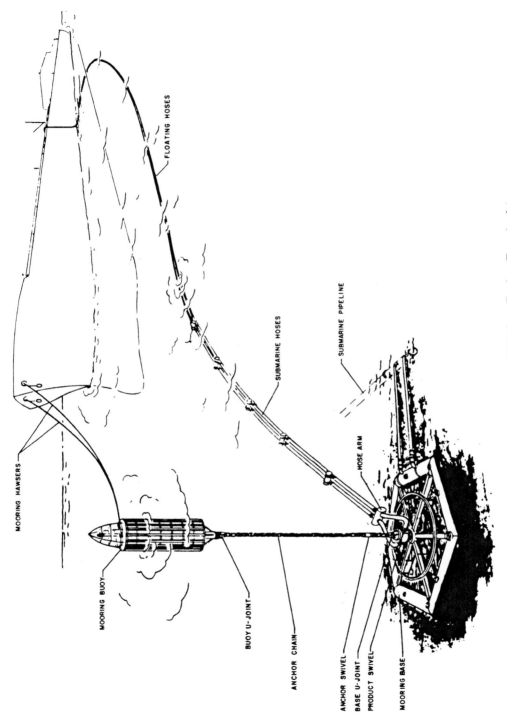

Fig. 8.19 Single-point SALM (reproduced by permission of SOFEC Inc. Tanker Terminals)

floating on the surface except for their rise from the cargo swivel to the surface. The swivel is turned by the hoses and not by the mooring line. This puts a relatively low limit on the permissible torque resistance of the swivel. This type of single-point mooring is especially cost effective in deeper water. It requires slightly larger minimum water depths than CALM due to the reduction of water depth above the anchor cargo swivel combination.

The tower mooring consists of either a single flexible pipe pile driven into the seabed or a trussed tower. The cargo swivel is usually above the water surface. This type of installation is especially well suited for vessels equipped with bow manifolds, in which case, floating hoses can be avoided entirely. The principal drawback is that they are more prone to collision damage and the required repairs are likely to be very costly.

It is claimed that vessels can remain at single-point moorings during extreme wave conditions (up to $H_s = 4\,m$ or more depending on the specific installation). However, due to problems associated with maintaining the floating hose securely attached to the amidships manifold, transfer is usually limited to conditions with $H_s = 3.5\,m$. Mooring operations are limited to conditions with H_s less than 2 m to maintain safe line handling operations.

Even when a port provides suitable locations and water depths for conventional berths it may be worthwhile to consider a single-point mooring in the open sea outside the port. While the risk of small oil spills is somewhat larger at a single-point mooring, the risk of a catastrophe is significantly reduced due to the fact that vessels do not have to navigate in the confines of the port among other vessels or near grounding depths. The required water depth in the port may perhaps be reduced, if tankers are accommodated elsewhere.

The utilisation of single-point moorings for the transfer of liquefied gas has been considered, but appears to be feasible only for gases with relatively high boiling points such as propane. The principal difficulties are the use of rubber hoses, which may not remain flexible at low temperatures, thermal insulation and the motion restrictions for certain LNG tankers, when they are only partly loaded (see 'Conventional terminals for refrigerated and/or compressed liquid gases', earlier in this chapter).

Dry-bulk terminals

General considerations

A large variety of goods is carried by ships as dry bulk. These goods can be divided roughly into four main categories:

(a)   minerals, such as iron ore;
(b)   coal;

(c)   foodstuffs such as grain; and

(d)   other commodities, such as cement.

During recent years, coal and iron ore respectively, accounted for about 38 per cent and 36 per cent of major dry-bulk commodity shipments, grain for about 18 per cent, and bauxite/alumina and phosphate rock for about 4 per cent each.

Depending on the volume being transferred at a particular port, one or more berths may be dedicated exclusively to dry bulk or perhaps even reserved for one particular commodity. The decision whether to have berths exclusively dedicated to dry bulk is an economic one, which involves a variety of factors; the most important of which is the forecast annual or seasonal volume. However, the location of storage facilities, contamination, exposure to the elements and safety hazards may also play significant roles in making this decision. In general, when a berth is dedicated exclusively to dry bulk, it is possible to employ high-capacity transfer devices to effect quick turnaround of the vessels. When the berth is used for a variety of purposes, it is usually only possible to use mobile transfer equipment of low capacity.

Dry bulk is frequently transferred between the loading or discharge device and the storage by means of belt conveyors. While these conveyors are usually not located at quay level, the supporting structures do obstruct horizontal handling of general cargo, vehicle traffic, etc. In general, it is desirable that dry bulk storage takes place fairly close to the berths. However, stockpiles may cause special geotechnical problems because of the high-unit loads they exert on surfaces close to berths.

Depending on the type of commodity and throughput, the storage is usually one of four basic types:

(a)   open storage yard, which is normally used for commodities which do not suffer serious degradation by being exposed to the elements;

(b)   shed, which is normally used for commodities which would suffer degradation if exposed to rain;

(c)   silo, which is used for storage of grain, cement and other commodities, which must be protected from the elements. Silos are normally equipped with efficient materials-handling equipment; and

(d)   slurry pond, which is, of course, only feasible for materials which are not damaged by water.

The choice between covered sheds or silos is usually based on economics. Silos would normally be preferred when the storage time is short, and for commodities which are fine powders, for dust-control reasons.

Bulk cargoes vary widely in regard to stowage factor, angle of repose, generation of dust, resistance to degradation by mechanical handling

313

Table 8.6 Bulk-cargo properties

Commodity	Stowage factor (m^3/tonne)	Angle of repose degrees
Coal	1.08–1.39	30–45
Iron ore	0.33–0.47	30–50
Bauxite	0.84–0.92	28–49
Alumina	0.70–0.84	35
Cement	0.61–0.64	
Phosphate	0.92–0.98	30–35
Potash	0.94–1.04	32–35
Super phosphate	0.84–1.00	30–40
Maize	1.25–1.41	30–40
Rye	1.40	30
Wheat	1.18–1.34	25–30
Sugar	1.11–1.25	40
Soya beans	1.23–1.28	30

and hazardous properties such as toxicity, susceptibility to fire and spontaneous ignition. For stowage factors and angles of repose see Table 8.6. In addition, there are large differences in regard to their corrosive and abrasive properties, both of which affect the handling equipment, which is used.

The required storage capacity for bulk materials is dealt with in Chapter 2. However, the following rules of thumb are commonly used for preliminary analysis. For open storage the capacity should be four to six times the largest shipload of each commodity, for covered storage three to four times the largest shipload and for silos two to four times the largest shipload.

When choosing materials-handling equipment, manufacturers, who specialise in installations for the commodity in question, should be consulted, as there is considerable flux in the state of art of materials handling. An efficient and reliable link between the vessel and the storage facility is frequently of paramount importance for a particular installation. For many bulk commodities, shipping and handling costs constitute a large percentage of the total delivered cost.

Conventional export terminals

This section deals only with high-volume export terminals, where one or more berths are dedicated exclusively to dry-bulk commodities and usually only to one commodity at each berth. For ore, coal and other minerals the terminal is usually located at a suitable site, close to the mine, or at an existing nearby port which is connected to the mine by rail and/or road.

The type of berth structure which is selected for dry-bulk export depends upon local conditions and the selected loading equipment. Due to the relatively high costs of conveyors there is less flexibility in the choice of suitable berth location. Therefore both continuous structures located close to the storage facility and platform berths, similar to the type which is common for liquid bulk facilities, are used.

Various types of loading arrangements are used depending on the commodity in question and also on the type of wharf which is used. Figure 8.20 shows four options for high-capacity ship loaders, one of which is normally used.

All the loaders shown are suitable for berths in rather exposed locations. There need not be direct contact between the loading device and the vessel, as the bulk material usually drops the last few metres into the hold, and for this reason even quite considerable ship motion can be tolerated without interruption of loading. The travelling ship loader is normally used when both sides of a pier are used. The other three systems are shown in reverse order of their time of introduction in the market, with the twin-orbiting ship loader being the most recent system.

The simplest, and often the lowest-cost, loading arrangement is a single spout at a fixed location. Its use means the berth has to be longer than normal, as the vessel has to be moved along the berth to shift the loading from one hold to another. This arrangement is only feasible for low throughputs. It is not suitable for exposed locations due to the difficulty of shifting the vessel under adverse weather conditions.

Another technique for loading of the vessels is to bring the material to an elevated point alongside the vessel and load it into the hold by means of a series of spouts. This technique is quite common in grain terminals. An overhead conveyor belt is provided along the berth and a travelling device trips the bulk material down into inclined spouts located at each hatch. Another more modern version is a gantry device travelling under the tripping device, which spreads bulk material over the hatch by means of a slewing conveyor with an adjustable spout at the end. In this manner, it is easier to accommodate a wider range of ship sizes and it is also much easier to control dust emission since open free-fall of the material is avoided. A typical grain export facility is shown in Fig. 8.21.

Typical loading rates for a conveyor with chutes are 1000 to 7000 tonnes per hour and for fixed chutes about 500 tonnes per hour.

For material stored in the open, it is common to provide combined stacking and reclaiming devices. A typical stacker–reclaimer device is shown on Fig. 8.22. It reclaims from the storage pile at rates of 1000 to 3000 tonnes per hour by means of a digging wheel which feeds directly onto the conveyor system which carries the material to the ship loader. In the

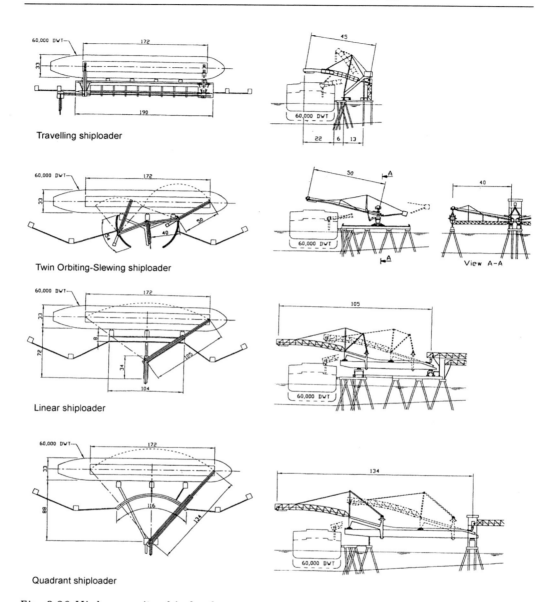

Fig. 8.20 High-capacity ship loaders

stacking mode the boom conveyor is reversed and the material falls onto the pile from the end.

Another method of bringing the material to the conveyor is by means of earth-moving equipment such as bulldozers, which scrape the material into fixed, below-ground-level hoppers, which deliver the material to conveyor belts. This method is frequently used in covered storage.

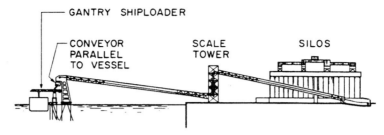

Fig. 8.21 Typical grain export facility (reproduced by permission of Dravo Van Houten Inc.)

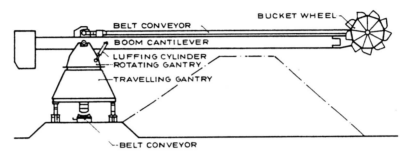

Fig. 8.22 Stacker–reclaimer unit (reproduced by permission of Trans. Tech. Publications from Stacking, Blending, Reclaiming of Bulk Materials by R. H. Wöhlbier)

Silos have built-in handling equipment. The design of silos is beyond the scope of this book.

Conventional import terminals

This section deals only with berths, which are designed exclusively for import of one or more bulk commodities. Such berths are typically encountered at power plants for import of coal, at steel mills for import of iron ore and coal, at chemical factories receiving raw materials and processed chemicals, and for import of grain.

Import terminals usually have lower throughputs than export terminals. It is more difficult to discharge the materials from vessels than to load them. For these reasons a larger variety of equipment is encountered at import terminals than at export terminals.

It is unavoidable that equipment used for discharging the vessel comes in contact with the vessel. Therefore, much calmer conditions are required at berths used for import than at those used for export. Pitch and heave, motions, especially, are detrimental to discharge operations due to the difficulty of maintaining a proper position between the unloading device and the pile of bulk material in the hold of the vessel. The following types of discharge devices are used:

(a) crane-mounted grabs removing a certain amount of material (up to 50 tonnes) in each cycle;
(b) continuous mechanical devices such as bucket wheels, chain buckets or screw conveyors; and
(c) pneumatic devices.

The crane-mounted grab is the most common discharge device for bulk materials. The crane dumps the material directly in a storage area immediately behind the wharf, or alternatively, dumps it into a hopper feeding a conveyor which carries the material to storage. Typically, the handling rate for a grab will be from a few hundred tonnes up to 1500 tonnes per hour if the crane revolves and up to 4000 tonnes per hour if the crane does not revolve. These rates are the peak rates obtainable. The average transfer rate would usually be 60 per cent of the peak rate.

Continuous mechanical devices are normally used for high-capacity terminals with frequent vessel arrivals. Such a device is shown in Fig. 8.23. Bucket elevators may have handling rates which are as high as 1000 to 5000 tonnes per hour.

Pneumatic devices are normally used for discharge of grain, cement and other similar materials. The pneumatic devices require a high energy

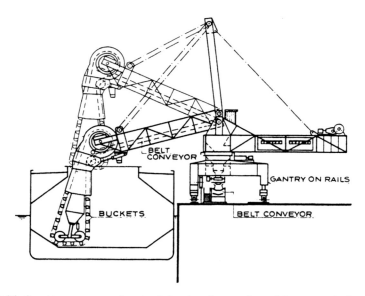

Fig. 8.23 Continuous mechanical device (reproduced by permission of Trans. Tech. Publications from Stacking, Blending, Reclaiming of Bulk Materials by R. H. Wöhlbier)

input but are easy to handle and provide good dust control. Their handling rates are typically a few hundred tonnes per hour.

The selection of type and capacity of material-handling devices for any particular import terminal is a complex problem, which involves the economy of the operation, the suitability of various types of conveying systems for a particular material and the susceptibility of the material to degradation by the equipment. The selection process requires inputs from manufacturers of material-handling equipment in order to arrive at the optimum solution.

Front-end loaders and bulldozers are commonly used to clean up the ship's hold and bring the last of the bulk material to the discharge device.

Environmental considerations

Most bulk materials are handled by conveyors, stacker-reclaimers, ship loaders and ship un-loaders. Their operation tends to create dust, which may be carried by the wind and contaminate the vicinity of the materials-handling equipment. To eliminate or minimise this is an important consideration in equipment selection.

To minimise the amount of dust escaping from the materials-handling operations, the equipment should be enclosed to the maximum extent possible and it should be maintained in top-notch mechanical condition, such that it does not leak or spill the bulk material. In many cases the material is handled by manual control, for example, when a crane with a grab is used. When this is the case, proper training of operators and proper operational procedures are of the utmost importance. For example, if the operator attempts to operate the equipment beyond its capacity, overfilling and spilling of the material results. Using excessive operational speed has similar effects.

The consequence of trying to operate, at speeds beyond those it was designed for, is degradation of the material by reduction of its particle sizes. This, in turn, increases the propensity for dust generation.

Even when proper equipment and proper operations procedures are used, the generation of some dust is usually inevitable. Therefore, steps must be taken to remove the dust once it is generated. A range of technologies exists to suppress or collect the dust generated by bulk-handling equipment. There are basically two alternatives, the dry method or the wet method.

The dry method consists of aspirating the air that contains the dust, and then recovering the dust by various means such as cyclones, filter bags, and electrostatic precipitation.

The wet method consists of spraying water onto the surfaces of the material. This method is most effective in open storage piles and at conveyor transfer stations. Addition of surfactants to the water enhances the efficiency of this system.

Inquiries should be made with the manufacturers of equipment in order to provide the most efficient and cost-effective solution for a given installation.

Multiple-use berths

When import or export, of a particular bulk commodity, results in very low occupancy of a berth, it is uneconomical to provide a berth exclusively for this purpose. In this case general cargo berths with sufficient land areas are used for transfer of one or more bulk cargoes. The bulk material is usually transferred by crane-mounted grabs or a combination of grabs and small mobile conveyors. Outbound bulk material is usually brought to the wharf by trucks, which may discharge into mobile hoppers from which the material is loaded aboard the ship by means of conveyors. Inbound bulk material is frequently placed in a small temporary storage pile on the wharf by mobile cranes equipped with grabs, loaded into trucks by front-end loaders and taken to storage areas. The material may also be discharged directly from the ship into railway wagons or trucks. In either case, the installations required to load or discharge the bulk material are mobile. They should be used in such a manner that they do not interfere with the normal general cargo use of the berths. In this respect, floating discharge devices are advantageous.

The discharge or loading of bulk vessels at a multiple-use berth is usually slow and relatively inefficient. However, due to the high capital investment required for alternative, efficient handling facilities, this mode of operation may be more economical. Any operation of this nature should be re-evaluated from time to time to determine whether a more efficient operation can be achieved through reasonable investment in improved facilities or whether the operation should be shifted to a specialised berth.

Offshore terminals for slurried bulk

Certain bulk materials can be pumped through pipelines as slurry, when mixed with water. This is the case for most minerals and most solid fuels. At locations where conventional port facilities are very expensive, terminals for export of dry bulk in slurry have been patterned on the single-point mooring facilities, which are common for petroleum terminals (see 'Offshore terminals', earlier in this chapter).

A major economic and technical problem is the quantity of water required to pump the slurry. Typically, more than 50 per cent (by volume) of the slurry consists of water. If this water had to be transported by the vessel a considerable extra cost would result. Therefore, the water is decanted off the holds of the vessel prior to departure and may be discharged overboard or sent back to shore depending on the allowable water contamination. In any event, significant quantities (about 10 per cent) of water cannot be removed and remains in the hold of the vessel, which increases the transportation cost for the bulk material. At the receiving port the material usually has to be slurried again and devices for doing this are available. It should be noted that material, which was loaded as dry bulk, may be discharged as slurry.

Slurry-handling problems are, to a large extent, similar to liquid-handling problems. However, there are some additional technical problems such as choking of pipelines and considerable wear and tear of pipelines and pumps. It should also be noted that the material may have to be dried before use at a considerable cost. This is, for example, the case for finely ground coal. Handling of dry bulk as slurry is the ultimate solution to dust control problems.

Fishing ports

General considerations

A fishing port represents a specialised type of port in the sense that it deals with one commodity exclusively. Compared with general cargo or container ports, fishing ports are generally much smaller, as the length of a fishing vessel usually does not exceed some 60 m.

While other ports from a logistics point of view represent facilities enabling a more or less rapid change of mode of transport, fishing ports normally serve three purposes. The fish does not only change its mode of transport, but a substantial part of it serves as raw material for industrial plants within the port's premises. Further, the port has to accommodate the fishing vessels until they leave on their next trip. Being a highly specialised port, it is crucial for its existence to maintain a commercially viable flow of the catch of fish. In reality, a fishing port's only alternative use, without substantial reconstruction, is for accommodation of pleasure craft.

In recent years, the increased focus on sustainable yield aspects and environmental considerations has led to severe catch restrictions. Such measures have been imposed all over the world, especially in waters near large concentrations of consumers, such as is the case in the North Sea area. Here a distinct decline in the consumer fish catch has taken place, while the catch of industrial fish has increased considerably.

In the planning process for a fishing port it is therefore of paramount importance to include an assessment of the available and sustainable resources, which can be exploited most successfully by vessels from the port in question and not by vessels from other ports. With this aspect as

321

a point of departure the main items in the planning process may be summarised as follows:

(a)   determination of present and forecast of future sustainable yields for the different species;

(b)   determination of corresponding fleet requirements;

(c)   selection of site(s) for port construction and/or extension;

(d)   determination of port facilities requirements broken down into the following main components:

   (i)   unloading facilities;

   (ii)   auction or other sales facilities;

   (iii)   fish processing plants;

   (iv)   quays and basins;

   (v)   service and repair facilities; and

   (vi)   ancillary facilities for supply of ice, water, electric power, rescue at sea, pilotage, tug assistance, waste collection and management, recreation and sanitation, etc.

(e)   forecasts of market demands for fresh and processed fish.

The planning process for the above components is highly iterative and the optimum solution is a result of several loops of modifications of earlier findings and estimates. A new fishing port or a major extension of an existing one should be subjected to a full fledged, technical, economic and financial feasibility study (see Chapter 4).

The physical planning processes take their point of departure in analysing and making provisions to ensure smooth operations, to avoid bottlenecks and conflicting activities in the same areas. Therefore substantial efforts should be made to separate activities to the extent possible. In this context, attention is drawn to the ever-existing conflict between nearby residential areas and port activities, which cause noise, odours, dust, etc. Therefore, in the general layout, a buffer zone between the port and the city should be provided. The buffer zone may contain offices, shops, restaurants, etc., but very limited housing. Further, activities associated with disagreeable environmental effects should in general be located as far from residential areas as possible. Their possible harmful effects on fish handling and processing should be prevented.

In the following six sections, general planning principles (PIANC, 1998) are presented. 'An example of a modern fishing port', later in this chapter, has been devoted to a description of the main features of the modern fishing port in Thyborøn, Denmark. It must be emphasised that such a description cannot be directly used for design purposes at other locations, but its parameters and components may serve as checklists and points of departure for experienced planners and engineers.

The catch and the fleet

Obviously, the vessels and their gear have to correspond to the species and their quantities. It should be borne in mind that almost all species exhibit seasonal variations, meaning that annual requirement calculations have to include peak values in order to ensure sufficient capacity. This may leave specialised fishing vessels with idle capacity during off-season(s). Therefore, many vessels change gear during the year in order to match the available catch. Other vessels may temporarily leave their home base in order to exploit available quotas elsewhere.

Landing and processing

For operational reasons, separate facilities for unloading of consumer fish and industrial fish respectively should be provided. Vessels with consumer fish arrive at the unloading facilities and deliver the catch in iced boxes into the sorting area of the auction or other sales facilities by means of their own gear and/or conveyor belts. The unloading rates vary according to the equipment employed. Rates in the range from 5 to 10 t/h may be expected.

In the sorting area the fish is washed and sorted according to species and size, placed in iced boxes and stored in a chilled area, where the temperature is kept at about zero centigrade. The total area requirements for these purposes vary according to mechanisation and supply patterns. With pronounced peak season(s), the area requirements increase. The total area may be expected to be in the range corresponding to 5 to 10 t/m^2 per year. Further, an additional chilled storage area must be provided, for the fish which is not sold the same day it is landed. From a quality point of view, such an arrangement should be reduced to a minimum.

It should be noted that an increasing number of vessels prepare and sort the fish onboard, so that it can be sold directly upon arrival in the port. This should be taken into account when determining the area requirements described above.

After the fish has been sold, it is prepared for delivery to the buyers in an area reserved for this purpose. It should be of rectangular shape, thus allowing simultaneous loading of as many trucks as possible. An area corresponding to an annual throughput of at least 10 t/m^2 should be adequate.

From the auction/sales facilities, the fish is transported to the buyers' premises for further processing. Part of the catch of certain species may be processed (marinated, smoked, canned, etc.) in plants situated between the buffer zone and the auction/sales facilities within the port premises.

Landing of industrial fish shall take place at special facilities near the processing plants in the port area. Great efforts to prevent emission of disagreeable odours should be made. The fish is brought into the processing plant by fish pumps or bucket elevators or enclosed conveyor belts. The rates of unloading depend on the equipment, but more than 100 t/h is easily achieved.

The industrial fish serve as raw material for production of fish oil and fishmeal. The products leave the factory on trucks or by rail.

Berthing facilities

Provisions must be made for quays and/or piers to be used while the vessels stay in port, except when being unloaded, serviced and repaired, etc. The necessary water depth obviously corresponds to the unloaded draft of the largest vessel to be accommodated, considering tidal variations, wind effects on the water level, siltation, etc.

The required total quay length depends on the number and length of vessels to be accommodated and the number of vessels, which can be moored side by side. Some 20 years ago, when the vessels were smaller, four or five vessels, side by side, were common. Such arrangements are not possible today due to the increased size of the vessels, which would make it hazardous for inside vessels to leave. Further, modern vessels have covered weather decks, making access to the outside vessels and handling of moorings difficult. More than two modern vessels side by side are not recommended.

Service and repair facilities

Provisions for hauling vessels on land for service and repairs once or twice a year should be made. Often, such facilities comprise a slipway with rail-mounted cradles and a side transfer system, operated by a system of steel wires and winches. The number of bays shall correspond to the demand for dry hull repair days. A normal overhaul would take 5 to 10 days. In this context it should be borne in mind that the demand for service will vary throughout the year, corresponding to periods with reduced catch possibilities.

The traditional slipway system described above represents a safe and reliable overhaul facility. However, its flexibility in terms of numbers of service and repair bays is limited. Many efforts have been made to improve the flexibility. An example of such a concept is a ship elevator system, with a carrying capacity corresponding to the vessel's docking weight, consisting of a steel platform with a cradle. It is equipped with skirts opening downwards. The platform sits on the sea bottom. Once the vessel is in position over the cradle, it is raised above the water table by means of compressed air pumped into the platform hull. Next the cradle with the vessel is transported onto land on air cushions. The concept of lateral transport by means of air cushions represents an extremely flexible land arrangement solution. However, it requires a smooth, tight and resistant, concrete surface. Also other lifting devices such as a system of wires and winches instead of compressed air have been developed and are operating successfully.

The environmental impact, associated with the ship repair facilities, should be given particular attention. Service and repair activities are in many countries subject to severe regulations by law due to their potential adverse effects on the environment. All activities and measures to reduce or prevent pollution are to be approved by the authorities before start of operations. In particular, this applies to cleaning and painting. All shot-blast material has to be collected for safe disposal and measures taken to prevent paint spills, due to their content of heavy metals and other poisonous substances. Restrictions on emission of dust particles from shot blasting are likewise imposed.

Ice supply

Supply of ice is essential for the proper functioning of a fishing port. On departure, vessels may require a volume of ice approximately equal to the expected catch. The required ice quantity is somewhat difficult to esti-mate and additional fuel would be required to carry excess ice from the port to the fishing grounds. Therefore, an increasing number of vessels have their own ice-making plant onboard.

Pollution control

In addition to the measures described above, systems for oil-spill control, liquid and solid waste management, shall be incorporated in the planning. From an environmental point of view, efforts should be made to recycle as much as possible. This calls for a detailed and systematic approach. An example of this is the development of a special net shredder, that enables burning of net material, with high calorific content, in public incineration plants.

Ancillary facilities

Ancillary facilities comprise essentially:

(a)   supply of water and electricity along the quays;
(b)   rescue at sea assistance;
(c)   pilotage;
(d)   tug assistance;
(e)   waste collection and management;
(f)   recreational and sanitation facilities; and
(g)   a fisheries inspection vessel.

With a few minutes notice a fast rescue boat shall be operational. During very rough sea conditions a fisheries inspection vessel, if available, may participate in such rescue.

325

The city fire brigade would normally be responsible for fire fighting in the port.

An example of a modern fishing port

(a)   *General*: Thyborøn port in Denmark has been chosen as an example of a modern fishing port, because it almost exclusively serves fishing activities, and because it has been developed in a manner which reflects efficient and rational responses to growing and changing demands.

The port is situated on the west coast of Jutland, in lee of the Thyborøn Bar. The port dates back to 1917. The location was selected due to easy access through the Thyborøn Channel to the rich fishing grounds in the North Sea as well as low breakwater construction and maintenance dredging costs compared with a location on the open coast, where severe wave action and littoral drift problems exist.

The initial small port, including its basins, covered a total area of 4.2 ha. The development of larger fishing vessels with powerful and reliable engines made it possible to increase the exploitation of the North Sea fish resources. Hence, the demand for proper port facilities increased rapidly, and additional facilities to cope with these demands were provided in stages over the years. The present day port is shown in Figs. 8.24 and 8.25. The port is widely recognised

Fig. 8.24 Aerial photograph of Thyborøn fishing port

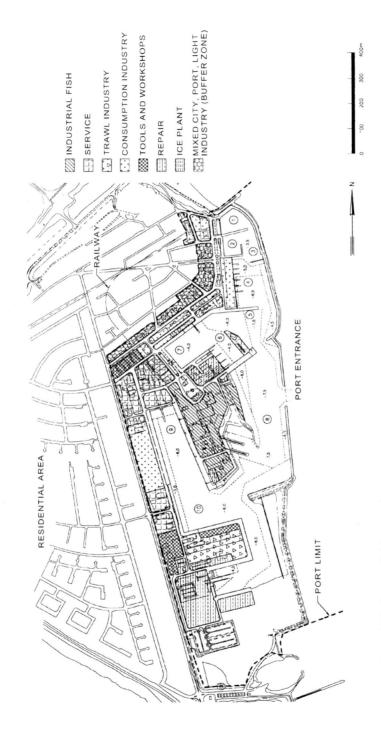

INDUSTRIAL FISH

SERVICE

TRAWL INDUSTRY

CONSUMPTION INDUSTRY

TOOLS AND WORKSHOPS

REPAIR

ICE PLANT

MIXED CITY, PORT, LIGHT
INDUSTRY (BUFFER ZONE)

RAILWAY

RESIDENTIAL AREA

PORT ENTRANCE

PORT LIMIT

N

Fig. 8.25 Plan of Thyborøn fishing port

as a signal illustration of port planning as a continuous, never ending and dynamic process, providing sufficient facilities, corresponding to the increasing demands.

The source of the following data is the State Port Authority of Denmark.

The total area occupied by the port today amounts to approximately 97 ha, of which approximately 63 ha are land and 34 ha are basins.

The port has approximately 3300 linear m of quay and 700 linear m of piers. The total length of breakwaters and revetments amounts to 1250 linear m. The catch of industrial fish has increased from about 100 000 t to about 350 000 t annually in recent years. However, the catch of consumer fish has decreased from a maximum of about 30 000 t in 1980 to about 15 000 t annually. This decrease is a result of severe catch restrictions imposed in order to sustain the species.

The recent throughput was achieved by 118 vessels, making about 7000 landings plus 300 landings by foreign vessels.

The number of vessels has remained fairly constant over the last 30 years, while their total tonnage has increased from about 8000 GRT to about 15 000 GRT.

(b)   *The fleet*: A total of 118 fishing vessels are based in the port. These vessels are exclusive traditional single-hull ships, except for one catamaran. Practically all vessels below 100 GRT are occupied with catching consumer fish. For the larger vessels the picture is more diffuse. However, the proportion of industrial fish vessels increases with increasing size. Seven large-beam trawlers and two large Danish seiner vessels are all used for catching consumer fish. Also, many of the vessels are used for catching both consumer fish and industrial fish.

(c)   *The catch*: The catch of consumer fish is rather unevenly distributed over the year. During December, the catch is only half the monthly average and in October it is some 70 per cent above the average. For industrial fish the peak month is May with a catch some 130 per cent above the monthly average. In December and February, the catch amounts to only some 40 per cent of the average.

(d)   *Port facilities*:
General:
As shown in Fig. 8.25 and described below, the main components of the port facilities are:
(i)   a consumer fish section in the northern part (Basins 1 through 4);
(ii)   a service and repair section for smaller vessels (Basins 6 and 7);
(iii) an industrial fish section (Basins 8 and 9); and
(iv)   a service and repair section for larger vessels (Basin 10).

Table 8.7 Land-use plan

1	Industrial fish	4.4 ha
2	Service	4.5 ha
3	Areas for tools and storage of equipment	2.0 ha
4	Repair activities	4.8 ha
5	Consumption industry	4.9 ha
6	Mixed city/port activities	2.4 ha
7	Trawl industry	1.9 ha
8	Ice plant	0.4 ha
	Total	25.3 ha

Basins and quay facilities:

At the seven piers, outside the auction facilities at Basin 4, 14 consumer fish vessels can be unloaded simultaneously. The unloading facilities for industrial fish in Basin 8 comprise four suction unloading plants and two fish pumps, allowing six vessels to be unloaded simultaneously.

The berthing facilities comprise a total of about 1500 linear m of quay. About 500 linear m are reserved for vessels below 20 GRT and 600 linear m for vessels above 600 GRT, while the remaining 400 linear m are available for all categories.

In Basins 6 and 7, a total quay length of 225 linear m is reserved for service and repairs of smaller vessels. In Basin 10, where the largest slipway is located, a total quay length of 900 linear m is available for service and repairs of larger vessels.

(e) *General land use plan*: Of the total land area of approximately 63 ha, about 38 are mainly used for public purposes, such as roads, railway, etc., and about 25 ha are used for a variety of activities shown in Table 8.7.

From an overall point of view the land-use plan reflects the efforts made during the port's development to separate conflicting activities. It should be noted that a buffer zone with mixed city and port activities has been placed between the port proper and the city.

Another essential feature is the location of service and repair as well as industrial fish activities placed far from the city. Thereby noise, odours and other undesirable effects are reduced. The prevailing wind directions are westerly, blowing away from the city toward the uninhabited fjord area.

(f) *Consumer fish*: A total of 14 vessels can unload their catch simultaneously at seven piers. The installations are operated 24-hours a day.

The catch is lifted onto conveyor belts by the vessel's own gear. The unloading rate is typically 5 to 8 t/h. The fish is transported to the auction sorting area on conveyor belts. After sorting and icing,

329

the fish in boxes, each weighing about 35 to 40 kg, is transported on battery driven forklift trucks to the chilled auction area. This area has a floor area of about 1200 m^2. The temperature is kept at about zero centigrade. During peak landings, an additional chilled auction area of about 1000 m^2 is used.

Industrial fish vessels often carry a small amount of consumer fish on deck. These vessels berth at the quay separating Basins 4 and 5 near the auction area. Vessels with chill-room facilities on board constitute an increasing proportion of the fleet.

After the auction, the fish boxes are transported to a sorting area where forklift trucks load them onto lorries.

The forklift trucks operating outdoors are not allowed in the chilled auction area for hygienic reasons. Similarly, used boxes are not allowed in the sorting and auctioning area until they have been cleaned. They are cleaned and washed in a separate room. The box cleaning and washing plant has a capacity of 500 boxes/h. The plant handles approximately 400 000 boxes per year during only one shift per day. Its capacity is utilised 100 per cent during three to four months of each year. The plant's capacity was designed to cope with earlier year's higher landings.

About 40 per cent of the auctioned fish are processed within the premises of the port.

(g) *Industrial fish*: Four suction plants and two fish pumps on two piers enable the unloading of six vessels simultaneously. The unloading takes place 24-hours a day at a rate of approximately 120 to 130 t/h. One to three men work in the ship's hull during unloading. Due to the severe physical conditions (hard manual work and risk of poisonous gas emissions), they are replaced every hour. During recent years, the fish quality has improved thereby reducing the risk of gas emissions.

The fish is transported to the processing areas through closed circuits for hygienic reasons and in order to reduce odour emission.

In the processing plants, fishmeal and fish oil are produced. The products are loaded into containers on train wagons and transported to buyers outside the city.

(h) *Ice plants*: Two ice plants operate in the port, a smaller one in the older part near the auction facilities and a larger one at the industrial fish basin. Their production capacity is approximately 350 t and 550 t per day, respectively. The vessels' ice requirements from external sources are diminishing as an increasing number of them have their own ice plant onboard.

(i) *Service and repairs*: The two service and repair facilities, shown in Fig. 8.25 are centred around slipways. The smaller slipway is situated in Basin 6. It comprises a traditional rail and wire pulled system with a side transfer system and cradles, enabling servicing and repair of 10 vessels simultaneously. The maximum docking weight is 80 t, enabling the facility to accommodate approximately the smallest 50 per cent of the home-based fleet.

The slipway in the southern part of the port is also a traditional one with side transfer and four service and repair bays. It can accommodate vessels with docking weights up to 500 t, meaning that all but the seven largest vessels may be hauled on land in the port.

Each vessel will normally use the slipway facility one or two times a year for maintenance and minor repairs. The duration on land is 5 to 10 days succeeded by approximately four days along the quay for work which does not require a dry hull. In case of major repairs or renewals, such as a new engine, additional time is of course required.

Although on average for the year the total capacity of the slipways is ample, it should be borne in mind that the overhaul demand is not evenly distributed over the year. Industrial fish vessels for instance, would not want to be out of operation during the peak season from April through September.

(j) *Pollution control*: In addition to the environmental impact control measures mentioned above, other important pollution prevention systems have been set up, which are as follows:
(i)   an oil-spill catch system; and
(ii) a system for collection, destruction and disposal of liquid and solid waste.

The oil-spill catch system consists of a traditional floating barrage system with pumps and collector containers. The system for collection, sorting and disposal of waste has been developed by and is operated by the port authority.

Fishing vessels approaching the port can request waste containers to be placed at their berth. Organic waste is put in plastic bags and other waste is put in the waste containers.

The waste is collected and transported to a paved sorting area within the port's premises. Here it is sorted into flammable and non-flammable waste as well as waste for dump disposal.

Some of the inflammable waste goes to the local incineration plant. Other materials, such as oil, air- and oil-filters, paint remnants and chemicals are transported to a central, national waste-treatment plant. Some items, such as fish boxes, are recycled.

Non-inflammable waste, such as lead, copper, brass, steel wires, electrical cables and fittings, bottles, batteries, concrete blocks, etc. are transported to different plants for processing and recycling.

Waste for controlled dump disposal comprises items, which do not fit into the above categories.

References

Barber, S. D. and Knapton, J. Port Pavement Loading, *The Dock & Harbour Authority*. London, 1979, April and 1980, March.

Frankel, E. G. (1980) 'Container Systems Selection', *The 2nd Terminal Operations Conference*, Cargo Systems Publications.

Fugl-Meyer, H. (1957) *The Modern Port*, Copenhagen, Danish Technical Press.

Jane's Yearbooks (various years) *Jane's Freight Containers*, London.

PIANC (1998) *Planning of fishing ports*, Report of Working Group No. 18, Supp. to Bulletin k 97.

State Port Authority of Denmark (various years) *Port statistics and port inventory*, Annual Reports.

Thomas, R. E. (1983) *Stowage. The Properties and Stowage of Cargoes*, Glasgow: Brown, Son & Ferguson.

CHAPTER 9

Berth and terminal structures design

Torben Ernst and Leif Runge-Schmidt

Introduction

The scope of this chapter is twofold. First, it is intended to provide guidance to engineers who do not have extensive experience in the design of berth and terminal structures. Second, the aim is to provide the port administrator and operator, who may not be very familiar with design aspects of harbour structures, with a background for assessing the range of available technical solutions and for understanding the rationale in the design process. Such a background will improve his ability to evaluate the work of his technical department or his consulting engineer.

The planning and design process for a berth or terminal structure may be considered to consist of the following six consecutive stages:

(a)   define needs;
(b)   translate needs into requirements;
(c)   determine alternative locations and develop general layouts;
(d)   execute site investigation;
(e)   select location and general layout; and
(f)   select type of structures and design.

The present chapter deals only with stage (f), but there is a strong interdependence between this and the previous planning and design stages. Site and model investigations, whether hydraulic or geotechnical, which are of paramount importance to the design and investigations necessary for identifying construction material sources or environmental conditions, have not been covered in this chapter.

Definitions

Below some of the terms used for describing fixed port and terminal facilities are defined:

> *Berth* – a facility where one vessel may be safely moored and load/unload cargo or let passengers or vehicles embark or disembark.

Quay – one or more berths continuously bordering on and in contact with a land or dock area. The quay apron reaches the quay front over the entire length of the berth(s). Often a quay is described as one or more berths, which are parallel to the shore, although the shore may not be well defined.

Wharf – same as quay, although there seems to be a tendency to use this term only for open structures on piles.

Pier or Jetty – a structure projecting into the water. Often piers and jetties will have berths on two sides and abut land over their full width. They may also be able to accommodate one or more vessels at the end. Piers and jetties may also be more or less parallel to the shore and connected to land by a trestle and/or a causeway.

The terms quay and wharf on the one side and the terms pier and jetty on the other side are mainly used to characterise berthing facilities with respect to their general arrangement relative to the shore line or adjacent land areas.

A berth which does not require an apron over its entire length but only a platform at mid-ship usually consists of a system of mooring and breasting dolphins. The individual dolphins will normally be connected by catwalks and a trestle bridge will connect the platform to land. Such a facility is usually also called a jetty.

Terminal – one or more berths for a limited purpose (such as oil, grain in bulk, containers) and/or limited to one or a few users.

Basis for design

Facilities requirements

The requirements are all related to the future use of the facilities and include dimensions, loads, utilities, etc. and location.

(a) Dimensions:
  (i)   water depth relative to datum or to a specified water level. The water depth should be specified not only along the berth to be designed but also in approach channels and port basins;
  (ii) height of structure above datum or other specified level;
  (iii) berth length; and
  (iv) apron width.
(b) Loads:
  (i)   live load from cargo on apron including quayside stockpiles;
  (ii) mobile and fixed cargo-handling equipment loads are not only for determination of the overall stability and structural design but also for design of apron surfaces;
  (iii) building loads;

(iv) berthing forces. Because these forces depend on fender types, the information obtained is normally data on vessels and approach velocities. Where the berth is located in a less sheltered area, the design of fenders may be governed by the movements of moored vessels subject to wave action. Requirements in this respect may almost exclusively be obtained through hydraulic-model tests; and

(v) bollard forces. The port operators may prescribe a certain minimum capacity of bollards and their approximate spacing. The engineer, however, should also get information on above-water hull areas and shapes of light vessels envisaged to use the berth, in order to be able to determine mooring forces from wind forces on the hull. For berths which are exposed to wave action, maximum mooring forces may occur as a result of combinations of wind forces on the hull and forces resulting from ship movements, the determination of which may require hydraulic model tests.

The various loads acting in disfavour of the stability of a structure should not indiscriminately be added together. Rather, they should be combined with regard to their probability.

(c) Utilities, etc:
   (i)   power supply to cargo-handling equipment and vessels;
   (ii)   lighting;
   (iii) telephone connections;
   (iv) water and bunker oil pipelines or barges; and
   (v)   fire-fighting equipment.

(d) Location: Design of berth facilities may include the determination of the optimal location within constraints established by a master plan. Optimisation of the general location is also a design matter but here it is assumed established during planning and design stages (c) and (e) (in the introduction to this chapter).

Site conditions

Information concerning actual site conditions is, of course, indispensable. Depending on the structures to be designed and the construction materials to be used, some of the items listed below may or may not be relevant. Some information, e.g. on topography and soils conditions may only be obtained from detailed site surveys or investigations, while other information is likely to exist and be available in a satisfactory form from other sources, e.g. information on climatic conditions.

(a) Topography: Topographic information both for the sea bottom and adjoining land areas may or may not be available in sufficient detail.

335

(b) Soils conditions: Full knowledge of soils conditions is of paramount importance and consequently careful subsoil investigations, both with respect to quality and area coverage, should be carried out. Usually such investigations are planned to yield adequate data on soil properties to:

(i) calculate earth pressures, bearing capacities and stability of structures;

(ii) analyse if subsidence of fills or structures may occur; and

(iii) assess the methods and costs of dredging, excavation and pile driving.

(c) Seismicity: Horizontal accelerations, which may occur in the region, have to be accounted for in the design.

(d) Water levels and wave conditions: Variations in water levels have to be known in order to determine the effective weight of soils and the hydrostatic pressure difference on walls with low or no permeability. Tide is normally of primary importance in this connection but also long-period waves should be considered.

(e) Climatic conditions: Data on wind speeds and directions are necessary for determining wind loads on moored vessels, buildings, cranes and other equipment.

Precipitation and temperature together with wind data will describe the climatic conditions under which the berth has to function and under which construction has to be carried out. The climate may have an impact on the design, e.g. the arrangement of storm-water run-off or on certain requirements with respect to curing of concrete in extreme high or low temperatures.

The design must consider the occurrence of ice as appropriate, either in the form of drifting ice which may hit the structure or as ice frozen to the structure and subsequently exposed to water-level variations which will introduce vertical forces due to the buoyancy or the weight of the ice.

(f) Corrosivity: Information is required, either from measurements on samples or from local experience, concerning corrosion by soil and/or seawater.

General design considerations

Least-cost solutions

When the facilities requirements have been established and sufficient information has been obtained on actual site conditions, the remaining part of the design activity is a matter of determining the least-cost structure, which will satisfy the requirements and at the same time be safe. Costs refer either to construction costs only or to construction costs plus maintenance costs. During the preliminary design stage, however, it will be sufficient to compare alternative solutions on the basis of construction costs only, because annual maintenance costs for well-designed berth structures are small compared with investment costs (0.5 to 2.0 per cent) and therefore their differences will have limited impact on the net present value of investment and maintenance costs. For two structures with the same construction costs, but with annual maintenance costs of 1 and 2 per cent respectively, the net present value of investment and maintenance for a 25-year life and an interest of 12 per cent would differ less than 8 per cent.

The least-cost criterion may also in certain situations be supplemented by a criterion concerning the expenditure of foreign currency so that a solution requiring a smaller amount of foreign currency may be preferred to a less expensive solution requiring a larger amount of foreign currency.

Another element that should be taken into account is that cost estimates for two different types of structures may often be significantly different with respect to their reliability. The reasons for this may be physical, environmental factors, such as sensitivity to variations in soil conditions or to wave conditions during construction. It may equally well be due to one type of structure being known in the country or location in question, while the other is virtually unknown and therefore prone to unrealistic costing by local contractors. It may also happen that a short construction time will be preferred at a higher construction cost, as savings would result from making the facilities in question operational earlier.

Availability of materials is often mentioned as an important factor, e.g. if suitable fill is available within a short-hauling distance for massive retaining structures or if sound aggregates for production of concrete are within reasonable reach. It is an important factor but it should not be considered as a separate issue, but rather an input to the common denominator, the cost analysis.

Two important site conditions

There is no doubt that two factors, water depth and soil conditions, have tremendous influence on the alternative solutions which ought to be investigated, as well as on the final result.

When a quay structure has to be constructed in shallow water or on existing land in connection with the dredging of a harbour basin, a structure which can be constructed through existing soils, e.g. a sheet-pile wall, is likely to be very competitive, while when the existing water depth is close to the desired one, other types, e.g. open structures on piles, may he more competitive.

The soil conditions are decisive, when hard layers prevent tension piles from penetrating to sufficient depth or make necessary dredging prohibitively expensive, and when weak and/or soft layers may endanger the overall stability or cause unacceptable settlements.

The presence of bedrock at or within a few metres below the design water depth tends to favour block walls and cellular bulkheads. A steel sheet-pile wall, however, may still be competitive even though trenching in the bedrock may be necessary.

Often the location of a new wharf or quay structure is fixed, such as in the case of extension of a marginal wharf, where neither bends nor offsets between the old and new stretches of quay are acceptable. In other cases, preliminary design work and investigations lead to an approximate location only. For the designer, the exact location becomes a parameter which is used for optimising the structure. Again, the existing water depth and subsoil conditions are likely to become the two main inputs for the optimisation process.

Vertical-face structures versus open structures

It is necessary to distinguish between vertical face or massive structures and open structures with a deck supported on piles, as they respond differently to wave action. Where planned facilities are located at a site which is not well sheltered, three-dimensional hydraulic-model tests are usually required to show that one of the types of structures will provide better mooring conditions and less operational down-time than the other.

Further, the open structure may introduce a fire hazard, which does not exist for the vertical-face structure. This is the case in harbours where debris, oil or other inflammable liquids may be washed underneath the structure and coat part of the substructure and the slope.

In situations where neither the hydraulic conditions nor the aspect of fire hazards are relevant, the choice of one of the two types should be governed by construction and maintenance costs only, except if the required construction time is different for the two alternatives and early completion is desirable.

Cost estimates

The actual design of a quay structure in terms of structural and geotechnical safety, is a fairly rigorous exercise. When it comes to measuring the result in terms of cost the accuracy is considerably lower. This is true in particular for 'absolute' cost estimates but also for comparative cost estimates for different types of structures. The usual practice is that alternative, preliminary designs with cost estimates are prepared and a solution for detailed design is selected on this basis. Not until tenders have been received or, in case of cost plus contracts, not until construction has been completed will the designer be able to check his estimate and use his experience for future jobs. The designer, however, hardly ever gets an effective feedback concerning whether his choice at the time of preliminary design was the right one, or if a solution, which at that time appeared to be more expensive, would in fact have a lower construction cost.

Cost estimates are usually based on unit prices obtained from previous projects of similar nature. Such unit rates are modified to reflect common price escalation and specific, actual conditions. It has been advocated that consultants should prepare their cost estimates in the same manner as contractors, i.e. by estimating separately the various components making up a unit price such as rent, write-off and maintenance costs for construction equipment, petrol, labour, purchase and freight of materials, administration, insurance and profit. This would require a detailed plan for the execution of the construction. We do not believe this is a move in the right direction. Designers are normally not construction experts and therefore would not improve the unit prices by estimating the above-mentioned components. Neither could it be expected that designers, in addition to their other skills, would be able to develop costing experience which compares with that of contractors. However, we think it would be in the interest of owners and clients, in cases where preliminary design indicates that two or more solutions are competitive within a relatively narrow margin, that both solutions be detailed and issued for tendering, leaving the pricing to the contractors. The additional cost of preparing detailed design of two solutions rather than one may be small, probably in the order of 1 per cent of the construction cost, while the savings may easily amount to five or ten times as much. If tender documents do not include alternative designs they should, at least, present complete design assumptions and invite alternatives from the contractors in order to take advantage of the possible availability of special equipment or experience favouring alternative solutions.

While the competent designer is not expected to have the experience to plan the construction in detail he should have a feel for what makes construction convenient and less expensive and what complicates matters.

In the general discussion of costs, it is worth noticing that when comparing the cost of a quay structure at one site with the cost of an apparently similar structure at another site the two unit costs may come out quite differently. Reasons may be differences in availability of materials, in mobilisation costs or in weather and sea conditions, which may require different methods of construction. Even when sites are similar, comparisons should be made with caution because quay structures are often constructed in conjunction with other works and such other works may have considerable influence on the cost of the quay itself, e.g. with respect to sharing mobilisation costs. The above reasons may explain why very little information has been published concerning comparison of actual costs for different structures. One investigation is, however, available

339

(National Ports Council, 1970). It covers a great number of structures and shows definite cost trends for the different types.

Quay structures

General considerations

In the following, a number of important types of quay structures will be dealt with. The number of types is not intended to be exhaustive and, moreover, the presentation goes in more detail with the sheet-pile solutions than with other types of structures. This is because a rational method for determining earth pressures acting on sheet-pile walls (flexible walls), developed by the late Professor J. Brinch Hansen (Brinch Hansen, 1953; Brinch Hansen and Lundgren, 1960) on the basis of theoretical work, laboratory model tests and practical experience, has made sheet-pile walls very competitive. Experience with this type of structure is available in abundance. The method has been incorporated in the Danish Code of Practice since 1965.

The traditional sheet-pile wall and the sheet-pile wall with a relieving platform are vertical face, massive structures. Gravity structures, like block walls, cellular-steel sheet-pile bulkheads and caissons belong to this group. The open structures are represented by pile-supported suspended decks or platforms.

The various types are shown in Fig. 9.1.

Traditional sheet-pile walls

(a)   *Principles and forces*: The normal arrangement of a sheet-pile wall is extremely simple and is shown in Fig. 9.2. A wall designed to absorb moments about a horizontal axis will retain fill or original soil and possibly also be exposed to hydrostatic pressures W. The earth pressure behind the wall is a function of the manner in which the wall is assumed to fail. This earth pressure E_a is mainly active and will increase with increasing live load p on the apron. The wall is laterally supported by tie rods at a level near the top and by the passive earth pressure E_p, which will develop in front of the wall, below the future seabed.

The necessary penetration depth z is determined as the depth at which the passive earth pressure E_p together with the anchor pull A is in a state of equilibrium with the active earth pressure E_a and the hydrostatic pressure W. Bollard forces B may either be distributed over a certain length of the wall and taken into consideration as a force acting on the wall or be picked up individually by separate anchors or in some cases by batter piles.

Berthing forces I from fenders are practically never a problem for these types of structures. They will relieve the anchor pull and

	TYPE OF STRUCTURE	Static principle	Construction materials	Remarks
VERTICAL FACE	 BLOCK WALL	Gravity structures	Blocks: mass concrete Bedding: stones, quarry run Backfill: stones	Reasonably good subsoil needed. Preloading of structure normally required. Some types susceptible to settlements
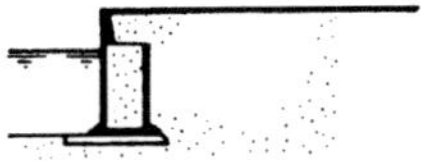	CAISSON		Caissons: reinforced concrete Bedding: stones, quarry run Capping: reinforced or mass concrete	Reasonably good subsoil needed. Calm sea while placing required. Somewhat sensitive to settlements
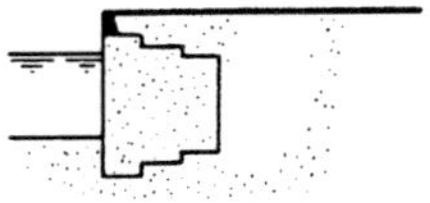	CELLULAR BULKHEAD		Cell: flat web-steel piles Fill: sand Capping: reinforced or mass concrete	Sensitive to horizontal loads until cell has been filled. Considerable settlements acceptable until *in situ* construction of capping wall
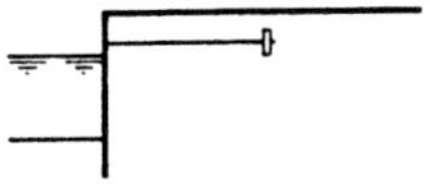	SHEET-PILE WALL	Flexible retaining structures. The wall distributes earth pressure by bending of sheet piles. The load is absorbed by tie rods anchored to slabs or batter piles and by passive earth pressure in front of the toe of the wall	Wall: steel, reinforced concrete Tie rods: steel Anchor slabs: reinforced concrete or steel Fill: sand	Structure itself absorbs settlements, which may however not be acceptable for quay apron. Often the least-cost solution
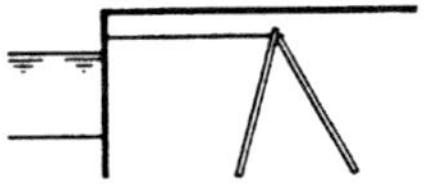	SHEET-PILE WALL		Wall: steel, reinforced concrete Platform: reinforced concrete Piles: reinforced concrete, prestressed concrete, steel	
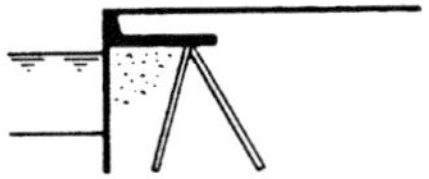	SHEET-PILE WALL WITH RELIEVING PLATFORM			
	SHEET-PILE WALL WITH RELIEVING PLATFORM			
OPEN	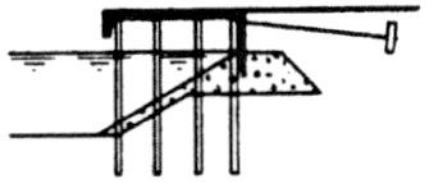 OPEN-PILED STRUCTURE	Vertical loads are carried by piles, while horizontal loads are absorbed by anchors or by batter piles	Deck: reinforced concrete, cast in situ or precast Piles: reinforced concrete, steel Slope protection: stones	
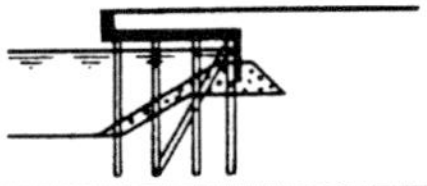	OPEN-PILED STRUCTURE WITH RELIEVING PLATFORM			

Fig. 9.1 Types of structures

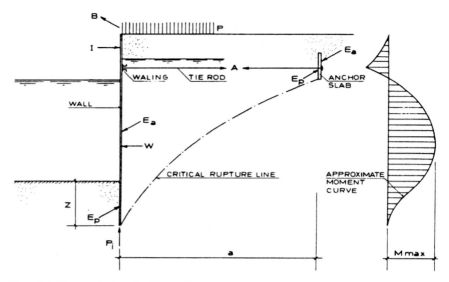

Fig. 9.2 Normal sheet-pile wall

develop local, passive earth pressure behind the top of the wall and for this purpose it may be necessary to arrange structural members on the rear of the wall which will ensure that the necessary length of wall is activated.

The resulting vertical component of E_a and E_p will normally be a downward force, which will be absorbed as toe resistance by the soil below the toe of the sheet-pile wall. For reasonable soil conditions the bearing capacity of the toe will considerably exceed the resulting downward force.

The anchor force is absorbed as passive earth pressure in front of an anchor slab for each tie rod (Ovesen, 1964). The distance a from the sheet pile wall to the anchor slab is determined as the minimum distance for which the overall stability of the structure can be obtained. Usually the critical rupture line will pass through the bottom of the anchor slab and the toe of the wall.

An alternative sheet pile wall is shown in Fig. 9.3. Here the wall penetrates to such a depth that it is fixed in the ground. For otherwise similar conditions, the maximum bending moment and the anchor pull will be less than for the wall on Fig. 9.2 and a lighter wall profile may be used. The optimum, of course, is when the moment from fixity equals the moment on the free part of the wall. This type of wall is suitable for large water depths or in case the backfill is of rather poor quality.

(b) *Bending moments in sheet-pile walls*: Figures 9.2 and 9.3 also show the approximate moment distribution on the two sheet-pile walls.

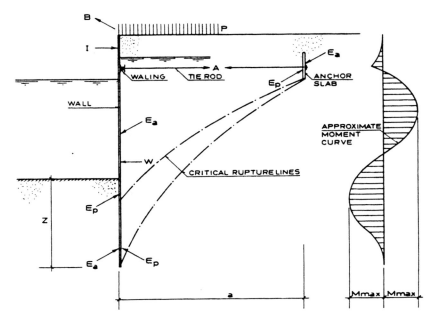

Fig. 9.3 Fixed-sheet pile wall

Where there is little or no tide, the tie rods will normally be connected to the wall slightly above the mean water level for convenient construction. In such cases and when the water depth is 8 to 12 m, the numerical value of the moment in the wall at tie-rod level may be only one-third to two-thirds of the mid-wall maximum moment. For steel sheet-pile walls, this lower value of the moment and its zero value at a lower level is a great advantage because initially a strength reserve is thus available in the zone of maximum corrosion.

It should be noted that sheet piles made of rolled-steel profiles provide the same bending strength over their entire length, but this strength is required only in one or two places. This fact has, in some cases, been exploited by using steel sheet-piles of insufficient strength and reinforced where needed by welding steel plates to the flanges. Similarly, for concrete-sheet piles the reinforcement may be designed to suit the bending moments if driving requirements are not governing.

In the wall on Fig. 9.2, the moment decreases from a maximum at mid-wall to zero at the toe. Therefore, the full section of the sheet-pile wall is not required at the lower end; every second pile could be stopped at a higher level. This would reduce the passive earth pressure on the part of the wall in question, but due to arching between the remaining staggered piles only by 5 to 20 per cent. Savings in material and driving costs will result, but may only be marginal. The aspect, however, is particularly important if driving is difficult and some of the sheet piles do not penetrate as required, because the loss of passive earth pressure may then be easily recovered by a small additional penetration of adjacent piles.

(c) *Anchors for sheet pile walls*: In Figs 9.2 and 9.3 the anchor slabs are shown at a certain distance from the sheet-pile wall. In some cases every second or third anchor slab could be placed closer to the wall without jeopardising the safety of the structure and thus result in savings for the total length of tie rods. However, the savings in materials may not always come out as a saving in total cost, because the contractor may find it inconvenient to install two or more different lengths of tie rods, or the variation in position of the anchor slabs may be inconvenient for other reasons.

Where soil conditions are poor it may not be possible to achieve the required overall stability with reasonable anchor lengths and it could be necessary to replace these by bents of batter piles. Absorbing anchor forces in bents of batter piles should also be considered where the piles could be utilised for other purposes, e.g. the support of crane-rail footings.

Instead of supporting sheet-pile walls by nearly horizontal tie rods, inclined anchor piles may be used (Schmidt *et al.*, 1964), see Fig. 9.4. This method may be preferable, e.g. in case a new wall is installed in front of an old quay wall and where it is found inconvenient and uneconomical to break-up large areas of existing

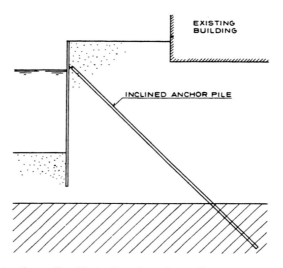

Fig. 9.4 Sheet-pile wall with inclined anchor piles

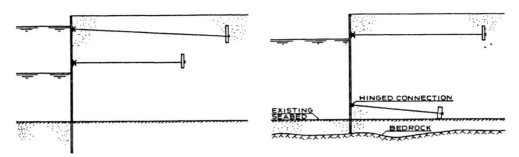

Fig. 9.5 Sheet-pile walls with anchors at two levels

pavement to be able to install tie rods and anchor slabs. The inclined anchor piles, however, will cause a considerable vertical, downward force on the sheet-pile wall, which must be accounted for.

Where conditions, e.g. a considerable tidal range, favour the installation of anchors at two levels (Fig. 9.5), substantial reductions in the bending moments in the sheet-pile wall may be obtained due to the more direct transfer of lateral loads, as axial load in tie rods rather than as bending in the wall. Also the necessary penetration depth of the wall will be reduced, because less passive earth pressure in front of the wall is required. The reduction of the passive earth pressure will increase the resulting downward force on the wall, and bearing capacity at the toe may be a problem. This aspect should always be checked when using anchors at two levels. Anchors at a very low level, as shown in Fig. 9.5, may also be of interest in cases where sufficient passive earth pressure cannot be mobilised, e.g. if penetration into rock is too costly.

(d) *Weak cohesive soils under toe*: Soils like sand and gravel are excellent for construction of sheet-pile walls. Also walls which have been constructed in alternating layers of granular and cohesive soils are performing well. It is, of course, not possible to discuss the performance of sheet-pile walls for all possible soil conditions. However, there are three important aspects to keep in mind for a sheet-pile wall located in 'sound' soils, except for a layer of weak or relatively weak cohesive soil immediately at or a little below the toe of the wall, as shown in Fig. 9.6.

First, it should be checked whether sufficient bearing capacity is present at the toe of the wall to absorb the downward force resulting from friction between the wall and the surrounding soils (resulting vertical components of E_a and E_p). Second, the calculation of the overall stability should not be restricted to curved failure lines as shown in Figs 9.2 and 9.3, but should also include horizontal

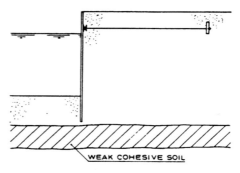

Fig. 9.6 Sheet-pile wall on weak cohesive soil

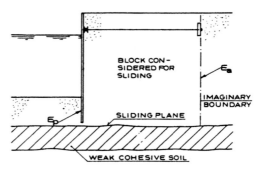

Fig. 9.7 Sliding of sheet-pile wall

sliding along the top of the weak cohesive soil layer, as shown in Fig. 9.7.

The third aspect is concerned with calculation of settlements arising from consolidation of cohesive soil under the toe of the wall, as illustrated in Fig. 9.8. Such settlements depend on the changes of effective, vertical stresses in the cohesive soil, which will decrease by dredging in front of the wall and increase by the placing of fill behind the wall. The change of effective vertical stress $\Delta p = \frac{1}{2}(h_3\gamma_3 + h_2\gamma_2 - h_1\gamma_1)$, where γ is the effective unit weight and h is the thickness of soil layers. Location of the wall at a certain existing water depth will make the two contributions cancel each other out, and there will be no settlement of the wall. Naturally, as the landward distance from the wall increases, the relieving effect of the dredging will decrease, and depending upon the properties of the cohesive soil, settlements may develop at a distance behind the wall.

(e) *Differential water pressure*: A hydrostatic or differential water pressure on a sheet-pile wall is a load which has to be supported by the structure. If it can be reduced by inexpensive means overall savings result.

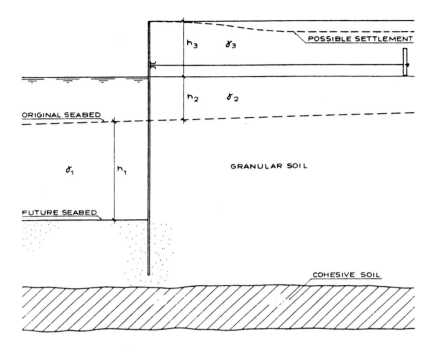

Fig. 9.8 Sheet-pile wall settlements

Differential water pressures develop when the water level in the harbour basin is lowered and remain until the ground-water level behind the wall has reached the same level. Lowering of the water level may be caused by tidal variations, sudden disappearance of wind set-ups or by long-period waves, as well as short-period waves. The magnitude of the differential water pressure depends mainly on the magnitude and velocity of the lowering of the water level, the permeability and specific yield (i.e. drained water volume relative to total volume) of the backfill and whether special drains have been provided.

Recommendations are available concerning the magnitude of differential water pressures at waterfront structures (Committee for Waterfront Structures, 1996). Different cases are dealt with on what appears to be an empirical basis. A more scientific approach (Carlsen, 1966) presents formulae and diagrams from which differential water pressures may be found. The presentation is related to different types of drains and it is found that a horizontal drain blanket is superior to a horizontal line drain and a vertical drain curtain, both immediately behind the wall.

It is important to realise that the effect of differential water pressures is not confined to the back of the wall itself but must also be considered

347

in rupture zones and lines. Consequently, it is not a matter of draining only the back of the wall but the entire rupture zone.

A higher water table behind the wall than in front will also cause flow around the toe of the wall, the gradient of which will influence the soil in the active and passive zones. Generally, the active earth pressure will increase and the passive earth pressure will diminish. Quantitatively, this effect can be investigated by tracing a flow net.

A special case of differential water pressure will develop if an impermeable wall is back-filled hydraulically. In that case, a free water table is created behind the wall and unless discharge by pumping is provided for, this temporary situation may be governing for the design of the wall.

(f) *Computer programs*: Computer programs for the Brinch Hansen theory are available for determination of necessary penetration depth, maximum moment in wall and, if relevant, anchor pull. The programs are able to cope with up to ten different layers of soil (cohesive and non-cohesive) on either side of the wall where surfaces may be sloping, while boundaries between layers must be horizontal. Input data comprise levels of top of wall, tie rods, water tables as well as boundaries, unit weight, internal angle of friction and cohesion for each layer. Also among input data are live loads and vertical components of gradients caused by different water tables on the two sides of the wall and a choice of the mode in which the wall is assumed to fail. This choice does not influence the safety of the wall but only its economy, and therefore more than one mode of failure may have to be investigated. The computer programs provide several possible modes of failure for walls with one, two or no yield hinges and walls fixed in a superstructure, e.g. as mentioned in the following section. These programs make the basic design of sheet-pile walls a very simple matter.

Sheet-pile wall with relieving platform

A sheet-pile wall with relieving platform may be an attractive alternative to the traditional sheet-pile wall in case the soil behind the wall is weak and therefore develops relative high-active earth pressures or unacceptable settlements, in case of extremely high unit loads on the quay apron, e.g. storage of bulk materials or metal ingots or in case it is desirable for reasons of economy or foreign exchange savings to increase the quantity of concrete and reduce the quantity of steel in the structure.

This type of structure was developed by Christiani & Nielsen (Brinch Hansen, 1946). The basic idea is that the weight of part of the fill immediately behind the wall is carried by piles directly to firm strata whereby the active earth pressure will be reduced and the wall relieved.

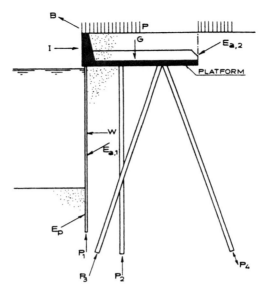

Fig. 9.9 Sheet-pile wall with relieving platform

Two designs of this type of structure are shown in Figs 9.9 and 9.10. In Fig. 9.9 the concrete superstructure is placed directly on fill, whereas in Fig. 9.10 the platform spans a sloping fill which will exert less pressure on the sheet-pile wall than the fill shown in Fig. 9.9.

The earth pressures E_a and E_p acting on the wall are functions of the mode in which the wall is assumed to fail, as for the traditional

Fig. 9.10 Alternative sheet-pile wall with relieving platform

349

sheet-pile wall. For this structure E_a is independent of the magnitude of the live load p. The sheet-pile wall is laterally supported where it joins the concrete superstructure and lateral forces are transmitted to the two batter piles acting as an anchor. These piles also support the lateral components of bollard loads B, berthing forces I and earth pressure $E_{a.2}$ acting on an imaginary plane through the back of the concrete superstructure.

The structural weight, the weight of the soil above the platform and the live load are supported by the piles. If the sheet-pile wall reaches firm soil strata it may be utilised as a row of piles; if not, then the second pile row in Fig. 9.9 should be moved closer to the wall or an additional row of piles just behind the sheet-piles should be introduced. To the extent sufficient support of the piles can be established no settlements of the apron above the concrete superstructure will occur.

The pronounced cantilever of the concrete-deck slab behind the two batter piles is intended for two reasons. First, because vertical load will be added to the batter piles whereby tension in the rear pile will be reduced or eliminated. This is of importance where the rear pile is fully embedded in granular material and therefore may not be able to develop substantial surface resistance. Second, because the cantilevered deck, to a considerable extent, will shield the rear batter piles from the effects of subsiding fill, which would cause shear forces in the piles and tend to break them.

It is possible to fix the sheet-pile wall in the concrete superstructure so that moments are transmitted and it is also possible to assume a mode of failure of the wall so that the fixing moment equals the mid-wall moment, whereby the capacity of the wall will be utilised to the maximum extent. The location of the fixing moment, however, will usually coincide with the zone of maximum corrosion. This aspect is of little concern if the sheet-pile wall is made of reinforced concrete, but steel sheet-pile walls require protection against corrosion. Otherwise, it is necessary to select a mode of failure which will reduce or eliminate the fixing moment at the cost of an increase of the mid-wall moment.

The construction of a sheet-pile wall with relieving platform in an area where the water depth is close to the desired depth requires installation of temporary anchors between the wall and the bents of batter piles before fill can be placed and the superstructure cast supported on the fill. Such temporary tiebacks are not necessary if the quay wall is constructed on existing or temporary land or in shallow water, and the harbour basin is dredged afterwards.

For the alternative design, shown in Fig. 9.10, it is necessary to cast the concrete-deck slab on forms supported by scaffolding, none of which may be recovered, or alternatively use precast concrete members.

Gravity quay wall

(a)   *Block wall*: Block walls consist of a number of concrete blocks placed on top of each other and capped with a mass concrete wall. Today, block walls are not normally considered to be economical structures, but over the years many have been constructed in various parts of the world. The chief advantages of block walls are high durability and simplicity in construction. This combination makes them an attractive alternative in certain circumstances, in spite of their normally high cost.

The static principle in block walls and in other gravity walls is to give the structure such a weight that the resultant weight of the wall and earth pressure will intersect the base, at a distance from the front edge whereby sufficient effective base width for transmitting the load to the ground will be obtained (see Fig. 9.11). Further, it is necessary that the resultant lateral forces can be supported by friction forces at the base plus possible passive earth pressure on the front face of the base. A block wall will normally have a toe, which will increase the effective width almost without adding weight or cost. Also, normally, one of the lower blocks will be protruding at the rear to pick up vertical load from the fill.

It is preferable to use rubble as back-fill because it has a relatively high angle of internal friction, low-active earth pressure and also will reduce differential water pressure through favourable drainage characteristics.

A block wall should be founded on even stone bedding placed on the top of a rubble base or on underwater mass concrete. The weight of the blocks may range from 40 to 125 t, the decisive factor being the

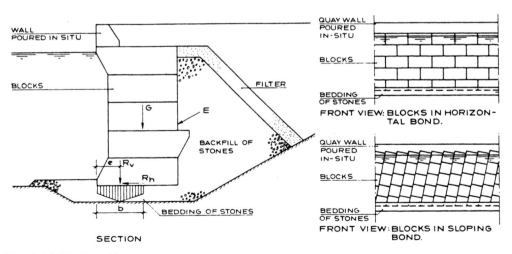

Fig. 9.11 Block wall

capacity of available plant. Larger blocks will lead to a smaller number of placing operations, an activity, which requires skilled divers and extensive supervision.

Because of their weight, block walls require reasonably good sub-soil conditions and it is therefore necessary to substitute soft soils with sand or other suitable material. However, despite the presence of good soil conditions, it is common to experience considerable settlements originating from compression of stone bedding and subsoil and therefore pre-loading of the wall prior to casting of the capping should be reckoned with. Pre-loading can be made with blocks to be placed later in other parts of the wall, or if the blocks are placed in a land-based operation, the transport can be arranged over sections of wall already in place. It is not uncommon that the top of the wall exhibits a seaward movement when being back filled. It is therefore preferable to postpone casting of the capping until back-fill has been placed to final grade, and perhaps to place the blocks with a small backward slope.

Walls with the blocks in horizontal bond are well suited if founded on bedrock or other hard ground, which will not cause settlements. Uneven settlement may cause cracking of blocks. Where settlements are expected it is preferable to use walls with sloping bonds because each inclined column of blocks can settle individually.

(b) *Caissons*: These may be used for construction of ordinary quays, although more frequently they are used in breakwaters and, in particular, where berths are located along the inside of a breakwater.

A quay constructed from caissons may be considered an extreme case of a block wall with each cross section consisting of only one block, not in the form of solid concrete, but in the form of a sand-filled concrete box. The static principle therefore is the same as for block walls. A typical caisson solution is shown in Fig. 9.12.

The caissons are placed on a bedding of stones, the surface of which shall be thoroughly levelled by divers. Sand is normally used as back-fill and therefore the joints have to be sand tight. If rubble, with a higher angle of internal friction, is used as back-fill the width of the caissons may be reduced due to the lower earth pressure. This reduction would only save concrete in the diaphragm walls and the bottom of the caissons, whereas, the total amount of fill inside and behind the caissons would be unchanged. Normally this saving is not sufficient to pay for the additional cost of using rubble as back-fill.

The construction of the lower part of caissons will have to take place on a slipway, in a dry dock, in a floating dock or in a shallow-water area where the water table has been temporarily lowered.

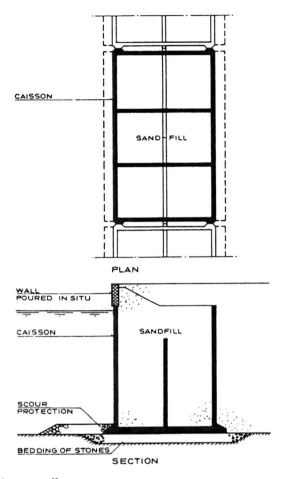

Fig. 9.12 Caisson wall

The upper part can be constructed directly on the floating caisson if it is in a sheltered area.

Whether a caisson solution shall be considered depends very much on actual conditions because, in some cases, it is a virtue that the caissons can be constructed at a different location, e.g. where materials and labour are available, and then towed to the site and placed during a calm period; whereas, in other cases, it is a problem that special facilities and requirements to water depths are necessary within reasonable towing distance.

Quay walls made of caissons require reasonably good sub-soils. If uneven settlements occur, joints between individual caissons could open and let back-fill escape and could also cause cracks in the concrete wall usually cast *in situ* on top of caissons.

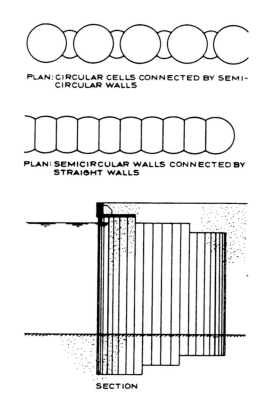

PLAN: CIRCULAR CELLS CONNECTED BY SEMI-
CIRCULAR WALLS

PLAN: SEMICIRCULAR WALLS CONNECTED BY
STRAIGHT WALLS

SECTION

Fig. 9.13 Cellular bulkhead wall

(c)   *Cellular bulkheads*: These consist of a number of sand-filled, circular
or semicircular cells constructed of flat-web steel-sheet piles. Two
arrangements of the sheet piles in plan are shown in Fig. 9.13. The
most common is the one where a row of circular cells is inter-
connected by semicircles, whereas in the other arrangement, two
rows of semicircles are connected by straight walls. On top of the
cells at the front is constructed a reinforced concrete wall.

The special type of sheet piles used for cellular bulkheads is
characterised by heavy locks able to sustain the ring loads developed by the fill inside the cells. These loads increase with depth
below the surface and consequently the steel is only lightly stressed
in the splash and tidal zone where maximum corrosion occurs.

The diameter of the cells or the width of the wall depends on the
water depth and the live load on fill behind the wall. The amount
of steel per unit of length of quay is almost independent of the
diameter and therefore the economy of the structure is tied to the
cost of placing fill in the cells and of the bracing of the cells during
construction.

The static principle for supporting the loads acting on a cellular bulkhead is the same as for the block wall or the caisson. However, when investigating the stability of a cellular bulkhead, possible ruptures penetrating through the fill from the toe of the front wall to the toe of the rear wall must be also analysed. When considering a cellular bulkhead, it is important to realise that during construction it is necessary to place all the piles for a cell, lock-in-lock, before driving can commence. During this operation and until the cell has been partly or completely filled, it must be adequately braced against horizontal loads from waves or currents.

Cellular bulkheads can be located on any subsoil, which is able to carry their load. During the period of construction considerable settlements and horizontal deformations may be accepted. However, settlements should not develop once the reinforced concrete wall on top has been cast. For this reason, it is preferable to drive the piles at the front to strata, which will not cause settlements.

For more detailed information on cellular bulkheads reference is made to Leimdorfer (1972–73) and Ovesen (1962).

Open structures

The open-piled structure with suspended deck is widely used. This type of structure offers a great freedom with respect to choice of materials. The piles can be of steel, reinforced or prestressed concrete or of timber. The deck will normally be constructed of reinforced concrete either cast *in situ* or partly pre-cast and partly cast *in situ*. The centre-to-centre distance between the rows of piles is a matter of least cost for deck plus piles. The optimum distance depends heavily on the magnitude of live load and the depth to strata where piles can be founded. Further, the pattern of piles may be governed by the location of crane rails or supports for other cargo-handling equipment.

Underneath the deck the slope should be as steep as possible. If the wharf is constructed in shallow water and the existing soil does not need to be replaced, it is natural to dredge this slope in the existing soil and to cover it with one or more layers of stones if protection against scour is necessary. In case the wharf is constructed at a water depth close to the required one or if it turns out that a slope dredged in the existing seabed cannot be made sufficiently steep, it is common practice to place a stone dyke to retain reclaimed fill. In general, the slope underneath the structure, whether established by dredging or by filling, should be constructed prior to pile driving to avoid lateral forces on the pile shafts developing as a result of lateral soil movements generated by construction of the slope.

At the rear of the deck where the slope reaches its highest level it is necessary to install a low bulkhead in the form of sheeting or a concrete

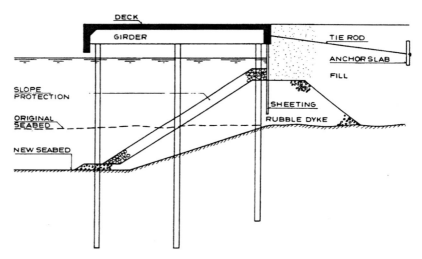

Fig. 9.14 Anchored open structure

structure to retain the fill. Where wave action occurs, this corner is a critical point in the structure requiring special attention to avoid damage.

Unless the deck slab has to support certain facilities, e.g. crane rails at fixed positions, its width is the choice of the designer; but again it is a

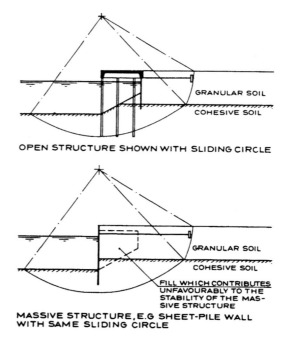

Fig. 9.15 Comparison of sliding stability for open structure and massive structure

matter of economy. If the width is reduced to save cost then the height of the sheeting or the concrete structure at the rear increases and adds cost and vice versa. A typical cross section is shown in Fig. 9.14.

Compared with massive structures like steel-sheet-pile walls or gravity walls, the open structure has more favourable overall stability properties because of the lack of fill near its front, (see Fig. 9.15). In many cases, however, this point has no bearing because either type of structure would be stable, but for certain conditions the open structure may come out safe where the massive structure would not be safe.

Vertical live loads are supported by the deck and transmitted to the piles. Horizontal forces from ships' impacts are transmitted through the deck and absorbed as earth pressure behind the low bulkhead. Mooring forces and active earth pressure at the rear of the low bulkhead are supported either by batter piles or by tie rods and anchor slabs as shown.

It is a fact that a number of open wharves have been constructed with fill placed on top of the deck as shown in Fig. 9.16. Both batter piles and tie rods plus anchor slabs have been used for picking up horizontal loads. In the former case the weight of the fill is likely to pre-compress the vertical pile adjoining the batter pile so that tension will be less or will not develop for horizontal loads such as mooring forces. When tie rods and anchor slabs are used, the fill on top of the deck only adds unnecessary load to the deck and the piles.

Often it is necessary to provide the edge of the deck with a curtain or take other measures, which will ensure that small vessels with low freeboard are not trapped underneath the deck at low water.

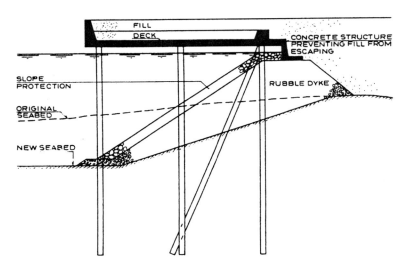

Fig. 9.16 Open structure with batter piles

Piers

General considerations

Piers may he divided into three types: (a) piers consisting mainly of reclaimed land and bordered by quays, (b) piled piers, and (c) floating piers. The first type is merely a particular arrangement of a quay structure, which is dealt with in the previous section. The other two types are dealt with below.

The layout of piers in the plan also varies. Finger piers abut the shore at one end while T-head and L-head piers are connected to land by a trestle and/or a causeway with an access road.

Piled piers

Piled piers may be chosen rather than massive piers because of cost, e.g. if the pier is located at considerable water depth, if the massive pier may develop unacceptable settlements or if a non-reflecting structure is preferred for wave disturbance reasons.

Piled piers simply consist of a deck supported on piles, but the pattern of piling, the type of piles and the particular design of the deck may be varied and combined almost infinitely. Piles may be of steel, reinforced or prestressed concrete or timber.

When steel piles are used, the fundamental choice is between an H-pile and a pile with a hollow section (tubular pile). The H-pile is usually less costly but offers less column bearing capacity for same weight, due to the smaller moment of inertia, in particular if the pile is long. Further, the pile with the hollow section is more corrosion resistant because corrosion will take place almost only on the outside surface. Hollow steel piles are often concrete filled. If this is done for the purpose of completely avoiding corrosion from inside, it is probably an expensive measure compared to allowing for some extra steel in the pile itself. The deck will normally be of reinforced or post-tensioned concrete either with a plain slab or with girders in one or two directions. Cast *in situ* concrete is widely used, but also pre-cast deck slabs or combinations of precast and *in situ* concrete are often used.

The most economic choice in a particular case depends mainly on the type of live load, water depth, soil conditions, availability of plant and materials, experience of potential contractors and whether the new pier is located in a sheltered or non-sheltered area. The static principle of a pier is illustrated in Fig. 9.17. Structural weight and vertical live loads are carried by the deck and distributed to the piles. Horizontal loads from ships' impacts and mooring forces are normally very decisive for piled piers. These loads have to be transmitted to the ground via the piles. Depending on the number, stiffness, capacity and fixity in the ground of the piles this may be possible with only vertical piles if the corresponding

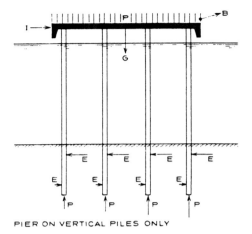

PIER ON VERTICAL PILES ONLY

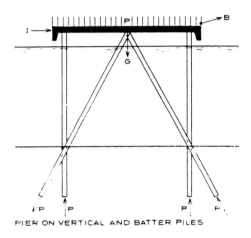

PIER ON VERTICAL AND BATTER PILES

Fig. 9.17 Pier on piles

deflections are acceptable. More frequently, however, batter piles are used to transmit horizontal loads. If sufficient pulling resistance cannot be obtained due to the ground conditions, it may be necessary to introduce extra weight in the deck either by concrete or by sand fill.

Because of the importance of the horizontal loads on piled piers, one should realise that mooring forces resulting from wind pressures on vessels cannot be reduced once the maximum profile of vessel is established. However, the berthing impact forces are functions of the type and characteristics of the fendering chosen for the pier and other parameters related to vessels and their approach manoeuvres.

Thus, for piled piers the fendering becomes an important integral part of the structure. This is contrary to the quay structures where berthing impacts are fairly easily transferred to the ground behind the structure.

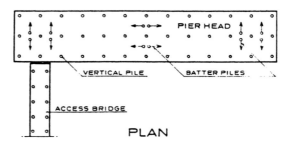

Fig. 9.18 Location of batter piles

The resilience of piers on vertical piles has in a number of cases been utilised to absorb impacts from berthing vessels. Deflections of the pier of up to 15 cm have been allowed and proper fendering has either been omitted or greatly reduced. It is necessary to ensure that pipelines leading to the pier are sufficiently flexible and that a flexible joint is provided between access bridge and pier. Further, it should be realised that the piles are subject to maximum utilisation (axial load and bending) at the soffit of the deck, which is a region with a harsh environment because of alternating wetting and drying. Finally, the ability of a pier to absorb berthing impacts by elastic deformation of the structure itself is considerably lower at the ends than in the central section, whereas normally the ends of piers, in particular, are exposed to impacts. In a rigid pier it is preferable to have transverse batter piles at the ends to absorb berthing impacts, which would otherwise cause torsion of the entire pier structure. Batter piles in the longitudinal direction may be located near the centre of the pier (see Fig. 9.18).

Whether the pier is rigid or resilient the deck will act as a beam deflecting in a horizontal plane and will distribute horizontal loads to the piles.

Often it is necessary to provide the edge of the deck with a curtain or take other measures which will ensure that small vessels with low freeboards are not trapped underneath the deck at low water.

Floating piers
A floating pier is a type of structure, which is completely different from the types of structures previously described because it is a vessel and not a fixed structure. A floating pier or pontoon could be an alternative to a fixed structure where soil conditions are so bad or water depths so excessive that the foundation, normally piles, becomes prohibitively expensive or in case a very short construction period is essential. A short construction period is possible because the floating part of the structure is constructed in a shipyard which will not require a time-consuming mobilisation of plant and equipment and because the structure, in fact, is a very simple

vessel with minimum equipment and therefore can be constructed very quickly if shipyard capacity is available. To the actual construction time shall be added time for towing to the site and time for installation of anchorage. Floating piers have been dealt with by Tsinker and Vernigora (1980).

A floating pier consists of a pontoon, an anchoring system and an access bridge connected to an abutment at the shore.

The pier itself is normally box-shaped and constructed as a ship's hull with longitudinal and transverse bulkheads. In addition to providing strength, the bulkheads serve to subdivide the pontoon into compartments where machinery, pumps and ballast tanks are placed. By changing the ballast, the draught and freeboard can be adjusted.

A floating pier would normally be built in accordance with rules and regulations of a ship classification society.

The width of a floating pier depends not only on the space required for cargo-handling operations, but also on the requirements to the stability with respect to uneven distribution of live load. In this respect, the width should not be less than 15 m. The length of the pier depends on the vessels to be accommodated. However, for reasons of wave conditions, not only at the actual site but in particular during the towing operations, it may he necessary to build the pier in sections, which are hinged together in place. Normal dock fenders are usually provided at the berthing faces.

Floating piers constructed and installed in a location in the Middle East were 180 m long, 40 m wide and had a draught of 2 to 3 m and a total height of 4.5 m. They were divided into three hinged, 60 m long, sections. Commonly floating piers are kept in position by anchor chains. The chains are fastened in wells, which go through the pontoons. From the deck, there is access to the wells for adjustment of the anchor-chain length. In order to obtain sufficient holding power of the anchors, it is necessary to place these at some distance from the pontoon so that the pull is near horizontal. On the other hand, it is also necessary that the anchor chains do not interfere with the berthing vessels so therefore sinkers are often fixed to the chains and the anchors are placed opposite the side where the chain is fixed to the pontoon.

Theoretically, four anchors are sufficient, but it is common to use six anchors, four for transverse forces and two for longitudinal forces.

The lengths of the anchor chains are so adjusted that they include slack to allow for tidal variations. When exposed to current and/or wind, the moored pontoon will move. The pier movements depend on the above-mentioned slack, but even if there is no tide, and therefore no particular slack in the chains, surge and sway in the order of 0.5 to 1.5 m should be allowed for.

If the bed is sandy or consists of firm clay, conventional steel anchors may be used. In case of very soft clay, which may be the reason for selecting a floating pier rather than a fixed structure, special mud anchors or specially designed anchors, which develop resistance against pull as a result of digging-in, can be used.

Access from the shore to a floating pier can be arranged in different ways but most commonly is obtained either by way of a trestle with a link span, which allows for motion of the pontoon, or an access pontoon bridge. The latter could be connected to the floating pier in a manner so that only relative movements in the longitudinal direction of the access bridge shall be accounted for.

Vessels berthing at a floating pier should not be moored partly to the pier and partly to fixed points, e.g. dolphins, because the moorings to the fixed points, if too tight, may have to absorb the combined motion of vessel and pier and thus be overloaded.

A floating pier is to be considered a moored vessel and, as such, there is a risk that it may come loose. Insurance companies have assessed this risk and charge an annual premium in the order of 0.5 to 1.5 per cent of the value depending on the extent of coverage.

Jetties

The characteristics of a jetty are that it consists of a number of individual structures each of which support a special type of load. The mooring dolphins pick up the pull from the hawsers, the breasting dolphins support fenders which absorb berthing impacts, and on-shore wind loads on the moored vessel and loading platform support special loading or unloading equipment, but normally no horizontal forces apart from wind loads on the equipment.

A typical layout of a jetty with one berth, only, is shown in Fig. 9.19. Catwalks which connect the various dolphins and platforms are indicated.

The location of the mooring and breasting dolphins, relative to each other, is an important issue in the design of jetties. Parameters are the

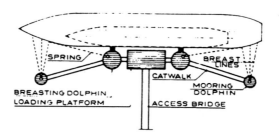

Fig. 9.19 Typical single-berth jetty

dimensions of the maximum and the minimum size of vessels to be served. The breasting dolphins should be placed as wide apart as possible but the distance should neither exceed the length of the straight side of the smallest vessel nor be less than approximately one-third of the maximum length of vessel, assuming that the vessels do not have to be shifted while loading or unloading. In addition to providing support for fenders, the breasting dolphins may also support bollards for spring lines.

Mooring dolphins for breast lines shall be located at bow and stern at a distance (about the beam of a ship) from the berth line, which will not make the moorings too steep. If the jetty shall serve vessels of different lengths, the location has to be a compromise, unless two or more sets of mooring dolphins are provided.

It is common, and in accordance with many captains' wishes, to also provide mooring dolphins for bow and stern lines because 'this is the way it has always been' and because they provide extra safety in case one of the other lines should burst. However, vessels can be safely moored with only breast and spring lines if the mooring points are properly located and if each mooring point allows for several hawsers. Further, under unfavourable conditions, bow and stern lines could introduce undesirable transverse forces in combination with surge.

That breast and spring lines are sufficient for safe mooring, at jetties where breast lines can be fixed well behind the berth line, has been confirmed at several jetties which do, in fact, serve vessels of considerably greater length than intended. In these cases, the dolphins meant for bow and stern lines are actually used for breast lines.

Unless the loading or unloading platform is combined with the two breasting dolphins, it is normally located at some distance behind the berthing line in order to avoid horizontal forces on the platform as a result of direct contact with the vessel.

If the jetty has two berths, one on either side as shown in Fig. 9.20, it may be an advantage to combine the two sets of breasting islands into one set

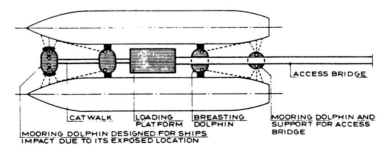

Fig. 9.20 Typical two-berth jetty

which could be used from both sides. Mooring dolphins then have to be placed (in the centreline) between the two berth lines and to avoid excessively wide breasting islands, it may be necessary to compromise on the distance from berth line to mooring point.

Mooring and breasting dolphins are almost entirely exposed to horizontal forces. If constructed with piles (vertical and batter), several of these will have to take pull because of the lack of downward, vertical loads, which could neutralise the pull. Sufficient pulling resistance may he obtained in clayey soils, but only with difficulty, if at all, in granular soils. Also gravity structures, e.g. in the form of caissons or individual, sand-filled sheet-pile cells, are widely used for dolphins in jetties. As for the gravity quay walls, the principle is that the dolphin is given a weight and the base is given width so that the resultant force intersects the base so far from the edge that the pressure under the dolphin will not exceed the bearing capacity of the subsoil.

It is often economical to give an individual caisson a circular, cylindrical shape. The hoop tension in the shell will make inside diaphragm walls redundant but a deck, which distributes the concentrated horizontal load to the walls, is required. The cylindrical shape compared with square or rectangular shapes will also reduce current forces and bed erosion effects.

Because the governing loads on a breasting dolphin are impacts from berthing ships, the fenders become integral elements in the design and they cannot be considered just items to be fitted at the end. The loading platform and its access way, if independent of the breasting dolphins, may be considered a pier and reference is made to 'Piers', earlier in this chapter.

Jetties are frequently located in open water outside harbours. If this is the case in regions where drift ice occurs from time to time, then the designer should realise that impact of drifting ice floes could exceed that of berthing vessels or could result in loads on the exposed mooring dolphins which may exceed the mooring forces. In areas with drifting ice of more than insignificant magnitude, piled structures should be avoided unless they are effectively sheltered by massive structures. Further, it may be necessary to stiffen thin-wall caissons and sheet-pile cells with inside concrete slabs buried in the fill at water level.

Fenders

It is a difficult task to select a fender for a quay or a jetty, because many parameters are involved and many options are available. Consequently, it is not possible to make general recommendations. In this section, some of the more important parameters are discussed and some of the options are presented.

Fenders may be installed for one or more of the following reasons:

(a) to prevent direct contact between quay and vessel while the vessel is moored alongside the berth (rubbing or protection fenders);

(b) to reduce the risk of a moored vessel moving in resonance with waves; and

(c) to absorb berthing impact energy.

Rubbing or protection fenders

These types of fenders are installed to prevent minor damage to quay and moored vessels as a result of the movements of the vessel caused by wind, water level variations, waves generated by wind or passing vessels, etc.

Rubbing fendering is the original type of fendering, so to speak, and commonly consists of vertical and horizontal timbers fixed directly on the quay wall or of timber piles supported at the top by the quay wall either directly or by a wale. Also small cylindrical hollow rubber fenders draped along the wall may be included in this category.

Energy absorbing fenders also function as rubbing fenders.

Fenders which improve mooring conditions

A vessel with its mooring lines and fenders constitutes a physical system with a natural frequency perhaps close to the frequency of waves or wave groups reaching the berth in question. If so, the vessel will move in resonance and, because of the increased amplitudes, loading or unloading will be impeded or made impossible and mooring and fender forces may reach unacceptable values. In such cases, it may be possible to change conditions radically by changing the characteristics of one or both of the elastic elements (fenders and moorings), e.g. fender deformation characteristics (ratio between force and deflection). Alternatively fenders, which dissipate energy rather than storing it and giving it back to the vessel, may be used. It is possible to determine, with some degree of accuracy, the requirements to fenders relative to movements of moored vessels not only in physical, hydraulic models but also by means of advanced simulation. The problems described above have grown in importance in step with more and more berths being located in less sheltered areas. The introduction of certain elasticity requirements does not necessarily mean that the fender cannot be used for energy absorbtion, but it will limit the options.

Energy absorbing fenders

The most important purpose of fenders is, no doubt, to absorb kinetic energy from berthing impacts. In the past, ships were smaller and comparatively much more sturdy and therefore less prone to damage during berthing. For today's average vessel of considerable tonnage, time lost in a dry dock is very costly. Therefore it is good economy to invest in effective fendering to reduce down-time and repair costs. Port owners and operators used to compare the cost of fendering with reduction of damage to quays only and therefore had less incentive for the investment than ship owners with their own terminals. The interest in fendering grew as oil companies acquired their own terminals.

Kinetic energy

Formulas for calculating the energy to be absorbed by a fender when hit by a berthing vessel have been presented by several authors but in particular, Vasco Costa (1964) has treated the subject thoroughly. The formulae developed by Vasco Costa and Ueda *et al.* (2001), for the virtual mass factor, are presented in PIANC (2002). For vessels with more than one hull, such as catamarans, the formulae cannot be used without modification. The formulae shall not be repeated here, but it shall only be recalled that energy to be absorbed is proportional to the displacement tonnage of the vessel and to the second power of the velocity. It also depends on the distance from the centre of gravity of the vessel to the contact point between vessel and fender and on the angle between the line connecting these two points and the velocity vector of the vessel. The energy increases with reduction both of distance and of angle, as illustrated in Fig. 9.21. The energy depends solely on the vessel and the manner in which it approaches the berth and the fenders. In particular, velocity is important. These parameters again depend on such conditions as current, waves (shelter), wind, loading condition of vessel, water depth, maneuvring space, number of tugs used, bow thrusters, single or double propeller and competence of pilot or captain.

Proper fender design would require determination of all these sub-parameters and their precise influence on the basic parameters. This is not possible. How may the competence and reactions of human beings

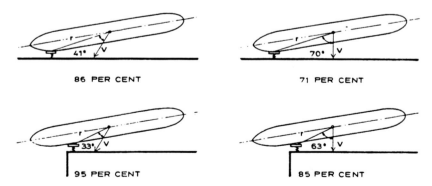

Fig. 9.21 Kinetic energy to be absorbed by fenders

involved in the process be predicted? However, it should be noted that efforts to develop computer simulation models are being made. In consequence hereof, the problem might more appropriately be treated as a statistical one. A guide to the determination of the energy to be absorbed by the fenders may be found from analyses of integrated computer simulations used by pilots and captains to familiarise themselves with new ships or ships at new ports.

PIANC (2002) contains a great deal of information, rules of thumb for preliminary design and selection of fender systems and other interface structures such as mooring dolphins, properties of materials, workmanship and inspection as well as maintenance. However, it seems not to have considered the overlapping areas covered in PIANC (1995) and there are contradictions between the two PIANC publications.

PIANC (2002) does not treat physical model determination of fender and mooring forces, which is recommended by PIANC (1995) to provide more detailed information about the most promising options and to optimise the final design. During the seven years between the two publications, sufficient further development of mathematical models have not taken place. The references are few and to a large extent limited to publications by PIANC and its committee members. The important subject of hull stress determination is dealt with summarily, saying only that if the side plating, longitudinal stiffeners and side traverses are known, it is possible to calculate the permissible hull pressure. This calculation is presented in 'Hull strength', chapter 5, with appropriate references.

Various textbooks and manufacturers of fenders provide data on approach velocities. Such data may be used but should not be relied on exclusively. They should rather be used as a point of departure for a dialogue with local pilots, who are going to be the users of the fenders. Quay fenders known by the pilots should be used for demonstrating the results of theoretical calculation. It is important that the two parties, during the discussions, agree on which are considered normal approaches and accident approaches, respectively, and to which extent the fenders shall provide protection. In this context the above-mentioned simulations constitute an extremely useful tool.

Fender reactions
Having thus determined the approximate energy required to be absorbed, the next step would be to establish acceptable fender reactions. For the vessels the total reaction is less important than the hull pressure (see 'Hull strength', chapter 5), except for such vessels as ferries and certain ro/ro vessels with a distinct fender list. If the fender is on a massive or open quay structure, normally there will be no problems in absorbing the reaction, which is transmitted to back-fill (see 'Quay structures',

earlier in this chapter). However, for open piers and jetties and on breast-ing dolphins, fender reactions are likely to be the governing horizontal forces for design of the entire structure. So the engineering problem is to work out if it pays off to invest in fenders with low reaction per absorbed unit of energy and save on supporting structure. Further, it is important to consider the fender performance beyond the rated energy capacity because fenders with a swift increase in reaction provide very little margin for overloading, which for these structures may have serious con-sequences. In this context it is noted that the buckling fenders (π shape or similar) seem to draw increasing attention. This type is associated with a load/deflection curve, where the maximum load is achieved at relatively small deflections and remains rather constant for the scheduled deflection range. As one consequence of this, the fender design safety factor should be applied to the ship's impact energy and the fender selected on this basis. Hence, the maximum force will almost certainly occur but it is well-defined (see below).

Types of fenders

The types of fenders available off-the-shelf and the types of fenders designed and constructed for ports and offshore terminals are numerous. Below are brief descriptions of only a few of the most common ones.

Rubber fenders are by far the most common. These fenders utilize the fact that rubber can deflect considerably under load and return to its original shape after unloading.

In pneumatic fenders air is contained in a bag of rubber. The air pres-sure will increase when the bag is squeezed between a quay and the side of a vessel. The fender will deflect under load and absorb energy.

Other types of fenders have been used but are not common, e.g. gravity fenders, where a heavy weight is lifted by the berthing vessel, or torsion-bar fenders, which absorb energy through elastic deformation of steel.

In all cases, the kinetic energy of the vessel (or part thereof) is tempora-rily stored in the fender and more or less given back to the ship when the fender returns to its original shape or position.

Rubber fenders

Depending on the shape of a rubber fender and the manner in which it is loaded the rubber will be exposed to compression, shear, bending, etc. Curves in Fig. 9.22 illustrate this. Fenders with approximately equal energy capacity have been chosen for comparison, except for shear fen-ders where the curve is valid for three fenders.

Deflection is a linear function of the load for fenders exposed to com-pression or shear (curves 2 and 3). For the side-loaded hollow cylindrical fenders the load increases progressively with the deflection (curve 1).

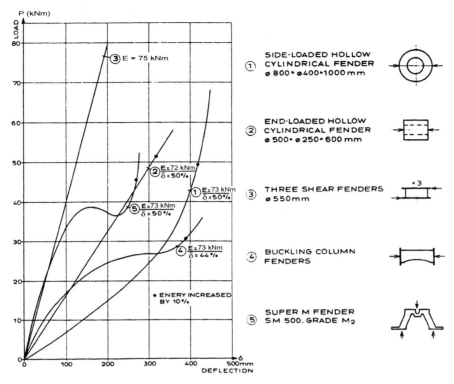

Fig. 9.22 Load deflection curves for different fenders

This means that for impact energy below the rated energy the fender will be relatively soft and accordingly accommodate smaller vessels. The buckling fenders have swift increases in loads but as the fenders are further deflected the loads are more or less maintained until the rated deflection is reached (curves 4 and 5). Consequently, a low ratio of load over energy (P/E) is reached for these fenders.

For structures sensitive to horizontal loads (open-piled piers and breasting dolphins), it is important to realise how fenders perform beyond the rated energy. A 10 per cent increase in energy (marked on the curves), which roughly corresponds to a 5 per cent increase in berthing velocity, leads to the following increases in loads: 16 per cent on curve 1, 6 per cent on curve 2, 9 per cent on curve 4 and 19 per cent on curve 5. The Super M fender (curve 5) could be improved in this regard by downrating the fender from 50 to 45 per cent deflection, which is indicated in the catalogue. For the side-loaded hollow cylindrical fender a downrating does not have the same effect because of the steepness of the curve.

The various manufacturers of rubber fenders provide data (usually graphs) on the performance of their products. Usually ±10 per cent

369

Table 9.1 *Characteristics of side-loaded hollow cylindrical fenders from two manufacturers 'V' and 'S'*

Fender No.	1	2	3	4	5	6	7	8	9	10	11
Manufacturer	V	V	V	V	V	S	S	S	V	S	S
Outside diameter D (mm)	700	800	1800	1850	1850	1800	1800	1800	2800	3000	3000
Inside diameter d (mm)	400	400	900	1000	1050	900	900	900	1500	1500	1500
Ratio D/d	1.75	2	2	1.85	1.76	2	2	2	1.87	2	2
Rubber compound	–	–	–	–	–	S0	S2	S4	–	S0	S4
Load P (kN/m)	290	390	870	850	740	1250	490	240	1240	2070	410
Energy E (kNm/m)	55	72	360	370	360	390	210	110	860	1070	300
Ratio P/E	5.3	5.4	2.4	2.3	2.1	3.2	2.3	2.2	1.4	1.9	1.4
Contact pressure (kN/m^2)	450	620	620	520	450	890	350	170	520	880	170

Note: Load and energy correspond to 50 per cent deflection.

inaccuracy should be allowed for, independent of whether this is stated in the catalogues or not. Some manufacturers produce their fenders in up to four different rubber compounds. Thus it is insufficient to specify a fender solely on the basis of geometry.

(a)   *Hollow cylindrical fenders*: Hollow cylindrical fenders are manufactured with outside diameters ranging from 200 to 3000 mm. For diameters up to 600 mm, the fenders are extruded, while the larger fenders are laminated. Table 9.1 presents characteristics of a number of fenders from two manufacturers 'V' and 'S'. While manufacturer 'V' produces his standard fenders with different ratios of outside to inside diameter, but uses only one rubber compound, 'S' varies only the rubber compound to achieve different characteristics. Fenders 3, 4 and 5 develop almost the same energy but the reaction for 5 is less than for 3 and 4. In addition, the contact pressure is smaller and fender 5 weighs about 5 per cent less than the other two. It is therefore difficult to see how fenders 3 and 4 would be preferred to 5. The influence of the rubber compound is indicated by comparing fenders 6, 7 and 8. With a softer compound, the rated energy capacity and the reaction force decreases together with the contact pressure, all of which are important parameters.

Fenders 1, 2, 9, 10 and 11 have been included merely to indicate the range of possibilities. The load/energy ratio (P/E) decreases from 5.4 to approximately 1.4 with increase in fender size.

It is common to install hollow cylindrical fenders on the quay wall supported by brackets and bars or by frames suspended in chains and let the vessels moor directly against them, see Fig. 9.23. If water level variations are so that small vessels will be trapped under the fenders, or if the fenders serve vessels equipped with

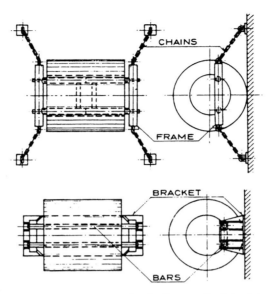

Fig. 9.23 Installation arrangement for hollow cylindrical fenders

horizontal fenders, e.g. ferries, it might be expedient to place pile-supported panels in front of the rubber fenders as indicated in Fig. 9.24. Flexible piles fixed in the seabed will assist in absorbing energy.

Where the contact pressure between fender and vessel is not acceptable, softer hollow cylindrical rubber fenders may be used or panels of steel and/or UHMW-PE surfaced steel frames may be used to increase the contact area. Such panels, which will also reduce the friction force between fender and vessel, could be supported on piles or suspended in chains as indicated in Fig. 9.25. Axial loading of hollow, cylindrical rubber fenders is possible, but only if they are equipped with front panels. Characteristics of such fenders are, of course, completely different from those of side-loaded fenders. This

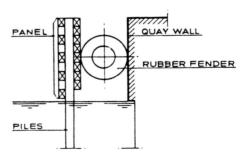

Fig. 9.24 Hollow cylindrical fenders with panel for smaller vessels

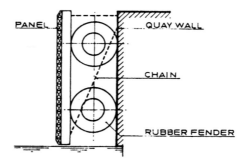

Fig. 9.25 Hollow cylindrical fenders with panel for increase of contact area

is demonstrated in Fig. 9.26 where the results of loading exactly the same piece of rubber in the two directions are shown. In this case, five side-loaded fenders would absorb almost the same energy and force as one end-loaded fender.

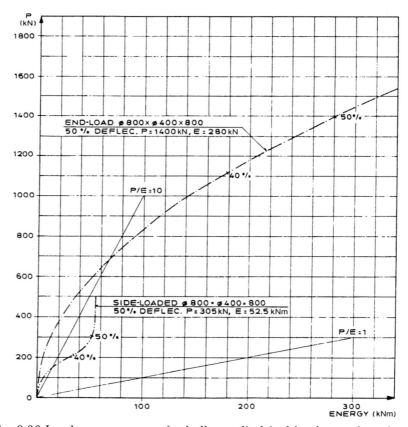

Fig. 9.26 Load-energy curves for hollow cylindrical fender, end- and side-loading

Table 9.2 Characteristics of buckling rubber fenders

	V-fender			V-fender[1]			Super M fender			Phi-fender		H-fender[2]		
Size H (mm)	200	600	1300	200	600	1300	250	600	600	1300	2000	400	1400	2500
Load P (kN/m)[3]	150	450	980	70	410	900	170	410	410	900	1380	190	670	1800
Energy E (kNm/m)[3]	10	90	420	10	90	420	15	89	90	420	1000	28	340	1640
Ratio P/E	15	5	2.3	7	4.6	2.1	11	4.6	4.6	2.1	1.4	6.8	2	1.1
Contact pressure (kN/m²)[4]	900–1000			560	1100	1100	580		200–400	250–450	280–550	200	270	400

Notes: [1] Low friction synthetic resin board. [2] Front panel. [3] Load and energy correspond to 45 per cent deflection and are for standard rubber compound. For other compounds values may vary ±30 per cent. [4] Contact pressure for fenders with front panels depends upon panel area and vary beyond values shown.

(b) *Buckling fenders*: the buckling column fender (see Fig. 9.22) consists of a section of solid rubber with one flat and one arched side. The rubber is bonded to two steel plates, one of which is attached to the quay and one to a pile-supported front plate. Under load the fender will buckle and, as can be seen from the load deflection curve, the reaction force increases swiftly and then more or less maintains its magnitude leading to a favorable energy characteristic. Further, the fender dissipates a considerable amount of the absorbed energy, which results in reduced rebound forces. This prevents vessels from bouncing away after impact.

The principle of buckling is utilised in a number of rectilinear fenders such as the V-fender, the Super M fender, the Phi (π)-fender and the H-fender.

Characteristics of these are shown in Table 9.2. A 1300 mm high V- or Phi-fender would absorb approximately 420 kNm which is slightly more than that of an 1800 mm diameter side-loaded hollow cylindrical fender (see Table 9.1) and could therefore be used instead, if the somewhat higher load is acceptable. A saving of 0.5 m in projection in front of the quay wall would increase the effective reach of quay cranes and ships' gear correspondingly.

Pneumatic fenders

Pneumatic fenders were originally developed as ship-to-ship fenders but due to their low load/energy ratio (P/E) and their low hull pressures they have been used on fixed structures in increasing numbers. The pneumatic fender is a sausage-shaped, inflated rubber bag of 0.5 to 4.5 m in diameter and 1.0 to 12.0 m in length. Usually the fender bag is protected by wire or chain net with tyres or rubber sleeves. The air pressure in the unloaded fender is normally 50 to 80 kN/m^2. For 55 per cent deflection, the pressure increases to 100 to 150 kN/m^2 corresponding to a hull pressure of some magnitude. Characteristics for a few of the existing sizes are shown in Table 9.3.

When used for a fixed structure the pneumatic fender will be floating and connected to the structure by chains in a manner which allows it to

Table 9.3 Characteristics of selected pneumatic fenders

Diameter and length (m)	0.5×1.0	1.0×2.0	2.0×3.5	4.5×12.0	4.5×12.0
Initial pressure (kN/m^2)	50	50	50	50	80
Load P (kN)	38	152	518	5000	6810
Energy E (kNm/m)	3.2	26	174	3690	5320
Ratio P/E	11.9	5.8	3	1.4	1.3
Contact pressure (kN/m^2)	100	100	100	110	150

Note: Load and energy correspond to 50 per cent deflection.

follow the water-level variations. It is thus required that the fixed structure has a vertical face which can support the fender at all required levels. Therefore it may be difficult to arrange pneumatic fenders at open piers and quays.

References

Brinch Hansen, J. (1946) 'Development of the C & N wharf type', *C & N Bulletin 56*, Copenhagen.

Brinch Hansen, J. (1953) *Earth Pressure Calculation*, Copenhagen: The Danish Technical Press.

Brinch Hansen, J. and Lundgren, H. (1960) *Hauptprobleme in Bodenmechanik*, Berlin: Springer-Verlag.

Carlsen, N. A. (1966) 'Drainage of quay walls, Calculations of overpressure behind drained and undrained quay walls', *Acta Polytechnica Scandinavica, Civil Engineering and Building Construction Series No. 40*, Copenhagen.

Committee for Waterfront Structures. (1996) *Recommendations of the Committee for Waterfront Structures*, EAU, Berlin: W. Ernst & Sohn.

Leimdorfer, P. (1972–73) 'The development of cellular bulkheads in maritime structures', *Acier, Stahl, Steel*, 1972, October and December, 1973, February.

National Ports Council (1970) *Port Structures. An Analysis of Costs and Design of Quay Walls, Locks and Transit Sheds*, Vols 1 and 2, prepared by Bertlin and Partners.

Ovesen, N. K. (1962) 'Cellular cofferdams, calculation methods and model tests', *Bulletin 14*, Copenhagen: Danish Geotechnical Institute.

Ovesen, N. K. (1964) 'Anchor slabs, calculation methods and model tests', *Bulletin 16*, Copenhagen: Danish Geotechnical Institute.

PIANC (2002) *Recommendations of the Committee for Waterfront Structures. Guidelines for the design of fender systems*. Report of Working Group 33 of the Maritime Navigation Commission.

PIANC (1995) *Criteria for Movements of Moored Ships in Harbours. A Practical Guide*. Report of Working Group 24. Supplement to *Bulletin No. 88*.

Schmidt, B. *et al.* (1964) 'The pulling resistance of inclined anchor piles in sand', *Bulletin 18*, Copenhagen: Danish Geotechnical Institute.

Tsinker, G. and Vernigora, E. (1980) 'Floating piers', *Proceedings of the Specialty Conference on Ports '80*, ASCE.

Ueda, S. *et al.* (2001) 'Statistical design of fenders', *Proceedings of the International Offshore and Polar Engineering Conference*, June.

Vasco Costa, F. (1964) *The Berthing Ship*, The Dock & Harbour Authority. Published as a booklet by Foxlow Publications Ltd, London.

CHAPTER 10

Environmental considerations

Hans Agerschou

Introduction

Port and coastal engineering was one of the first disciplines which had to deal with adverse environmental effects in the form of lee-side erosion of coasts as well as associated accumulation of sediment in access channels and port basins and along up-drift coasts.

Lee-side erosion has often resulted in loss of valuable land and buildings. Various, more or less effective, measures, such as groins, sea walls, sediment by-passing plants and artificial nourishment of beaches and near-shore sea bottom, were developed to stop or retard erosion and accumulation of sediment.

The following sections deal mainly with the broad outlines of pollution of soil, water and air as well as noise and vibration pollution. However, dredging and disposal of contaminated sediment, as well as the International Convention on Marine Pollution (MARPOL), we believe deserve a more detailed treatment. These subjects are dealt with in the following two chapters.

Adverse environmental effects resulting from construction and practical remedial measures

Effects of capital dredging and reclamation on the flora and fauna in seawater, on and below the sea bottom

Dredging will unavoidably disturb and/or destroy the flora and fauna on and below the sea bottom. However, these adverse effects may be only temporary. Also the increased water depth may enhance the habitat for certain species. The dredged material has to be disposed of either by dumping on the sea bottom or reclaimed as landfill. Dumping on the sea bottom will cover the existing bottom and create a different, more shallow habitat for the flora and fauna. Whether this creates an overall adverse effect or an improvement may vary from case to case. Reclamation eliminates the sea bottom and water habitats completely in the area in question. However, when the area and volume in question are small compared with the total local habitat, the impact may be quite limited or perhaps insignificant. Where reclamation is used to raise the level of land surfaces, coastal flora and fauna may be eliminated or disturbed.

Water pollution

Dredging may cause water pollution by putting silt and clay particles in suspension. This is a temporary effect. The particles are going to settle on the bottom.

Port basins may cause stagnation of the water, which could lead to organic pollution, in particular if municipal or industrial sewage flows into them. The latter should be avoided, as they are potentially much more serious pollution sources than port operations.

Air pollution

A certain limited degree of on-site air pollution, resulting from the use of diesel or petrol-fuelled construction equipment, is unavoidable. However, it can and should be limited to acceptable levels by proper engine maintenance. Air pollution outside the construction site from this source would normally not be a problem.

Earth moving on site may cause dust, which may be carried by wind over considerable distances if not controlled or eliminated by surface watering.

Noise and vibration pollution

Noise pollution would be generated by mechanical construction equipment, but should be limited by proper mufflers for all equipment as well as proper preventive maintenance. Further the areas surrounding the site may be protected against noise by temporary or permanent walls of various types. Pile driving, in particular, in addition to being noisy, may also generate ground vibrations, which may be felt at considerable distances from the construction site. If this cannot be reduced to acceptable levels, pile driving should not be allowed at night if it is a nuisance in nearby residential areas.

Adverse environmental effects resulting from port operation and practical remedial measures

Maintenance dredging

Maintenance dredging causes basically the same problems as capital dredging and not much can be done to remedy its adverse effects other than selection of less sensitive disposal areas. Reduction of the maintenance dredging rate should be attempted by selecting dumping areas from which the dredged material is not transported back to the dredged areas by littoral drift or other natural mechanisms.

Water pollution caused by cargo handling

General cargo if exposed to careless handling, e.g. by the use of hooks on bags of fertiliser, cement, etc., may be spilled in the water between the quay and the ship. This may be eliminated completely by the use of tarpaulins extending from the deck of the ship to the quay apron.

Dry as well as liquid bulk handling by modern equipment does not cause water pollution, if the equipment is operated correctly. The latter should be ensured by proper training of staff and by the presence of fail-safe measures.

Air pollution caused by cargo handling and storage

Coal and ore handling as well as storage may cause dust pollution. This may be reduced or eliminated by using closed conveyor systems and by surface watering of the stacks as well as of the cargo to be unloaded from ships. Cement in bulk is handled pneumatically through pipes and hoses and stored in closed silos and thus should not cause any air pollution.

Grain in bulk is handled pneumatically through pipes and hoses and stored in silos or sheds. However, some limited air pollution may result if the holds of a loading ship are not covered with tarpaulins or otherwise.

Fertiliser in bulk is stored in sheds. It is handled by clamshells and conveyors, which should be enclosed. Due to its moisture content it would not usually cause any air pollution.

Noise pollution caused by ships, land transport and cargo handling

Ships, lorries, trains and cargo-handling equipment do, unavoidably, generate some noise. As it is usually necessary to be able to carry out cargo handling 24-hours a day, noise abatement measures have to be employed if nearby residential areas are affected. Such measures would include walls of various types and/or green belts.

Pollution caused by dirty cargo

Dirty cargoes are normally of the following categories:

(a)   coal;
(b)   iron ore;
(c)   chemicals;
(d)   oil and petroleum products;
(e)   toxic substances;
(f)   infectious substances; and
(g)   radio-active substances.

Coal and iron ore, which are stored in the open, may cause dust pollution of the air. This is dealt with above.

Rainfall may cause soil, as well as water, pollution if proper remedial measures are not taken. The only way of preventing soil pollution is paving of the storage areas, which may be excessively expensive. Water pollution should be prevented by providing drainage around the storage areas and settling ponds behind the quay structure, where the solids would accumulate and be removed from when required.

All other dirty cargoes may cause soil, water and air pollution, when they escape from the cargo units, such as drums. Strict rules for their handling, storage and monitoring should be part of the regulations for any port and should be enforced at a zero tolerance level. The staff should be trained in emergency procedures, which should be rehearsed frequently.

Fires, explosions and other serious accidents caused by 'dangerous' cargo

Dangerous cargoes are normally of the following categories:

(a) explosives;
(b) gases (compressed, liquefied or dissolved under pressure);
(c) flammable liquids with low, intermediate and high flash-points;
(d) flammable solids;
(e) substances liable to spontaneous combustion;
(f) substances, which, in contact with water, emit flammable gases;
(g) oxidising substances;
(h) organic peroxides;
(i) corrosives; and
(j) e, f and g in the above section.

All of the above may cause serious damage to property, injury to people and loss of life if not handled and stored with utmost care. Strict rules for their handling, storage and monitoring should be part of all port regulations and should be enforced at a zero tolerance level. The staff should be trained in emergency procedures, which should be rehearsed frequently.

Water pollution caused by discharge of solid and liquid waste from ships

All ships' garbage and waste should be stored in suitable containers and only be disposed of when in port according to applicable port regulations.

The same is the case for bilge water containing oil and other impurities. The London Dumping Convention, which was adopted by an Inter-governmental *Conference on the Convention on the Dumping of Wastes at Sea* in 1972, governs this. It has been ratified by a number of countries.

Water, sea bottom and shore pollution caused by grounding or other damage to ships

The worst examples of pollution caused by damage to ships have been leaking crude oil tankers, which have in a number of cases resulted in large-scale damage to beaches, wetlands, mangroves, fish, shellfish, sea birds and other animals, as well as to the sea bottom and land flora.

This has resulted in mandatory double hulls for new tankers and forced retirement of older tankers, which are simply not allowed to call at ports and terminals in a number of countries.

Some ports are equipped and staffed for rapid containment of oil and other dangerous spills of liquid waste.

Environmental impact assessment

Feasibility studies for port projects should not only be concerned with technical, economic and financial feasibility, but also with environmental feasibility. The environmental assessment should cover positive as well as negative effects resulting from the project.

A number of countries and some international organisations have established checklists for environmental impact assessment. Most of these contain mandatory requirements. There is unavoidably a good deal of similarity in the substance of these checklists, whereas their format and detail vary considerably.

The World Bank (WB) checklist (Davis *et al.*, 1990) goes in great detail in regard to water- and land-related impacts, but is very brief in regard to air-related impacts and hazardous cargoes. It deals summarily with socio-cultural impacts, a review of existing and proposed regulations affecting the proposed port development, and the need for construction and facility operation environmental monitoring.

The Asian Development Bank (ADB) checklist (United Nations ESCAP, 1992) separates the causes, the adverse effects and the remedial measures in a systematic manner for eleven different categories of causes.

The checklist prepared by the International Association of Ports and Harbours (IAPH) (United Nations ESCAP, 1992) contains a number of less relevant or insignificant causes and effects. As the IAPH does not participate in the financing of port and terminal projects, which many countries as well as the WB and the ADB do, the use of its checklist is not compulsory.

References

Davis, J. D. *et al.* (1990) 'Environmental considerations for port and harbor developments'. *World Bank Technical Paper No 126*, Washington, D.C.
United Nations ESCAP (1992) *Assessment of the Environmental Impact of Port Development*, Bangkok.

CHAPTER II

Dredging and disposal of contaminated sediments

John M. Land

Introduction

Many ports and waterways are characterised by the presence of contaminated bed sediments. These may have to be dredged during capital or, more commonly, maintenance works. They may also be removed as part of a remedial scheme to improve sediment and water quality. The dredging, transport and disposal of such materials are, in themselves, environmental impacts and have the potential, if not undertaken carefully, to cause damage to the environment. Dredging and disposal of contaminated sediments are topics which have received considerable attention during the last 30 years. Techniques of dredging, disposal, treatment and use of dredged materials are evolving rapidly in response to general public concern and national and international regulation. This rapid evolution, in combination with the complexity of the topic, prevents a detailed review here. Readers requiring more than a brief overview are referred to the numerous literature sources of information, particularly the *Environmental Aspects of Dredging*, a series of seven guides published by the International Association of Dredging Companies (IADC) and the Central Dredging Association (CEDA).

The criteria that are used to assess contamination are complex and vary widely around the world and this topic is not considered here. However, most countries have a regulatory authority that is responsible for the establishment of contamination criteria and the permitting and regulation of operations involving dredging and disposal of contaminated sediments. This is usually done within the framework of the Convention on the Prevention of Marine Pollution by Dumping of Wastes and Other Matter, 1972 (the 'London Convention') and the Oslo and Paris Convention (the 'OSPAR Convention') which are principally concerned with marine disposal. The actual dredging operations and onshore disposal are generally regulated by means of National, rather than International, regulation or guidelines. Guidance on these aspects is provided in IADC (1997a, 1997b). The remainder of this section is written on the assumption that the sediments are deemed to be contaminated by the relevant authority and that special care must be taken during dredging and disposal.

The main topics of dredging, transport and of treatment, use and disposal of dredged material are separately reviewed below. However, it must be stressed that these topics are closely related and that when contaminated sediments are to be dredged, very careful attention must be given to the interfacing between these activities, i.e. the design of a contaminated sediment dredging operation should result in a comprehensive strategy in which all aspects of investigation, removal, monitoring, transport and ultimate disposal are considered in relation to each other with due regard to interfacing and the need to achieve maximum positive impact on the environment.

Dredging

General considerations

There are usually three main practical considerations when dredging contaminated sediments. They are the minimisation of:

(a) release of contaminants to the water column;
(b) the volume of material which is dredged; and
(c) bulking of the material during the dredging and transport processes.

In order to prevent the spread of contamination to adjacent areas and to avoid adverse impacts on water quality, special care must be taken to ensure that as little sediment as possible is released into the water column during dredging and transport to the disposal site.

It is often the case that a large proportion of the overall cost of dredging contaminated materials is expended on disposal, which may include some form of treatment. In order to reduce costs, and to conserve possibly valuable disposal resources, it is necessary to ensure that only the contaminated material is removed for disposal and that contaminated and uncontaminated materials are not mixed. This is achieved by a combination of detailed investigation of the extent of contamination, accurate definition of the materials to be dredged and the use of accurate dredging methods.

For the same reasons, it may be important in some cases to ensure that dilution and bulking during dredging is minimised although some types of treatment of contaminated sediments require the material to be delivered in a slurried condition. This is considered in 'Treatment of dredged materials', later in this chapter. In addition to these factors, careful attention must be given to the health and safety of those involved in work with contaminated materials. These aspects are reviewed in turn below.

Dredging and transport methods

Contaminated sediments most often comprise relatively soft fine-grained materials that can be dredged by most types of conventional dredging plant. However, the requirement to minimise sediment release during dredging, and to work accurately, has given rise to a range of equipment which is designed specifically to work in contaminated materials. These are often modified versions of conventional dredgers. The most important characteristics of the main types of conventional plant, with respect to working in contaminated sediments, are summarised below. Detailed descriptions of these dredgers and their manner of operation are given in Bray *et al.* (1997) and a more detailed account of their use specifically for dredging contaminated sediments is provided in *IADC (1998)*. A review of environmental impacts of dredging contaminated sediments, including several case histories of monitored trials, is given in Environment Canada (1994) and the IADC/CEDA guides.

Grab dredgers are essentially cranes that are mounted on a pontoon (Fig. 11.1) or, less frequently, a ship. They are equipped with a bucket and are able to excavate soft soils with minimal dilution. However, unless special precautions are taken, they can release relatively large amounts of sediment during excavation of the material and as the bucket is raised and lowered through the water column. Precautions include the use of watertight grab buckets and the deployment of silt curtains around the working area in order to minimise the spread of any sediment which is released. Further reductions of sediment loss can be achieved by using slow grab hoisting speeds and very careful loading of barges to avoid splashing and spillage. Following discharge, the grab can be washed in a large tank placed on board the barge in order to remove any adhering sediment.

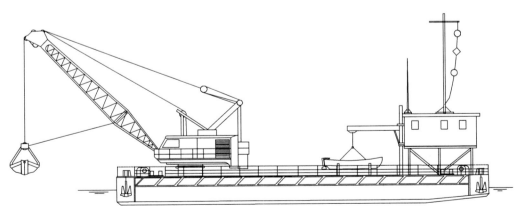

Fig. 11.1 A grab dredger (adapted from Bray et al., 1997)

383

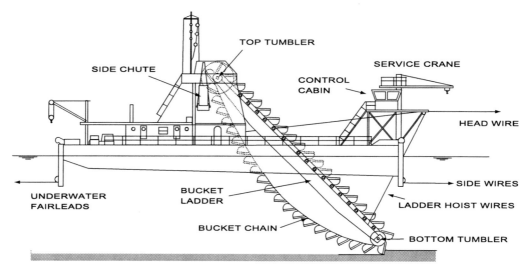

Fig. 11.2 A bucket dredger (adapted from Bray et al., 1997)

Conventional grab buckets are generally unable to remove material with a high degree of accuracy but special grabs have been developed which enable a smooth level cut to be achieved. Dredged material is normally transported to the disposal site in barges unless the grab is ship-mounted in which case it is placed in a hopper in the vessel for transport.

Bucket dredgers comprise large pontoons fitted with an endless chain of buckets (Fig. 11.2). The buckets move around a ladder that is hinged at the top of the dredger's superstructure so that its angle may be varied to dredge at different depths. The dredger swings from side to side along a working face as the bucket chain rotates. Material excavated by the buckets passes to the top of the ladder where it is discharged into chutes which load barges for transport to the disposal area.

Bucket dredgers tend to release relatively large amounts of sediment when working but do not greatly dilute the sediment during dredging. They can dredge with greater accuracy than grabs when working in calm waters but the working area cannot easily be enclosed with silt curtains. Sediment release to the water column may be minimised by:

(a)   enclosing the bucket ladder so that spillage is returned directly to the bed;

(b)   reducing the bucket and swing speeds;

(c)   installing one-way valves in the buckets to permit air to escape as the buckets descend to the working face; and

(d)   enclosing the barge-loading chutes and fitting splash screens to the barges to eliminate spillage during loading.

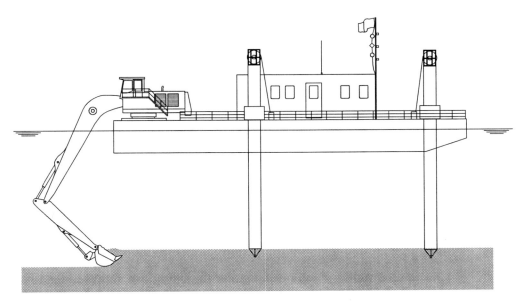

Fig. 11.3 A backhoe dredger (adapted from Bray et al., 1997)

Backhoe dredgers (Fig. 11.3) are usually only used for dredging contaminated sediments when small volumes are involved or when working in very confined areas. A large proportion of inland dredging is carried out by backhoe. They comprise conventional back-acting excavators mounted on a pontoon. They are also associated with relatively large sediment losses but can work with a high degree of accuracy and only slight dilution of the sediment. Buckets are available which close over the dredged sediment before being raised to the surface and it is possible to encircle the dredger with a silt curtain. As with most types of dredger, slow, careful working can also help to reduce sediment losses. The dredged material is usually transported to the disposal site in barges.

Cutter suction dredgers comprise pontoons with a ladder that is hinged at deck level, the lower end of which is fitted with a rotating cutterhead (Fig. 11.4). The cutterhead excavates the material, which passes into a suction inlet located within the cutterhead. The excavated materials are pumped into the hull of the dredger and thence, via a pipeline, to the disposal area. On rare occasions, it is loaded into barges for transport but great care must be taken to avoid spillage during barge loading. The dredger swings from side to side across the area to be dredged using a spud located at the back of the dredger. A few cutter dredgers are self-propelled ships.

Cutter-suction dredgers are associated with relatively small losses of sediment. These can be further reduced by maintaining high ratios

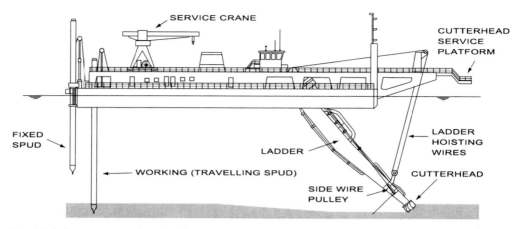

Fig. 11.4 A cutter-suction dredger (adapted from Bray et al., 1997)

between the suction inlet velocity and the swing and cutterhead rotation speeds, i.e. low swing speeds and low cutter rotation speeds are conducive to minimising sediment release. Cutters can dredge with a high degree of accuracy when working in calm waters but their use results in considerable dilution of the dredged materials. The degree of dilution will be increased if suction velocities are increased in order to reduce sediment losses.

Trailing suction hopper dredgers are sea-going ships with a large hold to contain the dredged material (Fig. 11.5). The sediments are pumped from the seabed through a suction pipe, which is fitted with a draghead designed to maximise the erosive forces of the pumping system. The

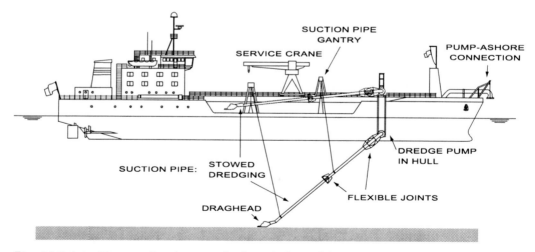

Fig. 11.5 A trailing suction hopper dredger (adapted from Bray et al., 1997)

vessel sails over the site when dredging. Most trailers are equipped with two suction pipes but some have only one. The latter are often designed to dredge soft fine-grained materials at high density.

With the exception of the softest of muds, which can be dredged at or close to their in situ density, materials dredged by trailing suction hopper dredgers are significantly diluted. However, this type of dredger is associated with low rates of sediment release when:

(a)   working in deep water where propeller scour does not put bed sediment into suspension; and

(b)   working without the use of overflow or Lean Mixture Overboard (LMOB) systems.

The accuracy of dredging is not great. Modern trailer dredgers with sophisticated monitoring and control systems can work to a vertical accuracy of about 0.2 m in good sea conditions but accuracy may be worse than 0.5 m for older vessels or when working in poor sea conditions. The dredged material is transported to the disposal site in the hopper of the ship and may be discharged by pumping ashore or by simple bottom-dumping through doors or valves. Some vessels can discharge into seabed pits using one of the suction pipes.

Hydrodynamic dredging covers two main types of working: water injection dredging and agitation dredging. Both result in the lateral displacement of the material, rather than its recovery to the surface. Water injection dredging involves controlled fluidisation of the bed materials with powerful water jets to form a density current which flows over the bed away from the dredging site under gravitational forces. The density current flow is only marginally affected by the ambient water current. Agitation dredging refers to the use of any one of several techniques that are designed to put the material into suspension so that it is carried away by water currents. This is usually achieved using rakes or harrows towed behind tugs. Alternatively, trailing suction hopper dredgers (or even cutter suction dredgers) can be used to raise the material to the surface where it is released back into the water. These techniques (particularly agitation dredging) are rarely used for work in contaminated sediments, and then only when such an approach can be shown to have acceptable environmental impacts.

In addition to conventional plant, which may be modified for work in contaminated sediments, there have been developed a number of dredgers, sometimes referred to as 'low-impact dredgers' (IADC, 1998) specifically designed for this type of work. These include:

(a)   the disc bottom dredger which uses a shielded cutting disc that rotates around a vertical axis;

(b)   the sweep dredger which is essentially a cutter suction dredger in which the cutterhead is replaced with a sweeping device that is either pulled or pushed into the material to be dredged;

(c)   the environmental auger has a shielded auger and is intended mainly for the removal of thin layers of material;

(d)   the pneumatic dredger (e.g. the Pneuma pump) which is able to dredge with very limited sediment release but has yet to be widely used because of limited efficiency in the relatively shallow water in which contaminated sediments tend to occur, sensitivity to debris and limited control; and

(e)   the scraper dredger which utilises a wide belt fitted with cutting blades that draw material up a sloping surface into a hopper; the material is enclosed during transport to the hopper and this type of dredger is well suited to the removal of thin layers of material with minimal dilution.

Impact mitigation

In addition to the careful selection of appropriate plant and working methods, which may include the use of silt curtains or restrictions on some of the dredger operating parameters, there are several measures that may be implemented to reduce potentially adverse environmental impacts. These include:

(a)   work can be restricted to certain seasons to avoid, for example, periods of fish breeding or migration, or so that the work is undertaken during periods of low water temperature when many contaminants are less able to pass into solution and when reductions of dissolved oxygen will be limited;

(b)   tidal working to ensure that any sediment which is released either settles in, or very close to, the working area or is transported away from particularly sensitive areas;

(c)   specification of maximum rates of working or the number of dredgers which are allowed to work at the same time; and

(d)   planning the works so as not to coincide with other nearby works which, in combination, may aggravate adverse impacts.

It should be noted that, in addition to sediment release, dredging operations may have other impacts on the environment which should be evaluated during the planning stage. These include noise and air pollution and potential interference with, or by, other marine traffic.

Monitoring of dredging operations

Monitoring of the dredging operation is required for several reasons:

(a) to verify predictions of potential effects made during the pre-contract studies;

(b) to control the works and ensure compliance with contract measures designed to minimise adverse impacts;

(c) to provide information on actual effects which may be used to modify, if necessary, operating procedures; and

(d) to increase knowledge.

In the context of working with contaminated sediments, monitoring combines the programmed measurement of environmental parameters of interest with detailed recording of all operational aspects of the works.

Environmental monitoring generally comprises measurement of water-quality parameters (e.g. suspended solids, dissolved oxygen, metals content) during a period before the works in order to establish baseline conditions. These measurements may also include temperature in cases where work can only proceed when the water temperature falls below a certain figure. The monitoring should continue during the work at pre-determined locations, which are chosen so that the effects of the work are measured and can be compared with background data from locations which are not impacted by the works. This is necessary because natural background conditions may often vary quite considerably for reasons that are quite unconnected with any dredging activity (e.g., seasonal or climatic changes of suspended sediment concentrations and the effects of storms).

Monitoring of the dredging operations should include recording of the locations where the dredger is working, the manner of its operation and the volumes of material that are dredged. Dredged material transport systems should be included within the scope of the monitoring. A more detailed review of monitoring is provided in IADC (1998).

Health and safety

Contaminated sediments inevitably pose a degree of risk to personnel working on dredging projects, the magnitude of which will depend on the type and degree of contamination. In all cases, the works should be planned and executed in a manner that minimises the exposure of personnel to the sediment, including fumes and aerosols. The site personnel must be instructed in safe working procedures. Each case should be examined and evaluated prior to the commencement of the work and appropriate measures put in place. When working with highly contaminated materials, safety precautions are similar to those employed in the chemicals industry and may include:

(a) full protective clothing with breathing apparatus;

(b)   the maintenance of a positive air pressure in the accommodation of the dredger; and

(c)   provision of comprehensive washing and disinfecting facilities.

Dredged material disposal

The primary objective of disposal is to place the material in a location, and in a manner, that ensures that contaminants are fully contained and isolated from the surrounding environment or that any loss of contaminants is at an acceptable level. In its broad sense, the term 'disposal' may include some form of treatment of the material in order to remove or immobilise contaminants or change them into a less harmful form. It also includes beneficial use of the contaminated material (or the products of treatment) in a manner that satisfies the requirement to protect the environment. In many countries, it is now a requirement for the proponents of dredging projects to seek out potential beneficial uses of dredged material and disposal licenses may only be granted if it can be demonstrated that there are no viable uses for the material.

Offshore and coastal disposal

There are two main types of offshore or coastal disposal of contaminated sediments: Contained Aquatic Disposal and Confined Disposal.

Contained Aquatic Disposal (CAD) refers to the placement of the dredged materials in submarine depressions (which may be natural or specially-dredged pits) followed by capping with a suitable clean material. The use of CAD facilities is usually restricted to the disposal of slightly or moderately contaminated materials. CAD has been adopted on a large scale in Hong Kong (Shaw *et al.*, 1998) where, since 1992, contaminated sediments have been placed in a series of 11 dredged pits located near the site of the new airport at Chek Lap Kok. Up to 40 million m^3 of material is expected to have been placed by 2003. In the USA (Palermo, 1997), CAD has been adopted in Los Angeles, Portland and Boston. The CAD facilities in Boston comprise a series of pits dredged in the main navigation channels. The pits have capacities of between $20\,000\,m^3$ and $230\,000\,m^3$. A 1.2 million m^3 CAD pit has been constructed in Newark Bay to accommodate material from the Port of New York and New Jersey and additional pits are under consideration. In Europe, CAD facilities have been used in the Netherlands in both marine and inland waters, including canals and lakes. Limited, small-scale CAD facilities have also been used in the UK and Germany.

The planning and design of CAD facilities must ensure effective containment of the contaminated material both during and after placement. The main issues which need to be addressed are summarised in Table 11.1 (after Palermo, 1997).

Table 11.1 Confined aquatic disposal – planning and design issues

Issue	Objective
Excavation	Selection of appropriate equipment; identification of disposal or re-use options for excavated material
Dredging and placement of contaminated material	Appropriate equipment and methods must be selected, fill scheduling must meet dredging requirements and the geotechnical properties of the sediments on completion of filling must allow for cap placement
Loss of material during placement and before capping	Releases of contaminants to the water column must meet applicable criteria; material must be placed and retained in the pit
Cap placement	Design thickness must be achieved without excessive mixing with or displacement of the contaminated materials
Cap effectiveness	Must contain the contaminated material in the long term and reduce flux of contaminants through the cap so that the sediment quality of the cap and the water quality criteria are met
Long-term cap integrity	The potential effects of bioturbation, consolidation, erosion and disturbance by storms/vessels must be controlled and minimised

A Confined Disposal Facility (CDF) is a site that is enclosed by bunds (or dykes) which extend above the high-tide level. The facility may or may not be lined, depending on geological and hydrological conditions and the degree of contamination of the dredged materials. Compared with CAD facilities, CDFs are relatively expensive but may be the only option for moderately and highly contaminated sediments. They may also be used as temporary storage facilities when the dredged material is to be treated. CDFs may be built as islands or constructed on the shoreline.

CDFs are a common solution for dredged material disposal and have been more widely used than CAD facilities for contaminated sediment. In the USA, contaminated sediments from Baltimore Harbour are placed hydraulically in the Hart Millar Island CDF, which was constructed using dredged sand to form the containment dykes. Large-scale CDFs have also been constructed in Mobile Harbour, Tampa Bay and in the Great Lakes, among others. Some of the most sophisticated CDFs have been constructed in the Netherlands and include the Slufter (150 million m^3 capacity), the Papegaaiebeck (1.5 million m^3 of the most highly contaminated materials) and the recently-constructed Isseloog facility in the Ketelmeer lake (15 million m^3).

The concept of CDFs can be extended so as to utilise the contaminated material as a construction material, incorporating it into reclamations and other structures, with appropriate isolating measures in place. For

Table 11.2 Confined disposal facilities – planning and design issues

Issue	Objective
Containment dykes	A range of structure types can be considered and the dyke construction costs will have a major influence on the overall cost of the option
Transport and offloading of material	The methods of transport and placement must be compatible with dredging methods and site conditions
Site geometry and sizing	The size of the CDF must account for the necessary degree of solids retention, bulking during placement, and long-term consolidation of the material
Rates and sequencing	The process by which materials are placed in the CDF must match likely offloading demands and sequencing of placement should be planned to minimise the potential for contaminant release
Contaminant pathways and controls	Each pathway of concern, potentially including effluent during filling, surface run-off, ground water leachate and dyke seepage, direct uptake, and volatilisation, should be separately evaluated and appropriate controls factored into the design
Dewatering and long-term management.	Once the dredged material fill-height rises above the mean tide level, further dewatering through drainage can be implemented and a cover can be placed

example, in Japan, contaminated sediments are placed in reclamation areas that incorporate collection and treatment facilities for leachate. The sediments are often stabilised during placement by mixing with cement or lime. In Vancouver, contaminated sediments are to be sealed within concrete in a berth structure for future shipyard operations. There are several examples of CDFs which have been converted into wild-life habitats, particularly in the USA and Canada. The Dutch Isseloog facility will, when filled, be landscaped and given over to recreational areas and a nature reserve.

The design and planning of CDFs has been widely reported (e.g. USACE, 1987; Laboyrie *et al.*, 1995; Palermo, 1996). The main considerations are summarised in Table 11.2.

Onshore disposal

Onshore or 'upland' disposal is often used for contaminated sediments. Depending on the degree of contamination, disposal may take place in a dedicated landfill (including abandoned mines), the material may be used to create a mounded, landscaped area or may be used to create or restore habitats such as wetlands. In many cases, it is necessary to isolate the material from the environment by lining and capping. Complex drainage arrangements may be required to collect (and usually treat) pore water, which is expelled from the sediments as they consolidate. Some dewatering of the material may be required before it is placed in the disposal area.

In many locations, the high cost of land and the limited availability of suitable sites mean that onshore disposal sites are a valuable commodity. It is therefore often the case that onshore disposal is carried out following treatment by dewatering and separation into contaminated and uncontaminated fractions in order to reduce the volume of the material. In Western Europe, in particular, it is common for the relatively uncontaminated products of such treatment to be used in some beneficial manner.

A notable example of an upland disposal facility is that of the Port of Hamburg, in Germany, where approximately 2 to 3 million m^3 of predominantly mixed sediments are dredged annually in the Port (Harms, 1995). Much of the fine fraction of this material is polluted with both organic and inorganic contaminants. The contaminated material is placed ashore and hydraulically transported to the METHA plant (Glindemann and Csiti, 1996) where it is treated to separate the coarse uncontaminated fraction from the fine contaminated fraction (see the following section). The fine fraction is disposed of at two 'silt hills' constructed on former hydraulic filling areas at Francop and Feldhofe. The silt hills act as impermeable covers over the previously deposited contaminated materials, and comprise alternating layers of contaminated material from the METHA plant and drainage layers. An edge structure, which includes collector drains, provides edge stability. The hills will eventually be covered with a root barrier and topsoil.

Two upland disposal sites for contaminated dredged material are located near Ghent in Belgium (de Vos et al., 1997). The Callemansputte started receiving dredged material from the Ghent–Terneuzen canal in 1981 and now contains more than 4 million m^3. The site is a pit excavated in sand to a depth of about 20 m with its base in a layer of clay. The pit is surrounded by a 0.8 m-thick bentonite slurry cut-off wall designed to intercept leachate and protect ground water. Observation wells and piezometers are installed inside and outside of the cut-off wall to monitor contaminant migration. The Geuzenhoek site is an experimental facility, constructed in 1990, to research the adsorption characteristics of natural materials and to establish whether they can be used as isolation materials. The site comprises eight bunded fields of 20×25 m that contain a 3 m thickness of dredged material. The fields are constructed using a variety of isolation materials including glauconite, bentonite, peat, chemically and mechanically dewatered sludge, plastic sheeting and mixed materials. A comprehensive monitoring system has been installed to observe migration of contaminants from the different fields.

In the UK, while many ports and inland waterways contain areas of contamination, these are generally localised and the volumes are relatively small. This, in combination with the scattered locations of contaminated materials, precludes the establishment of large centralised facilities dedicated to contaminated dredged materials. The most cost-effective onshore solution for the disposal of contaminated materials is to place them in sanitary landfills. There are numerous sites in the UK which are suitable for the small volumes of material. These are mostly municipal landfills but there are several private landfills, owned, for example, by chemical companies, which may accept such materials.

Treatment of dredged materials

Treatment of dredged material may be viewed as part of the disposal process. Treatment may be undertaken to:

(a)   reduce the volume of contaminated materials in order to conserve disposal resources; and
(b)   remove, immobilise or isolate the contaminants.

Volume reduction is often referred to as pre-treatment as it may be designed to reduce the cost of more expensive treatment options. Some or all of the products of treatment may be put to beneficial use.

Volume reduction

Volume reduction may involve two processes: separation of coarse materials from the finer materials to which the majority of the contaminants are strongly bound, and dewatering. The processes may be combined into a single continuous process train. It is usually the case that, where separation is a part of the process, the dredged material is required to be delivered as a slurry and this works to the advantage of hydraulic dredging methods.

Separation

Separation techniques usually commence with screening to remove coarse materials and debris following which the dredged slurry is usually pumped into a long lagoon in which coarse particles settle close to the inlet and the finer particles settle progressively further from the inlet; after drainage the materials are selectively excavated and may then pass to one or additional stages of separation. These may include use of the hydraulic classifier, spiral classifier, hydrocyclones, magnetic and electrostatic separation, and flotation. In some cases, e.g. when a mobile processing plant is being used, initial separation in settling lagoons is omitted.

An example of a permanent separation facility is the METHA plant in Hamburg, which is a high-efficiency separation and dewatering plant. The dredged material is pumped directly to the plant from the transport barges and passes through screens, which remove material greater than 80 mm, into a 300 000 m^3 storage basin. It is then passed through the main separation plant, comprising a rotary screen, hydrocyclones and fluidised bed classifier to remove the sand fraction. The fine fraction then passes through a thickening plant, where flocculants are added, and two stages of belt press filtration, following which it has a 55 per cent dry solids content. Water removed from the sediments during the separation and drying process passes through a treatment plant and is then discharged into the Elbe. The METHA plant has recently been upgraded to improve the sand–silt separation process (Detzner *et al.*, 1997). The original plant produced a residue that comprised particles up to 150 μ. The addition of a 'fine sand' plant comprising spiral concentrators separates particles at 20 μ diameter, thus further reducing the volume of contaminated materials for disposal.

Dewatering

The simplest and most effective method of dewatering is to use drying ponds. The material is placed in shallow bunded enclosures which are lined and equipped with under-drainage facilities and a weir arrangement to remove supernatant water with minimal fines loss. As the material drains, techniques such as progressive trenching can be applied to accelerate drainage and desiccation.

The North and South Blakeley drying-pond facilities in Mobile Harbour, Alabama, USA are examples of areas which have had their capacity increased by trenching and dewatering (Walker, 1989). The material dredged from Mobile Harbour is placed in such a manner as to ensure the maximum possible dewatering between dredging campaigns. Similar techniques have been used extensively in the Netherlands for many years.

Other, mechanical methods of dewatering can be used including belt filter presses, chamber filtration, vacuum filtration, solid bowl centrifuges, gravity thickening and thermal evaporation (Averett *et al.*, 1990). While using less space, these methods are relatively expensive and have a limited throughput. In particular, it should be noted that, with the exception of evaporation (or thermal dewatering), none of the alternative methods yield an end product with as high a solids content as the simple drying pond method.

Treatment methods

The treatment methods are designed to remove, destroy, isolate, insulate or immobilise the contaminants in the sediment. There are numerous

Table 11.3 Treatment times and space requirements for selected biological treatment processes (Ferdinanday-van Vlerken, 1997)

Method	Application	Space required per tonne dry solids (m^2)	Time required
Greenhouse farming after dewatering	Sandy material (60% solids)	1.27	1–3 months
Intensive land farming	Sandy material (45% solids)	1.7	1—3 years
Slurry decontamination process (SDP)	All material (30% solids)	2.5	8–12 days
Continuous bio-cascade (CBC)	Fines (20% solids)	4	8–16 days

methods available or under development and, while most are expensive, rapid advances are being made in developing cost-effective technology. It is important to note that the treatment process must be designed, often after some experimentation, to suit the particular sediment type and the type and degree of contamination. Treatment techniques fall into four main groups, biological, chemical, thermal and immobilisation.

Biological treatment

Biological treatment processes break down organic contaminants such as polycyclic aromatic hydrocarbons (PAHs) and polychlorinated biphenyls (PCBs) using micro-organic processes. In practical terms, they fall into two groups: those which require large work areas, and several months or even years of processing, such as land farming and greenhouse farming; and those which require a relatively small work area, and a period of several days, such as the slurry decontamination process (SDP) and continuous bio-cascade (CBC) systems. These processes, which involve mainly aerobic digestion processes, have been subjected to small scale (100 to 10 000 m^3) field trials and evaluated for applicability as well as space and time requirements (Table 11.3).

Chemical treatment

A wide range of chemical treatments can be used to extract, destroy or alter contaminants. They include wet oxidation and solvent extraction. Treatment processes can be designed to work on both organic and inorganic (i.e. metals) contaminants. Few, if any, very large scale chemical treatment facilities have been constructed for dredged material, mainly because the high cost of chemical treatment makes it suitable only for the most severely contaminated sediments, which generally comprise only small volumes. Recent feasibility and costing studies (Reinkes, 1997; Rulkens et al., 1997) focus on plants with throughputs of 10 to

15 t/h, suggesting that low-cost methods of treating very large volumes are unlikely to be developed in the near future.

Thermal treatment

Thermal treatment techniques are generally very effective (but they are expensive due to high energy consumption) and offer a high degree of security in achieving their objectives. The main techniques are thermal desorption, thermal reduction, incineration and vitrification. Thermal desorption removes organic and mercury contaminants, and thermal reduction reduces organic contaminants to less toxic forms.

Vitrification involves heating the sediment to a temperature sufficient to vaporise organic components and melt inorganic components. Upon cooling, the vitrified product can be used as a construction material (e.g. artificial gravel and building blocks). Prototype vitrification systems have been constructed in Russia, with a throughput of $9200\,m^3$/month (Anon., 1995) and in the Netherlands (Bolk *et al.*, 1997). Incineration operates on a similar principle but requires less energy and is thus generally a less expensive option.

During incineration organic contaminants are volatilised but, in contrast to vitrification, the inorganic contaminants in the resulting product are not immobilised and incinerated dredged material may require further treatment (e.g. immobilisation/ solidification) prior to disposal.

Immobilisation

Immobilisation treatments are designed to prevent the release of contaminants from the sediment particles by chemical binding (fixation) or physically preventing their migration (solidification) (PIANC, 1996). Chemical fixation can be achieved by increasing the pH in order to immobilise metals, or by the use of silica solutions to encapsulate the sediment/ contaminant agglomerations.

Solidification processes rely on the addition of cementing agents such as cement or lime. These are relatively inexpensive methods, which do not require a high-tech approach, and are commonly used for in situ treatment of sediments. The lime stabilisation technique has the advantage of both dewatering and binding inorganic contaminants in dredged material. It is noted, however, that the addition of lime may not be appropriate for stabilising materials with high levels of organic contaminants.

Use of dredged materials

The use of dredged materials, even when contaminated, is becoming increasingly common. Use reduces demand for other materials and requirements for often costly disposal facilities. There are several uses for contaminated dredged materials, in addition to their use as fill materials in reclamations such as those described above in Japan and Canada. These include using the products of treatment for making bricks (see below) and artificial aggregates (Smits and Sas, 1997). Some contaminated cohesive materials may be used to construct the cores of berms and dykes. They have been used as lining and capping materials in landfills and can even be used, in some cases, for habitat restoration although very careful consideration must be given to the nature of the contamination and its uptake by plants and animals.

In Hamburg, as part of the effort to reduce the volume of contaminated sediment which needs to be stored in the silt hills following treatment in the METHA plant described above, a brickmaking company worked in association with the Port Authority's engineering division and the Technical University of Hamburg to develop a process to use the material for brickmaking (H-SZ and Strom- und Hafenbau, 1997). A DM 20 million pilot plant commenced operations in 1996 and produces up to 5 million high-quality bricks per year. A large proportion of the conventional clay used for the brickmaking process is replaced with processed dredged material from the METHA plant, amounting to some 30 000 t/year. The pollutants, which become gaseous during the firing process, are completely removed from the flue gases and the contaminants which remain in the bricks (predominantly metals) are totally encapsulated. A full scale plant is planned to be constructed in the near future.

More detailed accounts of some potential uses of contaminated dredged materials are given in Brandon *et al.* (1992) (wetland creation); Tresselt *et al.* (1997) (use as barrier material); Leeuwen *et al.* (1997) (production of sand and clay).

References

Anon. (1995) 'Treatment and disposal of dredged material: an update', *Port Engineering Management*, July–August.

Averett, D. E. *et al.* (1990) 'Review of removal, containment and treatment technologies for remediation of contaminated sediment in the Great Lakes', *Miscellaneous Paper EL-90-25*, US Army Engineer Waterways Experiment Station, Vicksburg, Mississippi.

Bolk, H. J. N. A. *et al.* (1997) 'The processing of contaminated sediments into high-quality artificial basalt building material', *International Conference on Contaminated Sediments*, Rotterdam, The Netherlands, September.

Brandon, D. L. *et al.* (1992) 'Long-term evaluation of plants and animals colonizing contaminated estuarine dredged material placed in upland and wetland environments', *Misc. Paper D-91-5*, US Army Corps of Engineers Waterways Experiment Station, Long-Term Effects of Dredging Operations Program.

Bray, R. N. *et al.* (1997) *Dredging: A Handbook for Engineers*, London: Arnold Publishing.

Detzner, H. D. *et al.* (1997) 'New technology of mechanical treatment of dredged material from Hamburg Harbour', *International Conference on Contaminated Sediments*, Rotterdam, The Netherlands, Pre-prints pp. 267–74, September.

De Vos, K. *et al.* (1997) 'Environmental impact of the storage of contaminated dredged material – two case studies', *International Conference on Contaminated Sediments*, Rotterdam, The Netherlands. Pre-prints. pp. 981–91, September.

Environment Canada (1994) 'Environmental impacts of dredging and sediment disposal', prepared by *Les Consultants Jacques Bérubé Inc.*, for the Technology Development Section, Environmental Protection Branch, Environment Canada, Quebec and Ontario Regions. Cat. No. En 153-39/1994E.

Ferdinanday-Van Vlerken, M. M. A. (1997) 'Chances for biological techniques on sediment remediation', *International Conference on Contaminated Sediments*, Rotterdam, The Netherlands. Pre-prints, September.

Glindemann, H. and Csiti, A. (1996) 'Remediation of contaminated dredged material – the Hamburg experience', *CEDA Seminar on Environmental Aspects of Dredging*, Technical University of Gdansk, Poland, September.

Harms, C. (1995) 'Monitoring the Hamburg silt mound', *Proceedings of the 14th World Dredging Congress 1995*, Amsterdam, CEDA, The Netherlands. pp. 697–708.

H-SZ and Strom- und Hafenbau, (1997) 'Bricks made of dredged material', Hamburg: Commercial brochure, Hanseaten-Stein Ziegelei GmbH and Freie und Hansestadt Hamburg, Wirtschaftsbehörde, Strom- und Hafenbau.

IADC (1997a) *Environmental Aspects of Dredging – Guide 2a: Conventions Codes and Conditions: Marine Disposal*, Delft, The Netherlands: International Association of Dredging Companies, The Hague and the Central Dredging Association.

IADC (1997b) *Environmental Aspects of Dredging – Guide 2b: Conventions Codes and Conditions: Land Disposal*, Delft, The Netherlands: International Association of Dredging Companies, The Hague and the Central Dredging Association.

IADC (1998) *Environmental Aspects of Dredging – Guide 4: Machines, Methods and Mitigation*, Delft, The Netherlands: International Association of Dredging Companies, The Hague and the Central Dredging Association.

Laboyrie, H. P. *et al.* (1995) 'The construction of large scale disposal sites for contaminated dredged material', *Proceedings of the 14th World Dredging Congress*, Amsterdam, The Netherlands.

Leeuwen, J. L. M. *et al.* (1997) 'The production of sand and clay out of contaminated sediments', *International Conference on Contaminated Sediments*, Rotterdam, The Netherlands. Pre-prints. pp. 129–36, September.

Palermo, M. R. (1996) 'Design and operation of dredged material containment islands', *Proceedings of the Western Dredging Association 16th Technical Conference*, New Orleans, June.

Palermo, M. R. (1997) 'Contained aquatic disposal of contaminated sediments in subaqueous borrow pits', *Proceedings of the Western Dredging Association 18th Technical Conference and 30th Annual Texas A&M Dredging Seminar*, Charleston, SC, June–July.

PIANC (1996) Handling and treatment of contaminated dredged material from ports and inland waterways – CDM', *Report of Working Group No. 17 of the Permanent Technical Committee I*, Brussels, Belgium: *Permanent International Association of Navigation Congresses*.

Reinkes, J. (1997) 'Comparison of results for chemical and thermal treatment of contaminated dredged sediments', *International Conference on Contaminated Sediments*, Rotterdam, The Netherlands, September.

Rulkens, W. H. *et al.* (1997) 'Clean-up of PAH-polluted clayey sediments: the WAU-acetone process', *International Conference on Contaminated Sediments*, Rotterdam, The Netherlands, September.

Shaw, J. K. *et al.* (1998) 'Contaminated mud in Hong Kong: a case study of contained seabed disposal', *Proceedings of the 15th World Dredging Conference*, Las Vegas, Nevada. WEDA, Vancouver, WA.

Smits, J. and Sas, M. (1997) 'Maintenance dredging – the environmental approach', *Proceedings of the 2nd Asian and Australasian Ports and Harbours Conference*, Ho Chi Minh City, Vietnam. CEDA, Delft, The Netherlands.

Tresselt, K. *et al.* (1997) 'Harbour sludge material in landfill cover systems', *International Conference on Contaminated Sediments*, Rotterdam, The Netherlands, September.

USACE (1987) 'Confined disposal of dredged material', *US Army Corps of Engineers*. Engineer Manual 1110-2-5027, Office, Chief of Engineers, Washington, DC.

Walker, J. E. (1989) 'Conditioning and management of mobile harbour CDF's', *Proceedings of the XIIth World Dredging Congress*, Orlando, Florida.

CHAPTER 12

MARPOL

Tom McKay

Background

The second half of the last century saw significant moves by the international maritime community with regard to the protection of the
marine environment from pollution sources generated by shipping operations. Particular attention was paid to oily wastes arising from cargoes or
the operation of onboard machinery. Historically, such wastes were
dumped at sea in a largely uncontrolled manner giving rise to a range
of marine environmental impacts. During the 1960s and 1970s the worldwide growth in shipping movements, particularly those involving the
transport of large volumes of crude oils and hydrocarbon products
together with the aftermath of the Torrey Canyon disaster in 1967,
resulted in growing international concern with regard to the potential
pollution of the marine environment by ships.

As a consequence, the International Convention for the Prevention of
Marine Pollution from Ships (MARPOL) was adopted by the International
Maritime Organisation (IMO) in 1973 and subsequently amended by the
Protocol of 1978 *(IMO, 1991)*. The Convention, which is commonly
referred to as MARPOL 73/78, comprises the following five Annexes
defining regulations for the control of ship-generated pollution:

- I   Prevention of Pollution by Oil;
- II   Control of Pollution by Noxious Liquid Substances in Bulk;
- III   Control of Pollution by Harmful Substances in Packaged form;
- IV   Prevention of Pollution by Sewage from Ships; and
- V   Prevention of Pollution by Garbage from Ships.

Adoption of Annexes I and II is compulsory for Member State ratification of the Convention, which finally entered into force on 2 October
1983. Regulation 9 of the Convention addresses the control of oil discharge
from ships and includes severe restrictions on the discharge of oily wastes.
Regulation 12 defines the requirements for reception facilities to accept
waste oil from ships and states:

...the Government of each Party undertakes to ensure the provision at loading terminals, repair ports, and in other ports in which ships have oily residues to discharge, of facilities for the reception of such residues and oily mixtures as remain from tankers and other ships, adequate to meet the needs of ships using them without causing undue delay to ships.

MARPOL Annexes I and II are generally considered to be 'complete' with ratification by Member States representing well over 90 per cent of the world's fleet. The achievements of MARPOL Annex I have been endorsed by studies carried out by the National Research Council Marine Board of the United States and the United States National Academy of Science. MARPOL 73/78 has been credited by both these organisations as having made a substantial impact in decreasing the amount of oil that enters the sea. Studies showed that in 1973, some 2 million tons of oil entered the world's oceans as a result of shipping operations; this was reduced to around 1.5 million tons in 1980 and then further reduced to just over 500 000 tons in 1990. However, Member States continue to approach IMO where they feel there may be room for improvement and there is concern over the fact that a number of important oil producing and exporting nations have so far failed to ratify MARPOL. One of the key reasons for this is thought to be the costs associated with the provision of reception facilities for oily wastes as outlined above.

In 1997 a new Annex on the prevention of air pollution from ships was added and IMO's Marine Environment Protection Committee (MEPC) has drafted mandatory regulations covering the management of ballast water to prevent the spread of unwanted aquatic organisms and the banning of anti-fouling paints which are potentially harmful to the environment. The MEPC, which meets three times during every biennium, is an important forum for governments, inter-governmental and non-governmental organisations with an interest in protecting the marine environment from pollution from ships. MARPOL remains a 'living' document, which may be amended to meet changing needs.

Importantly, IMO is concentrating its efforts on full implementation of MARPOL requirements by all Flag States and Port States.

The present status of the MARPOL Annexes and the associated development of their regulations are set out in IMO (1998).

Aims and priorities

IMO and the European Union have taken the lead in carrying out studies and preparing guidelines for the handling of ships' wastes in order to best meet the needs of governments, the ports and shipping industries and the international community (EMARC Consortium, 1996; BMT *et al.*, 1997).

In order to determine the best way forward the following principles have evolved:

(a) the overall impact of ships' wastes on the environment must be reduced – not simply transferred from sea to land;

(b) mandatory discharge regulations and systems for charging should not unwittingly encourage dumping at sea;

(c) the production of waste on ships should be reduced – greater investment to meet this aim should be encouraged;

(d) legislation should be effective across all Member States;

(e) mandatory requirements should be written in such a way as to minimise exceptions;

(f) the cost of using facilities should be made clear in advance;

(g) existing patterns of shipping should not be disrupted;

(h) the costs of improved standards should be shared fairly by users and suppliers and should not preclude opportunities to derive commercial benefit;

(i) the overall costs to society should be broadly in proportion to the environmental gains which may be achieved;

(j) port waste management systems and facilities should not unduly delay ships nor create an excessive administrative burden; and

(k) a ship behaving reasonably and responsibly with regard to its waste management should not be penalised for the operational choices it makes.

The role of ports

Protection of the maritime environment and the need for greater understanding of pollution control measures related to the planning, development and operation of ports have been issues of growing concern over the past two decades. Those public or private sector bodies responsible for the management of port facilities are now required not only to take full account of environmental aspects but also to do so as part of a formal process in order to meet national and international regulations, directives and codes of practice.

Ports form the interface between land and sea and hence provide an essential link in the transport chain. Increasingly, they have had to adapt to take account of environmental issues related to their waterfronts, land use within the port, and access from land and sea. In order to take proper account of marine environmental issues, ports have had to develop extensive contingency plans for the prevention of marine pollution including the need to provide marine support facilities and emergency services booms, dispersants and other stockpiled equipment. Ports are well aware of the associated impacts on their capital and operating costs and the need for these to be recovered. However, if port pricing is not competitive, ship operators may decide to go elsewhere. The need to provide MARPOL reception facilities has clear impacts on the economics of port use and hence has a major influence on the decision making process of port management.

In order to address the requirements for the provision of such facilities it is necessary to take account of a wide range of issues affecting the operation and management of ports including the following:

(a) port location – coastal, estuarial, river, etc.;
(b) marine access – navigation channels, manoeuvring, locked access, etc.;
(c) port-related land use;
(d) adjacent land users;
(e) range of port operations;
(f) cargoes handled; and
(g) volume and mix of vessel traffic.

Studies carried out by international maritime bodies have identified the following key issues facing ports and their need to compete in national and international maritime markets (BMT *et al.*, 1997):

(a) the provision of adequate reception facilities which are capable of fast and efficient transfer and disposal operations; and
(b) the development of appropriate financing models which will require sea-going vessels to pay charges independently of their use of reception facilities (i.e. 'inclusive' or 'no special fee').

The following requirements have been identified as critical if the above aims are to be realised:

(a) the role of ports should not necessarily include policing of what ships do with their onboard wastes. For the sake of efficiency and bearing in mind the commercial relationship between ports and port users, policing should be the responsibility of the Member State through some form of regulatory control;
(b) costs should be met either by the ship directly or by fees included in port dues;
(c) adequacy and convenience of use of facilities and efficiency of port operations are more important than price;
(d) a key aim should be to avoid distortion in competition between ports and to attempt to achieve a 'neutral' position in this regard;
(e) ports should be able to receive any amount of waste from ships operating at their berths;
(f) ships operating at the port should have a minimum onboard pumping capacity below which there would be an additional charge;

(g)   ships engaged on short sea routes cannot be expected to discharge waste every time they enter a port. There must be systems in place, which allow monitoring of such operations in order to ensure that waste is properly deposited in appropriate locations; and

(h)   special regulations should apply to specific cases such as scheduled ferry operations.

Within the EU all the above topics have been under review by maritime bodies including the European Sea Ports Organisation, British Ports Association, UK Major Ports Group, and the UK Department of Transport Shipping Policy. In 1996 the EMARC Consortium brought together 12 environmentally conscious commercial and research organisations to carry out a research project funded by the European Commission to assess the effects of MARPOL regulations on the port environment throughout Europe and to investigate present and possible future systems for the management of ships' wastes (EMARC Consortium, 1996).

Marpol reception facilities

Waste-management planning

Under the UK's Merchant Shipping (Port Waste Reception Facilities) Regulations 1997, which came into force on 27 January 1998, there is a requirement for ports, harbours, terminals and marinas to prepare a Waste Management Plan with regard to the provision and use of waste reception facilities within its jurisdiction.

The Associated British Ports Group, which owns and operates 23 of the UK's ports has taken a lead role in the development of the mandatory waste management plans. It has drawn up the following process which may be followed regardless of the size, nature or function of the port:

(a)   review of port operations, facilities and waste management respon-sibilities including the numbers and types of different vessels calling at the port;

(b)   consultation with waste regulators, port users and other interested parties as part of a process which leads to improvement and greater utilisation of existing facilities;

(c)   analysis of the amounts and types of wastes generated and landed by ships using the port including the development of arrangements for collecting and recording data on waste;

(d)   consideration of the type and capacity of facilities needed to handle landed wastes;

(e)   consideration of the location and ease of using reception facilities;

(f)   consideration of the various methods of charging for facilities to determine a balance between the quality of facilities and their cost and to ensure that costs do not act as a disincentive to waste discharge at ports and harbours;

(g) development of a publicity strategy, which ensures that adequate and clear information, is available to ships' masters and other port/harbour users; and

(h) review the planning process regularly.

Implementation

The MARPOL Convention requires governments to ensure the provision of adequate port reception facilities capable of receiving shipboard residues and mixtures, containing oil, noxious liquids or garbage, without causing undue delay. In effect, depending on the size of the port, such a facility could comprise anything from a simple garbage skip and oil barrel to a large tank farm with storage and treatment facilities.

In technological terms there is little difference between the treatment, recycling or disposal of wastes generated on board ships or from land-based sources. If a viable waste management sector is in operation it will no doubt be prepared to invest – possibly in partnership with the port authority or government department – in providing facilities to deal with the additional wastes from ships if the stream of waste arising is seen to be reliable and an appropriate financing mechanism is in place. Where differences do occur is in the collection and handling of waste from ships bearing in mind the need to ensure that delays are minimised. Shipping operations and procedures within ports mean that simple systems for land based collection of industrial wastes cannot necessarily be applied.

Although the MARPOL Convention requires waste 'reception' facilities, it does not specify 'treatment' of the wastes prior to disposal. Where the local waste management industry already includes facilities for the treatment of oily wastes as a consequence of waste oil collection from land-based sources, commercial benefit may be obtained by the addition of oily wastes from ships.

In general, the implementation of a MARPOL waste reception facility will involve the following key requirements:

(a) systems for collection of wastes from ships;

(b) facilities for storage or treatment; and

(c) systems for transfer of treated wastes either for recycling or disposal.

Developments in the port and shipping industries have resulted in increasingly rapid turnaround times within ports which limit access to ships while in port. Where tanker trucks are commonly used to collect waste oils from ships in ports, access from the landside may be restricted by cargo loading and unloading operations to a very short timeframe.

Collection by self-propelled barges offers greater flexibility, as they are able to service ships at anchor or alongside a berth. Although such barges may be expensive to buy and operate they also offer economies of scale as a result of their carrying capacity and their ability to service a large number of vessels at different locations within a port or even, if appropriate, at a number of ports in close proximity.

Storage and treatment facilities may include a range of skips or holding tanks with associated treatment plant if appropriate. Transfer from a collection barge is likely to require a dedicated berth with a pumping facility and pipeline to the holding or treatment plant. Disposal or the recycling of treated waste may involve discharge to road tankers or transfer to an outfall to inshore waters. The definitive guide for the planning and implementation of reception facilities is IMO (1999).

Financing, control and regulation

Although the MARPOL Convention states that signatory countries should 'ensure' that reception facilities are provided it does not say how this aim may be achieved. Ports may identify local sources of capital investment or consider introducing a foreign investment partner to assist with the construction and operation of the waste-handling facilities. The problem, which has slowed the programme of development of such facilities world-wide, is generally linked to financing and regulatory enforcement and, in particular, how to ensure that the establishment and operation of reception facilities may be made to be commercially viable.

There are currently three main financing systems which have evolved worldwide with the introduction of MARPOL reception facilities. They have been the subject of ongoing study by international maritime bodies, and may be summarised as follows:

(a) *Free-of-charge*: disposal is provided from general governmental revenue sources at the expense of the taxpayer. This system was adopted for a trial period in Germany (1988 to 1991) and was found, perhaps surprisingly, to have only limited success. Only 15 per cent of ships calling at ports covered by the project opted to discharge their waste oils which implies that the associated impacts on their operations and turnaround times were significant factors.

(b) *No Special Fee*: this system includes the cost of waste reception facilities in the port dues. Ships are not charged separately for discharge of engine room waste. The system was introduced in Sweden in 1980 and has been maintained as it is considered by the Swedish Authorities to be the most effective way of bringing waste ashore. The system does have clear problems associated with the need to deal with waste, which should have been discharged at other ports prior to a ship's arrival. Sweden has been a strong supporter of a regional no-special fee system, which was proposed by the Convention for the Marine Environment in the Baltic Sea Area (HELCOM), in 1996. The HELCOM strategy attempts a regional mechanism, which will result in the greater availability of reception facilities, a harmonised no-special-fee system, uniform rules on discharging of waste and co-operation on detection of discharges and prosecution of offenders.

(c) *Direct charge:* the direct-charge system with a fee based on the quantity of waste discharged is the most widely used system throughout Europe and elsewhere in the world. In some cases the facilities are provided by the ports, but in a large majority of cases the service is provided and operated by private-sector waste management companies as an extension of their existing land based business of waste collection, transfer and treatment.

In order to examine typical levels of direct charges and their variation among European ports, British Maritime Technology (BMT *et al.*, 1997) obtained charge levels from various EU Member State ports for the following three types of vessel:

(i)   1600 GRT Coaster
(ii)   12 000 to 15 000 GRT General Cargo/ Handy Size Bulk Carrier
(iii) 60 000 to 80 000 GRT Panamax Bulker.

Each port was asked to quote the level of port fees for each ship type along with the costs of disposal of 17 t of Annex I waste and 2 t of Annex V waste. The results of the survey demonstrated a wide variation in charges with fees for the reception of 17 t of Annex I waste ranging from 275 to 1500 EUR representing an average of 50 EUR/t. Charges for the reception of 2 t of Annex V waste varied from 30 to 320 EUR with a mean figure of 148 EUR/t. Table 12.1 summarises these average costs as a proportion of average port dues. (Note: 1 EUR = US$ 0.90 at September 2001).

The table demonstrates that the relative proportion of waste charges compared with port dues decreases dramatically with ship size. However, these results, which are based on a single port visit do not provide a realistic comparison between port dues and reception fees as they fail to take account of differences in the rates of waste generation and, therefore, the frequency with which the various ship types may generate and discharge the quantities of waste under consideration. These additional factors, again based on work carried out by BMT *et al.*, are shown in Table 12.2.

Table 12.2 should be interpreted with some caution as it is limited to a specific data gathering exercise but it does provide a realistic

Table 12.1 Fees for reception of Annex I (17 t) and Annex V (2 t) wastes, as a percentage of total port costs (single port visit)

Ship type	Average port dues (EUR)	Average reception fee (EUR)	Reception fee as % of dues
Coaster	2827	Annex I–858 Annex V–126 Total–984	35
Handy-size bulker	18 496	Annex I–839 Annex V–148 Total–987	5
Panamax bulker	73 064	Annex I–861 Annex V–190 Total–1051	1

Source: Direct enquiry to ports (BMT, June 1997). Figures are based only on ports with direct charges for Annex I and V wastes.

Table 12.2 Fees for reception of Annex I (17 t) and Annex V (2 t) wastes, as a percentage of total port costs (averaged over all trips)

Ship type	Average port dues (EUR)	Average reception fee (EUR)	Waste generation per voyage (tonnes)	Voyages per waste delivery	Average fees amortised over all trips as % of dues
Coaster	2827	Annex I–858 Annex V–126 Total–984	Annex I–3.6 Annex V–0.04	Annex I–4.7 Annex V–50.0	Annex I–6.5 Annex V–0.1 Total–6.6
Handy-size bulker	18 496	Annex I–839 Annex V–148 Total–987	Annex I–35 Annex V–0.7	Annex I–0.5 Annex V–2.9	Annex I–9.1 Annex V–0.3 Total–9.4
Panamax bulker	73 064	Annex I–861 Annex V–190 Total–1051	Annex I–45 Annex V–0.7	Annex I–0.4 Annex V–2.9	Annex I–3.1 Annex V–0.1 Total–3.2

Source: Direct enquiry to ports (BMT, June 1997). Figures are based only on ports with direct charges for Annex I and V wastes. Crew sizes and voyage lengths (coaster: 9 crew, 2 days; handy-size: 25 crew, 14 days; Panamax: 25 crew, 14 days) were assumed for the purpose of this comparison, although they vary widely in practice. Waste generation rates, based on a review of several sources, will also vary in practice.

demonstration of the potential importance of waste reception charges for ships which discharge their wastes in a legal manner. While not very high in proportion to port dues, the charges are high enough for ship operators to take into consideration in their strategies for cost control and minimisation.

Safety during discharge and other handling of wastes

Ship generated wastes are hazardous to the marine environment and/or human health. As a consequence, it is important to ensure good practice with regard to safety issues when discharging or handling such wastes while in port. Safety regulations for the discharge and handling of wastes should be based on national legislation, international conventions and recommendations, accepted industry standards and guidelines.

IMO has published the following safety guidelines, which apply to the discharge and handling of hazardous wastes:

(a) recommendations on the safe transport, handling and storage of dangerous substances in port areas;

(b) international maritime dangerous goods code; and

(c) crude oil washing systems.

The following industry safety guides have also gained wide international recognition for operations within ports:

(a) *International Safety Guide for Oil Tankers and Terminals* (ISGOTT) published by the International Chamber of Shipping, the Oil Companies International Marine Forum and the International Association of Ports and Harbours.

(b) *Safety Guide for Terminals Handling Ships Carrying Liquefied Gases in Bulk*, published by the Oil Companies International Marine Forum.

(c) *Guidelines on Port Safety and Environmental Control* published by the International Association of Ports and Harbours.

Present status of MARPOL convention

(a) Annex I – prevention of pollution by oil: as discussed above, the Annexes are generally considered to be complete but there is concern with regard to the slow development and implementation of port waste reception facilities worldwide. IMO is presently addressing the problem of inadequate reception facilities and the MEPC is examining options for the financing of port reception facilities. IMO has also been involved in a number of technical co-operation projects to assist developing countries with the implementation of MARPOL requirements.

With the aim of improving the implementation of Annex I the MEPC has reviewed all the provisions of the Annex and prepared a General Action Plan for the Revision of Annexes I and II in 1995. The aims of the revised Annexes are to simplify present requirements, take account of technical progress and to identify and correct inconsistencies. The Revision is expected to be completed by 2002.

(b) Annex II – control of pollution by noxious liquid substances in bulk: this Annex addresses the transportation by sea of liquid chemicals in bulk form. Chemical tankers are generally smaller than oil tankers, ranging from around 500 to 40 000 GRT. They have become gradually more complex in their design and construction, being designed to carry many different substances at the same time all with different properties, storage characteristics and handling requirements. The main chemicals carried in bulk can generally be classed as heavy chemicals, molasses and alcohol, vegetable oils and animal fats, petrochemical products and coal tar products. The MEPC are presently looking at the categorisation of products related to their potential impacts on the marine environment and human health. It is envisaged that the whole of Annex II will be revised by 2002 including the re-evaluation and categorising of the hazard profiles for all substances falling within the Annex. The MEPC is also examining the issues of reception facilities and how to ensure adequate implementation within ports.

(c) Annex III – prevention of pollution by harmful substances carried by sea in packaged form: the main aims of this Annex have been the identification of harmful substances so that they may be packed, stowed and clearly marked in such a manner that possible accidental impacts to the marine environment may be kept to a minimum. Initially, this was hampered by the lack of clear definition of harmful substances but was improved by amendments to the International Maritime Goods Code (IMDG Code) to include marine pollutants.

The main changes likely to affect Annex III are now related to the IMDG Code, which is being reformatted for ease of use and understanding.

(d) Annex IV – prevention of pollution by sewage from ships: discharge of raw sewage into the sea can create health hazards and, within inshore coastal waters, may reduce oxygen levels leading to subsequent impacts on marine ecological resources. Associated visual pollution can also lead to very significant impacts on tourism. The Annex requires governments to ensure the provision of adequate reception facilities at ports for the reception of sewage.

Although the Annex has not yet come into force, many countries have introduced regulations in line with its requirements in order to reduce impacts on health and the marine environment brought about by the discharge of sewage from ships. In the meantime, IMO have been reviewing the regulations within the Annex in order to encourage further ratification and is also working on the harmonisation of IMO standards on sewage treatment with those being developed by the International Standards Organisation (ISO).

(e) Annex V – prevention of pollution by garbage from ships: garbage from ships may have severe impacts on the marine environment.

Annex V defines 'garbage' as 'all kinds of victual, domestic and operational waste, excluding fresh fish and parts thereof, generated during the normal operation of the ship and liable to be disposed of continuously or periodically except those substances which are defined or listed in the Annexes to the present Convention'.

Annex V totally prohibits the disposal of plastics anywhere in the sea and severely restricts discharges of other garbage from ships into coastal waters or 'Special Areas' with low water exchange or heavy volumes of maritime traffic. These special areas established under the Annex are the Mediterranean Sea, the Baltic Sea, the Black Sea, the Red Sea, the Gulf, the North Sea, the wider Caribbean region and the Antarctic area. The Annex also obliges governments to ensure the provision of facilities at ports for the reception of garbage. Such facilities may include receptacles for the reception and segregation of garbage, a regular collection service, recycling and/or final disposal of garbage. The Annex received sufficient ratification to enter into force in December 1988.

(f)   Annex VI – prevention of air pollution from ships: the Protocol and new Annex VI will enter into force after being accepted by 15 states with not less than 50 per cent of the world merchant shipping tonnage. When the Annex comes into force it will set limits on sulphur oxide and nitrogen oxide emissions from ship exhausts and prohibit deliberate emissions of ozone-depleting substances. The Annex also prohibits the incineration onboard ship of certain products such as contaminated packaging materials and PCBs.

Possible future development of MARPOL

IMOs Marine Environment Protection Committee is working on draft regulations related to the following two environmental issues which may result in additional Annexes to MARPOL 73/78, although it is possible that they may be dealt with by independent Conventions.

Unwanted aquatic organisms in ballast water

Ballast water is used to stabilise ships at sea, when they have discharged or partially discharged their cargoes and are in transit to their next port of call. In the past ships have carried large numbers of varying species of aquatic organisms across the oceans and discharged them into non-native marine habitats which, over time, can have significant impacts on local marine ecosystems. In 1990, the MEPC set up a working group on ballast water, which developed guidelines for addressing this problem. The guidelines, which were aimed at providing information on procedures for minimising risks from the introduction of unwanted species, were adopted in 1991.

It is likely that regulations will be developed which will make it compulsory for ships to choose between exchanging their ballast water in mid-ocean where they are less likely to pick up sea life or discharging ballast water into special reception facilities or using some method of killing off alien aquatic life forms carried in the ballast waters.

Toxic anti-fouling paints

Anti-fouling paints are used to coat the underside of ships' hulls to prevent sea-life, such as algae or molluscs, from attaching themselves to the hull thereby slowing down the ship or increasing fuel consumption. However, it is known that underwater marine life may be harmed by these products, which may also affect the food chain. One of the most effective anti-fouling paints contains organotin tributyl tin (TBT). In March 1998, the MEPC agreed to establish a Working Group to begin drafting mandatory regulations to ban TBT in anti-fouling systems with a view to the adoption at a conference after the year 2000.

MARPOL 73/78 – conclusions

Despite the number of years that it took for MARPOL to enter into force and the subsequent slow pace at which many Member States have set about the process of ratification and the implementation of associated facilities, it is well recognised that the 1973 Conference, which adopted the Convention and provide the basis for IMO's future work on environmental issues, has been of great significance to the international maritime community. MARPOL is accepted as the most important set of international regulations for the prevention of marine pollution by ships. Figures do show that there has been a significant decline in marine pollution since the Convention came into being. The implications for port development and operation will continue to grow as maritime governments, port authorities and operators will have to accept the need to comply with international guidelines and practices with regard to port safety, efficiency of operation and protection of the environment.

Other factors under continuing development and review by IMO, which contribute to the prevention of marine pollution, include the issue of Port State Control, the introduction of the International Safety Management Code (ISM Code) and the amendments to the International Convention on Standards of Training, Certification and Watchkeeping for Seafarers (STCW).

In particular, Port State Control sets out provision for ships to be inspected when they visit foreign ports to ensure that they comply with IMO requirements. This was originally intended to support Flag State implementation but the systems have been found to be particularly effective if organised on a regional basis. IMO has encouraged the establishment of Regional Port State Control in many parts of the world including Europe, North America, Asia and the Pacific, Latin America, the Indian Ocean, the Mediterranean and the Caribbean.

The ISM Code is aimed at ensuring that ships are properly managed and operated. Its stated objectives are to 'ensure safety at sea, prevention of human injury or loss of life and avoidance of damage to the environment, in particular to the marine environment and property'.

There are ongoing concerns, which remain with regard to marine pollution and its impacts on the world's oceans, but it is now widely accepted that the key to combating marine pollution is the implementation of IMO Conventions. IMO is continuing to focus its efforts to achieve these aims through its Committees and Sub-Committees and through its Technical Co-operation programme, which aims to assist Member States, where necessary, in the development of infrastructure and trained personnel required to achieve ratification and implementation of the international regulations.

References

British Maritime Technology and Prof. Glen Plant. (1997) *Study on the Feasibility of a Mandatory Discharge System of Ships Waste to Shore Reception Facilities in Ports and the Financing Systems of such Facilities*, European Commission Contract B96-B2 7020-SIN 6054-ETU.

EMARC Consortium (1996) *A European Union Fourth Framework Research Project Endeavouring to Quantify the Effectiveness of the United Nations Marine Pollution Convention.*

IMO (1991) MARPOL 73/78 (Consolidated Edition).

IMO (1998) Focus on IMO, MARPOL – 25 Years.

IMO (1999) *Comprehensive Manual on Port Reception Facilities.*

Further reading

Port Waste Management Planning – How to Do It, UK Department of the Environment, Transport and Planning, 1998.

Best Practice Guidelines for Waste Reception Facilities at Ports, Marinas and Boat Harbours in Australia and New Zealand, Australian and New Zealand Environment and Conservation Council and Australian Transport Council, 1997.

The Baltic Strategy for Reception Facilities for Ship-generated Waste, Helsinki Commission (HELCOM), 1998.

International Safety Guide for Oil Tankers and Terminals (ISGOTT) the International Chamber of Shipping, the Oil Companies International Marine Forum and the International Association of Ports and Harbours.

Safety Guide for Terminals Handling Ships Carrying Liquefied Gases in Bulk, the Oil Companies International Marine Forum.

Guidelines on Port Safety and Environmental Control, the International Association of Ports and Harbours.

IMO safety guidelines which apply to the discharge and handling of hazardous wastes are:

(a) recommendations on the safe transport, handling and storage of dangerous substances in port areas;

(b) international maritime dangerous goods code; and

(c) crude oil washing systems.

APPENDIX I

Mathematical models

Hanne L. Svendsen

The most important development in hydraulic engineering during the 1970s and 1980s is no doubt mathematical modelling, made possible by advanced digital computers (Abbott, 1979). Strictly speaking, the term mathematical modelling is not necessarily related to the use of computers. A mathematical model is, in principle, nothing more than a mathematical formulation of a physical process, such as, for instance, the Navier–Stokes equation. However, it was only after the arrival of large capacity digital computers that such mathematical expressions could be handled in the form of a discrete numerical model and used for sufficiently detailed practical application. The advantages of mathematical (i.e. numerical in the following) models over physical models are many and important.

First and foremost the mathematical model can take into account any physical phenomena that can be described in mathematical form. Out-standing examples of this are Coriolis' forces and wind effects. Wind effects have important impact on currents and water levels in shallow-water regions, and on ships manoeuvring at slow speed. In a physical model, it would be impossible or very costly to simulate wind effects in a realistic way for such purposes, while in a mathematical model it is a simple matter. Another important advantage is that scale effects as such, do not exist in mathematical models. Errors may of course be introduced by simplifications or approximations in the mathematical equations describing the natural processes or, and this might perhaps be called a scale effect, the model grid made too coarse thereby smoothing out important features of the region described. However, contrary to scale effects in physical models, this type of 'scale effect' is not a matter of principle; rather it is a matter of computer capacity or cost. Finally, the mathematical models have the obvious advantage that many different port layouts may relatively easily be tested, and the models be stored and re-mobilised at insignificant cost, while for a physical model the cost of maintaining its availability in most cases is prohibitive.

With the arrival of very powerful personal computers many mathematical and numerical programs have been developed during the last ten years. This has given engineers flexible tools in planning and design of ports, harbours and marine terminals. It is now possible to use three-dimensional models necessary in stratified water bodies, and models with 'unstructured' meshes (i.e. grid varying in size and shape within the model area), useful in problems where details are needed in some areas but not in others.

This, together with advanced two-dimensional models, means that nowadays mathematical and numerical modelling is one of the strongest and most important tools in port and coastal engineering. A broad range of modelling tools are available, among others models for determination of design waves, which is a very important factor in design of any man-made structure in the coastal zone. In the following these wave-modelling tools are further described.

Offshore waves

Data on offshore wind and wave conditions are needed to determine the design waves near and inside a port.

Measured data is often not available during periods long enough to allow statistical analysis for the establishment of sufficiently accurate estimates of extreme sea states for use in more local models. In this case, measured data (if any) can then be supplemented with hindcast data through the simulation of wave conditions during historical storms using mathematical models. Such data are nowadays available from a number of commercial sources.

Collections of simultaneously measured wave and wind data at a regional scale used in regional wind and wave models are also available nowadays for use as extreme offshore boundary data for more detailed local studies. Another new development for determining offshore sea states (design offshore wave and wind parameters) is the access to satellite wave and wind data calibrated against traditional measurements.

These models, and other various systems, can also be used to forecast waves, e.g. for ship voyage planning, etc. with meteorological weather forecasts as input.

Nearshore waves

Nearshore waves can be determined using the offshore wind and wave conditions by transforming these with mathematical wind–wave modelling of processes such as refraction, diffraction, shoaling, breaking, etc., resulting in the nearshore design waves just outside the port (Fig. A1.1).

417

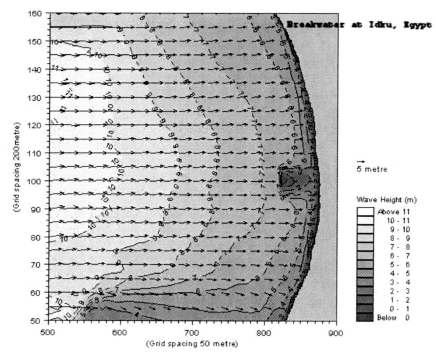

Fig. A1.1 Nearshore waves simulated with the Nearshore Spectral Wave Model Mike 21 NSW near the breakwater at the Idku LNG terminal in Egypt

Wave disturbance

Wave agitation inside ports is one of the key issues to be considered in the design of ports, harbours and marinas. The development of two-dimensional hydrodynamic models to incorporate the influence of vertical accelerations in short-wave motions made it possible to describe the combined effects of refraction, diffraction and reflection on short-period wave phenomena in an arbitrary nearshore bathymetry (Abbott et al., 1978). Advanced time-domain wave models based on the Boussinesq equations are thus available for determining the wave disturbance inside the port. These models use, for example, the nearshore design waves as input and are based on descriptions of the evolution of non-linear, frequency dispersive directional, irregular waves in water of variable depth (Fig. A1.2). The models simulate, in the time domain, all relevant wave processes in the port, including resonance. The wave reflection characteristics of structures with full reflection or partial reflection can be modelled too.

Accordingly, much more detailed description of the influence of wave conditions caused by local depth irregularities, for example severe wave

Halul Harbour

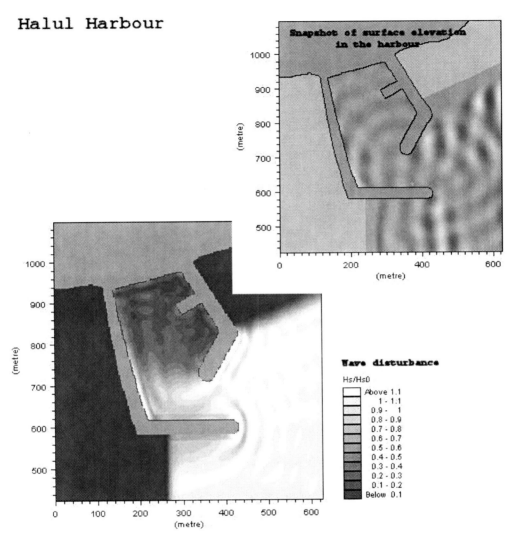

Fig. A1.2 Wave disturbance simulated with the Boussinesq Wave Model Mike 21 BW at Halul harbour in the Arabian Gulf

breaking, on wave disturbance in ports, may be obtained without having to construct a hydraulic scale model.

For large ports subject to the occurrence of long waves (seiching), due to the effects of wave grouping and wave breaking, physical models are still the most reliable tools. This is also the case for large vessels moored in a port exposed to severe long waves. For unprotected terminals, such as tanker jetties, mathematical models are available for assessing conditions at berth, i.e. vessel motions, mooring and fender forces, etc.

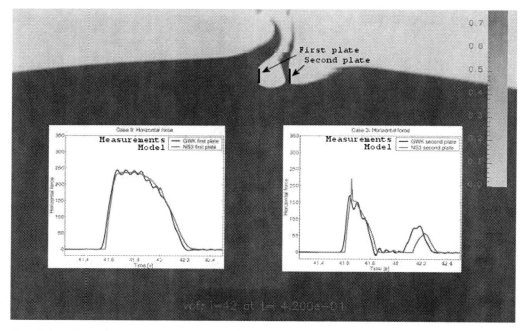

Fig. A1.3 Wave forces on structures simulated with a three dimensional Narier–Stokes Model developed by DHI Water and Environment (courtesy of DHI)

Wave forces on structures

Calculation or estimation of wave-induced forces on structures are also needed. Various design-code formulas or physical model tests have traditionally been the tool. The latest development in mathematical models means that these forces (Fig. A1.3) as well as scouring around structures can now be determined by modelling.

Further rapid development of mathematical and numerical modelling capabilities may be expected in coming years.

Water levels, currents and sedimentation

Besides the need for design waves in and outside a port and wave forces on structures, information regarding water levels, currents and sediment transport are also needed.

Today a large range of hydrodynamic models for predictions of water levels and currents are available (Fig. A1.4). By using these models the design engineer has powerful tools for the design of marine and coastal installations. Models for prediction of littoral transport, siltation in access channels and port basins and for water quality are available. Further, hydrodynamic models using weather forecasts as input are available for forecasting water levels and possible flooding in the coastal zone.

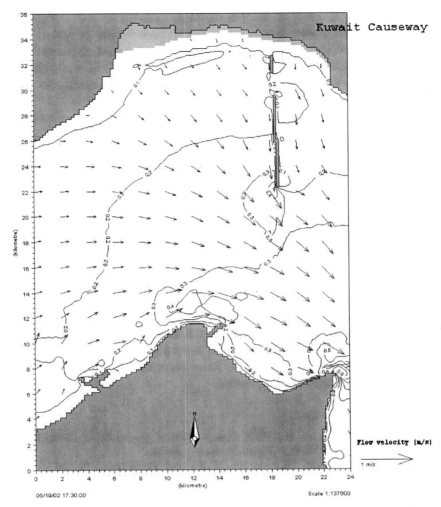

Fig. A1.4 Tidal current arrows and iso-velocity-lines at the proposed causeway between Kuwait City and Subiyah in Kuwait, simulated with the two dimensional Hydrodynamic Model Mike 21 HD

The limits of application of mathematical models are set by insignificant knowledge in terms of quantitative description of complex physical processes, inaccuracies in the numerical solution techniques, insufficient input data and by computer capacity and costs. Progress in all of these areas has been tremendous during the last ten years and has enlarged the applicability of mathematical modelling, so that it has rapidly gained in importance as a tool for solving hydraulic and related problems. Inaccuracies or numerical errors are present in the output of any numerical model. The art of computational hydraulics is concerned with developing numerical solutions techniques with insignificant, or at least acceptable inaccuracies, within the whole range of application for a particular model. Users of such models are cautioned to question their numerical accuracy and to thoroughly test and verify their models against measurements as well as physical phenomena for which analytical solutions are available.

The mathematical models enable execution of reliable studies in fields never adaptable to physical modelling. It is, however, not yet the case when it comes to breakwaters and coastal structures with loose elements such as rubble mounds, revetments, etc. An optimal approach for certain cases combines mathematical and physical modelling in situations, where different complex hydraulic phenomena, e.g. wave breaking, diffraction and reflection, exist at the same time. Further development of mathematical and numerical modelling will continue to replace physical models, giving engineers easy access to more reliable data to be used in the design process.

A broad range of mathematical and numerical modelling is available, giving hydrodynamic design inputs as described above. It is however, as it has always been, no trivial matter to understand the physical and environmental processes affecting man-made structures in the marine environment. Models will never replace knowledge and experience, but will become more and more important tools for elaboration of accurate and reliable design data.

References

Abbott, M. B. (1979) *Computational Hydraulics; Elements of the Theory of Free Surface Flows*, Pitman: London.

Abbott, M. B., Peterson, H. M. and Skowgaard, O. (1978) 'On the numerical modelling of short waves in shallow water', *Journal of Hydraulic Research*, 16: 3.

APPENDIX 2

Case studies concerned with new ports and port extensions

Hans Agerschou

General considerations

New ports or port extensions become necessary when cargo throughput increases or when entirely new cargoes have to be accommodated. In principle, if there is a choice, they should be located where they will minimise the relevant total sea and land transport costs to the economy of the country, considering the forecast origins and destinations. Obviously, this is not exclusively a matter of geographical location, but also a matter of different costs for different sites, depending upon foundation conditions, water depths, maintenance dredging requirements, etc.

Land transport costs should include an appropriate share of the cost of additional land transport links or of improvement of existing links. Sea transport costs should consider the effects of the different sizes of ships, which it may be technically and economically feasible to accommodate at different locations. This is likely to be particularly important for bulk cargo. The total sea and land transport cost in question is the sum of the discounted annual costs during the life of the project. Thus cargo throughput and ship traffic forecasts, with their inherent uncertainties, are unavoidably introduced into attempts to optimise port locations, in particular for new ports.

It is sometimes argued that transport projects generate development of production of commodities and/or of traffic, e.g. feeder roads or farm to market roads generate agricultural production over and above that required for local consumption. For port projects there are many indications to the contrary. A new port should not come into existence earlier than connecting land transport facilities and/or production and storage facilities, which are to be served by the port. However, the port has to be part of an economically feasible project containing the above components. The port alone may not be feasible, but may be an indispensable part of the project.

Factors, which are not quantifiable in economic terms, such as overall strategic planning for regional or demographic development, may play an important role in selection of general locations for ports.

The case studies below, which are presented as thumbnail sketches, are based on the author's personal experience during the past decades. Although they are presented in a non-commodity specific manner, this is not believed to make them less educational. Also the actual locations and the involved organisations are not mentioned, as the intended purpose is not to allocate blame, but to learn from insufficient or incorrect project analyses in the past and attempt to improve methodology for the future.

The new port on the other side of the peninsula

In one country, most industrial and agricultural production facilities as well as ports were located on one side of the peninsula, which is the mainland of the country. It was decided to construct a deep-water port on the other side of the peninsula, as part of the government's policy aimed at promoting economic development in the area. The new port is located approximately at the centre of the coast. Its distance from an existing major port, on the opposite side of the peninsula, is only about 250 km. Good roads connect the two ports. Another major port is located at the tip of the peninsula, about 300 km from the new port.

Based on forecasts of future production and consumption, the cargo throughputs were assigned to the three ports based mainly on official road transport user charges, the differences between which are largely proportional to the differences in distances by road, using the hinterland concept. In the determination of berth, storage and cargo-handling equipment requirements as well as for economic and financial evaluation of the project it was assumed, that upon completion, the new port, during its first year of operation, would achieve its full share of cargo throughput. Outbound cargo amounts to 67 to 75 per cent of the total.

The following important aspects were not considered:

(a)   the lack of processing facilities for exports at the new port;
(b)   the lack of storage facilities for exports at the new port;
(c)   established trade relationships; and
(d)   actual road-transport-user charges rather than official charges.

Various development projects for the hinterland were embarked upon. However they were largely unsuccessful. Within the port, large areas had been allocated for long-term lease to port-related industries and storage facilities. In the medium and long term this was to become the salvation for the project. The lack of processing and storage facilities combined with established trade relationships resulted in non-diversion of cargo throughput to the new port. Rebates granted on road-transport-user charges also contributed to this. Actual cargo throughputs during the two first years of port operation were only 29 and 35 per cent of appraisal forecasts, respectively. Liner service had not started seven years after the port started operations.

The port management has through its determined efforts been able to attract processors and operators of storage facilities to long-term leases of port areas. This means that seven or eight years after the port became fully operational actual throughput has exceeded all forecasts. More than 60 per cent of the throughput has its origin or destination within port limits. Forecast and even planned development of the hinterland has only materialised to a very limited extent.

The very large shortfall in cargo throughput during the early years of operation meant, despite its rather rapid growth since then, that the economic and financial returns from the project have been unsatisfactory. However, this does not detract from the fact that the port is now close to full utilisation and will no doubt require extension to accommodate its throughput in a few years' time.

The three ports, the two roads and the three lending agencies

The setting is the southern part of a major island, where one or more new ports and/or port extensions and improvements as well as road improvements were needed. Port A and Port B are situated on the east side of the island and Port C on its west side. First one lending agency agreed to finance upgrading of the road from Port B to Port C and, shortly after, a second lending agency agreed to finance upgrading of the road between Port A and Port C. The distances between the above mentioned ports are about 240 km in both cases, corresponding to three to four hours by truck. Before the port projects were embarked upon it was thus known to the three international and bilateral lending agencies, which were interested in financing the ports in question, that the two road projects were definitely going to be implemented.

One of the lending agencies agreed to finance a large extension of Port A. Before any decision was made concerning other port projects the two other lending agencies met to co-ordinate their efforts. One of them was apparently in doubt about whether to finance extension of Port B (for all practical purposes it was going to be a new port) or upgrading of the road between Ports A and B, a distance of less than 150 km. It was agreed that the needs of the island as a whole rather than those of isolated areas should be governing. However, this perspective was lost sight of soon after. One lending agency went ahead with the Port B project and not much later the other agency went ahead with Port C, a completely new facility. A national ports' authority, which could perhaps have prevented the developments described above from taking place, was not established until after the ports and the roads had been built. It had to take over operations of the three ports.

The result was very considerable excess port capacity for the southern part of the island. Nearly all of the cargo and passenger traffic for the island has its origin and destination, respectively, in the largest city of the country, which is located on another, distant island. The freight rates and the passenger fares, respectively, between the largest city and any one of the island ports were the same, but a call at the desired port was not promised. Truck transport for cargo, and bus transport for passengers, were included to or from any other port. The advantage from the point of view of the shipping lines was the freedom to reduce calls at ports from three to two or, on many occasions, only one. This also reduces ship-operating costs, because it is often not necessary to go around the island but only to one of its sides. It meant diversion of cargo and passenger throughput away from the smaller ports. Accordingly Port C, which had the smallest throughput of the three, and which became operational after the other two ports, suffered most from loss of traffic to the other ports. During its first year of operation actual cargo throughput reached only 31 per cent of the forecast tonnage and the prospects for the future were far from bright. Both the economic and financial returns on the port investment were below acceptable levels.

The pattern of shipping to and from the ports in question and the relevant user charges were apparently not considered in the planning and decision-making processes, which took place. Thus the problem of excess capacity for two of the ports was made even worse, in particular for the smallest port.

Port extension for vessels engaged in domestic traffic

A rather large port in an island country had a total of about 4100 m of quay for ocean-going ships and domestic carriers. About 500 m of quay exclusively for domestic carriers was added. The main economic justification was expected to be reduction of ships' waiting and service time.

The port was unusual, compared with most ports, in that cargo handling was very labour intensive. Very little mechanical equipment was used onshore, although some was available. It was like going back to the days before forklifts and tractor–trailer systems had been introduced in ports. Handcarts were used extensively. However, cargo-handling rates per gang hour for break bulk cargo were quite high. They were almost the same for international cargo as for domestic cargo. In the largest port in the country, which was highly mechanised, the corresponding handling rates were not much better. In both ports more than 50 per cent of the break bulk was uniform cargo in bags and bales. For domestic cargo the tonnage handled per alongside gang-day was far below that for international cargo. This was not understood until more detailed analysis of available data revealed that, while handling of international cargo took place during close to 20 h on average per 24-h day, it only took place during about 7 h on average for domestic cargo. This seemed unreasonable until it was established that there was very considerable excess carrying capacity in the domestic fleet and that the shipping lines, which operated these carriers, leased their own (one berth or a few berths) terminals. Congestion of the port was not part of the problem. Average waiting time to average service time ratios were low and would have remained low for a number of years.

The main official justification for providing the additional 500 m of quay for domestic carriers was the value to the economy of reduction of ships' waiting time. However, time had no real value for domestic carriers because of the excess carrying capacity of the fleet. Also there were no incentives to reduce the time spent alongside, such as steeply increasing daily charges for alongside accommodation.

The site for the container terminal extension

The major container terminal in the country was forecast to be beyond its optimum utilisation within a few years. A site selection study for a large, phased extension was undertaken. All the extension possibilities, except one, involved large-scale land reclamation using fill material, which would have to be transported from one or more distant sources. The one existing land area was immediately adjacent to the existing container terminal, its area was more than adequate, the port authority owned it and it was inside the port boundaries. It had been reclaimed more than 50 years ago. The fill which had been used was of acceptable quality. On average the fill layer was 5.5 m thick. Its weight had consolidated the lower, softer layers. However, a number of years ago it had been leased mainly for one-family homes on small plots of land. The lease contracts said clearly that the port authority had the right to discontinue the lease, after giving one year's notice, whenever it needed the land for port related purposes. The number of small homes was approximately 8400. The roads in the area were in a poor state of repair, the water supply was inadequate and there were no sewers at all, only septic tanks. The area was often flooded during high tide. Cost estimates, which included the costs of land, infrastructure and housing for resettlement, demonstrated conclusively that the least-cost reclamation alternative would be about twice as costly as the non-reclamation alternative. The port authority owned another area, outside the port boundaries, for which it had no use in the foreseeable future. Its area was sufficient for resettlement of the families which lived in the area, needed for extension of the container terminal.

Providing good infrastructure in the form of roads, water supply, electric-power supply and a sewer system as well as houses and/or plots for resettling the families, would only cost a fraction of the cost difference between the least-cost reclamation site and the non-reclamation site. However, the government decided at the time that the resettlement was not feasible for political reason because the resettlement scheme outlined above would establish a potentially dangerous precedent. Later, before implementation of the extension got under way, private business interests in the country in joint venture with a large foreign container-shipping line proposed to build and operate a private terminal in the existing land area adjacent to the public container terminal. Relatives of a high-level government official were involved in this joint venture. Suddenly it seemed not to be entirely impossible to use the area in question for the terminal. Several years later a decision had apparently still not been made.

The site for the first phase of the multi-purpose terminal

A large multi-purpose terminal as an extension to an existing river port was found to be economically and financially feasible. It was to be located immediately downstream of the existing port. Its first phase of about five berths was intended to be located in the area closest, but not immediately adjacent, to the existing port. The local planners and engineers had not given any reasons for this choice.

In the review carried out for an international financing agency, it became apparent that foundation conditions in the area chosen for the first phase were very inferior to conditions somewhat further downstream, in the area originally intended for the third phase of the terminal. The upper layers below the river bottom consisted of non-consolidated silt with organic content. They varied in thickness between 6 and 17 m, had SPT values of 0 to 2 and water contents which exceeded their liquid limits. Below were sand layers, followed in most places by other, non-consolidated layers with organic content. The surface of weathered rock along the quay line originally chosen for the first phase was below elevation -31 to -32 m. How far below was not known, as the borings which penetrated to the above elevations did not reach the weathered rock layer. Along the quay line intended for the third phase the weathered rock surface was at elevation -13 m on average. The geotechnical properties of the weathered rock were not known, as none of the borings penetrated into this layer.

Based on the existing, although insufficient, geotechnical data it was concluded that a deck supported on piles was likely to be the least-cost quay structure. The piles would have to be driven into the surface of the weathered rock layer. The reduction in length per pile for the third-phase site compared with the first-phase site would thus amount to more than 18 to 19 m. The total reduction of pile length for 1000 lin. m of quay structure required for the first phase was estimated to be at least 27 000 lin. m and the corresponding reduction for pile support for dry bulk stacker/reclaimer tracks and rail tracks to be at least 40 000 lin. m.

The main disadvantages associated with the much more favourable site, were earlier construction of 1150 lin. m of rail track and 600 lin. m of road, the additional discounted cost of which was only a small fraction of the very large reduction of pile foundation costs.

Thus the conclusion was that the first phase should be located at the designated third phase site.

Alongside port replacement of lighterage port

Very appreciable delays occurred during planning and design, before construction started. In the meantime the one export commodity, which amounted to more than 75 per cent of the total throughput, became increasingly containerised at the insistence of consignees, who wanted smaller and more frequent shipments. During the years of delay of the new alongside port the production of the export commodity in question increased steadily. The containerisation requirement meant that lighterage at the port in question was no longer suitable and the port lost nearly all its throughput to the largest port in the country and to a nearby foreign port. Road transport was employed to carry the commodity from its production areas to the ports.

The private operator, who had obtained a long-term lease from the government to operate the port from its completion, was thus faced with the difficult task of recapturing lost throughput as early as possible. During the first full calendar year of operation the actual throughput was only 20 per cent of the forecast. However, during the second year of operation the throughput nearly doubled and reached 36 per cent of the forecast. For the third year of operation (1991) indications were after the first five months that the total for the year would reach about 50 per cent of the forecast, corresponding to more than 500 000 metric tonnes. There are no reasons to believe that this growth will not continue, until the economic capacity of the existing facilities will be reached around 1996, at which time additional berth, storage and equipment facilities are likely to be required.

How was this accomplished? The private operator had a contractual obligation to hire expatriate expertise to facilitate the start of operations. He retained two of the expatriates who, in addition to their management and operations competence, have as their native language the language (which is not the same as the national language) of the shippers of the major commodities. The port management has succeeded in attracting the required increasing numbers of calls by container feeder-line ships to satisfy the requirements of consignees. Thus there are good reasons to believe that the future will be bright for the port in question.

The second deep-water port

In a country, which had only one deep-water port, an additional port was believed to be needed. The existing port was connected both by rail and by road to the economically most active regions of the country.

The country was blessed with several tidal inlets (in addition to the one in which the existing port was located), which are excellent natural port sites. The access to these port sites could conveniently be improved by a small quantity of channel dredging and required practically no maintenance dredging. However, none of the sites had rail connections.

The only site which was more than 100 km from the existing port had no land transport connections at all, it was farther from the economically active regions of the country than the existing port, and its hinterland, though vast, was not developed to any significant extent. Thus a port should not be constructed here except as part of a project involving land transport connections and agricultural or other development of its hinterland.

The other sites, relatively close to the existing port, were located at approximately the same distance from the economically active regions of the country, to which they were connected by road only. As rail transport was advantageous for many commodities, because of the considerable distances involved, a port at any of these sites would have to include a rail connection to the existing railway. However, the existing port could be extended considerably and the other sites presented no advantages, only the disadvantages sketched above. Thus the solution was, and remained for a number of years, to extend the existing port.

The transit sheds for the rehabilitated berths in the old colonial port

The old colonial port was only served by rail once or twice a week and had no road connection. Therefore the transit shed floor area per berth was a rather large 15 000 to 20 000 sq m, which is four to five times the area which is usually required at present. After the country became independent an all-weather road connecting the port to the largest city in the country, as well as roads to other population centres, were built. Road transport became the major mode, with an increasing share of land transport.

A rehabilitation project for some of the existing berths included construction of new transit sheds. The floor area per berth was not changed, although the transit storage requirements had been reduced to a small fraction of its earlier value, because of the road transport. In fact the overall covered storage requirements had been reduced so much for the port as a whole, that no transit sheds at all were required at the rehabilitated berths. Apparently those involved in project preparation had not paid any attention to this issue at all.

Lessons to be learned from the case studies

In the above cases standard economic and financial analyses, involving calculation of economic internal rate of return (EIRR) and financial internal rate of return (FIRR) or other suitable indicators were carried out to justify the projects. 'Standard' forecasts of cargo throughput and their split between competing ports, based among other factors on official land-transport-user charges, were employed. It was assumed that a new port would, during its first year of operation, achieve its full share of forecast throughput. Actually, only 20 to 30 per cent of forecasts were reached in three of the above cases. Disregarding developments which may not be considered foreseeable, such as long delays in project implementation and rapid containerisation of a commodity, which was previously not transported in containers, there are important areas where the methodology needs to be improved, in particular in regard to throughput forecasting for the early years of port operation. The likely timing of start of operations of processing facilities for outbound commodities has to be considered in the forecasts. To assume that they will automatically be available when the corresponding port facilities become operational is clearly not realistic. The same is true for specialised storage facilities both for outbound and inbound cargo. General storage facilities such as transit sheds would normally be part of the port project.

Long-standing trade relationships as well as the probability of land-transport-user charge rebates, considering the availability of back-haul cargo should not be neglected. In certain cultures personal loyalties are valued highly and not easily dispensed with. These factors may work very much in favour of existing ports, especially if a new port handles predominantly either outbound cargo or inbound cargo.

The case studies demonstrate clearly that standard forecasting methods for distribution of cargo throughput between different ports, based on a hinterland concept, minimisation of official land transport user charges etc., are quite misleading for new ports in new locations.

Shipping patterns, based on a uniform tariff for two or more ports, designed to reduce ship operating costs and including land transport as required, should be considered for cases involving competition between ports.

For cases, where major economic benefits are attributed to reduction of ships' waiting time and/or service time, a realistic assessment should be made of the value, if any, of ships' time.

The above recommendations are mainly qualitative. Quantitative recommendations for new ports in new locations are difficult to make, based on the limited number of case studies, which are presented above. One may, perhaps, suggest for discussion that only 25 per cent of the 'standard' forecast value should be used for the first year of operation, 35 per cent during the second year and perhaps 50 per cent during the third year. If the above, or a similar recommendation, is implemented it may become necessary to lower somewhat the cut-off EIRR value required for approval for financing of new ports in new locations. However, the appraisal process is going to become more realistic.

Another quantitative recommendation would be to include in projects, or otherwise make certain by means of loan conditions etc., that all required production and processing facilities as well as other facilities will be provided and will be ready, when the port becomes operational. However, this may not turn out to be realistic, because requirements may only be estimated, not known with any great degree of certainty.

Index

Page numbers in italics refer to illustrations, tables and diagrams.

Printed in the United Kingdom by
Lightning Source UK Ltd., Milton Keynes
137880UK00001B/15/A